T0235824

Thermal and Statistical Physics

Sandeep Sharma

Thermal and Statistical Physics

Concepts and Applications

Ane Books
Pvt. Ltd.

Springer

Sandeep Sharma
Department of Physics
Guru Nanak Dev University
Amritsar, Punjab, India

ISBN 978-3-031-07687-9   ISBN 978-3-031-07685-5 (eBook)
https://doi.org/10.1007/978-3-031-07685-5

Jointly published with ANE Books Pvt. Ltd.
In addition to this printed edition, there is a local printed edition of this work available via Ane Books in
South Asia (India, Pakistan, Sri Lanka, Bangladesh, Nepal and Bhutan) and Africa (all countries in the
African subcontinent).
ISBN of the Co-Publisher's edition: 9789390658268

This Springer imprint is published by the registered company Springer Nature Switzerland AG
The registered company address is: Gewerbestrasse 11, 6330 Cham, Switzerland

*Dedicated
to my
students and Shivansh*

Preface

This book is an outcome of lectures that I have delivered to bachelor and master students at **Guru Nanak Dev University, Amritsar, Punjab, India**. The content presented in the book is classroom tested and various students have benefited from it. The primary objective of this book is to introduce the fundamental concepts of thermal and statistical physics and their use in understanding various problems. The concepts are presented in a very simple way so that reader can build up a firm foundation in the subject.

In Chap. 1, starting with some historical facts we introduce the main subject of classical thermodynamics. The concept of thermodynamic limit, intensive and extensive variables, and perfect and imperfect differentials are introduced.

In Chap. 2, we introduce the Zeroth law of thermodynamics, concept of work and sign convention. This chapter also highlights and clarifies the use of different sign conventions, and emphasizes that both lead to the same results. The chapter includes a discussion of the first law of thermodynamics and ends with a description of heat and the idea of heat capacity.

Chapter 3 deals with the crucial second law of thermodynamics. The idea of a Carnot heat engine is introduced in this chapter. Various statements of the second law of thermodynamics which originate from the discussion of the heat engine are introduced. At the end of this chapter, a description of the thermodynamic scale of temperature and its equivalence with the ideal gas scale is presented.

Chapter 4 begins with the crucial concept of entropy. This is followed by various illustrations chosen from daily life. The important concept of free expansion is also included in this chapter. This chapter ends with a discussion on the third law of thermodynamics.

In Chap. 5, various thermodynamic potentials, such as the internal energy, enthalpy, Helmholtz function and Gibbs function, are presented. A qualitative discussion on free energy, Maxwell's relations, Clausius–Clapeyron equation, TdS equations and magneto caloric effects is also included in this chapter.

Chapter 6 presents the kinetic theory of gases along with the Maxwell–Boltzmann law of distribution of velocities. The Doppler broadening, and Zartman and Stern's experiments are included in this chapter. At the end of this chapter, we present degrees of freedom, evaluation of mean free path and discussion on various transport phenomena.

Chapters 7 and 8 deal with real gases. In Chap. 7, deviation of ideal gas behaviour is presented. This is followed by the Van der Waals equation of state, Andrew's experiment and evaluation of critical constants of a real gas. In Chap. 8, the Joule expansion and Joule–Kelvin expansions are discussed.

Chapter 9 is devoted to the theory of radiations. The chapter covers various aspects of Black body radiation spectra, Wien's displacement law, Newton's law of cooling, Rayleigh–Jeans law and Planck's law are covered in detail.

Chapter 10 of the book is devoted to the introduction of basic concepts in statistical mechanics. In this chapter, we introduce the idea of phase space, distinction between microstates and macrostates. Thermodynamic probability and its derivation for three different cases of Maxwell–Boltzmann, Fermi–Dirac and Bose–Einstein statistics are discussed.

Amritsar, India   Sandeep Sharma

UGC Approved Syllabus for Choice-Based Credit System

Laws of Thermodynamics

Thermodynamic Description of a system: Zeroth law of thermodynamics and temperature; First law and internal energy, conversion of heat into work, various thermodynamical processes; Applications of First Law: General Relation between C_P and C_V, Work Done during Isothermal and Adiabatic Processes, Compressibility and Expansion Coefficient, Reversible and Irreversible processes, Second law and Entropy, Carnot's cycle and Carnot's theorem, Entropy changes in reversible and irreversible processes, Entropy–Temperature diagrams, Third law of thermodynamics and Unattainability of absolute zero.

[22 Lectures]

Thermodynamic Potentials

Enthalpy, Gibbs, Helmholtz and Internal Energy functions, Maxwell's relations and applications, Joule–Thompson effect, Clausius–Clapeyron equation and Expression for $(C_P - C_V)$, C_P/C_V, TdS equations.

[10 Lectures]

Kinetic Theory of Gases

Derivation of Maxwell's law of distribution of velocities and its experimental verification, Mean free path (Zeroth order); Transport Phenomena: Viscosity, Conduction and Diffusion (for vertical case), Law of equipartition of energy (no derivation, qualitative description only) and its applications to specific heat of gases; monoatomic and diatomic gases.

[10 Lectures]

Theory of Radiation

Blackbody radiation, Spectral distribution, Concept of Energy Density, Derivation of Planck's law, Deduction of Wien's distribution law, Rayleigh–Jeans law, Stefan–Boltzmann law and Wien's displacement law from Planck's law.

[**6 Lectures**]

Statistical Mechanics

Maxwell–Boltzmann law, distribution of velocity, Quantum statistics, Phase space, Fermi–Dirac distribution law, electron gas, Bose–Einstein distribution law, photon gas and comparison of three statistics.

[**12 Lectures**]

Thermal and Statistical Physics—Concepts and Applications

Salient Features

According to the latest syllabus as per UGC guidelines;
   Inclusion of topics related to NET and GATE examinations;
   Inclusion of Numerical problems with solutions from the last 20 years of examinations (NET-JRF, GATE, JAM, JEST, etc.);
   Various illustrations from daily life;
   Enriched with various diagrams, illustrations and mnemonics;
   More than 380 solved problems including 240 MCQ type questions;
   Covers newly introduced topics like radiation, cooling of real gases.

Contents

About the Author

Sandeep Sharma Ph.D. (University of Groningen, The Netherlands) and M. Tech (IIT-Delhi), is an Asst. Professor at Department of Physics, GNDU Amritsar, Punjab. Prior to joining GNDU, he has worked as an engineer at Semiconductor Laboratory (Department of Space, India) Mohali (Punjab) and Chartered Semiconductor Manufacturing (now Global Foundaries), Singapore. Dr. Sharma has worked at various world renowned laboratories, such as MESA$^+$ Institute for Nanotechnology at University of Twente, The Netherlands. Thereafter, he shifted to Spintronics Research Center, at National Institute of Advanced Industrial Science and Technology (AIST) Tsukuba, Japan for his Ph.D. assignment.

With more that fourteen years of teaching and research experience in India and abroad, Dr. Sharma has co-authored more than forty five research papers in peer reviewed international journals including two in Nature. He is recepient of various awards like Merit fellowship (HPBSE Dharamshala), GATE-Fellowship (MHRD-New Delhi), prestigious FOM-Fellowship (The Netherlands). He is a life-time member of Indian Association of Physics Teachers (IAPT) and Punjab Science Congress. He is also serving as member Board of Studies (BOS) at GNDU-Amritsar.

Introduction

1

1.1 Origin of Thermodynamics

The industrial revolution during the nineteenth century involved the construction of machines like steam engines, which involves the conversion of heat energy into mechanical work. Study of these engines and the conversion of heat into other forms of energy led to the development of thermodynamics. Initial developments in this field occurred at a very rapid pace, and by 1900, the subject was well understood. In a broader sense, thermodynamics is applicable to all kinds of processes that involve heat and temperature. In this course, we will deal with applicability of underlying principles to a wide range of examples. Thermodynamics allows us to understand the phenomena such as thermal radiations on the one hand and low-temperature properties of a paramagnetic salt on the other hand. It is applicable not only to steam engines, but equally well to refrigeration and rocket propulsion also.

1.2 A Macroscopic Approach

In thermodynamics, the properties of a particular system can always be described by directly observable physical quantities. For instance, when the system under investigation is a gas confined to a certain volume V, then its properties can be understood in terms of its temperature T, volume V and pressure P that is exerted by it on the wall of container. Here, T, P and V are macroscopic measurable parameters. The laws of thermodynamics enable us to correlate the macroscopic physical quantities without bothering about microscopic description of the system. The great generic applicability of thermodynamics in understanding various problems is a direct consequence of this.

© The Author(s) 2022
S. Sharma, *Thermal and Statistical Physics*,
https://doi.org/10.1007/978-3-031-07685-5_1

1.3   The Thermodynamic Limit and Its Consequences

In this section, we will try to understand how the large numbers of particles in a typical thermodynamic system enable us to deal with average quantities. For this purpose, we will take an example. Assume a tiny hut with a flat roof. It is raining outside, and one can hear the occasional raindrop striking the roof. Each rain drop that hits the roof, transfers its momentum to the roof and exerts an impulse on it. If the mass of a raindrop and its terminal velocity is known, one can evaluate the force with which it hits the roof. If one plots force versus time, the graph will look like as shown in Fig. 1.1a.

Now suppose that catchment area is increased (let us say it is 100 times previous one). In this case, more raindrops will fall, and force versus time graph will look like Fig. 1.1b. Now scale up the area of the roof by a factor of one thousand, and the force versus time graph would look like that shown in Fig. 1.1c. Notice two key points about these graphs:

(i) The average force gets bigger as the catchment area of the roof gets bigger. This is because a roof with larger area catches more raindrops.
(ii) When area is quite large, the fluctuations in the force get smoothed out. It appears that the force stays much closer to its average value. In fact, in this case, the fluctuations are still large but, as the catchment area of the roof increases, they grow more slowly than the average force does.

As we just discussed, the force grows with an increase in area. Let us define another quantity, the pressure P as

$$P = \frac{\text{force}}{\text{area}}$$

As the catchment area is increased, the average pressure due to the falling raindrops will not change. The fluctuations in the pressure will decrease in this case. Under the condition, that the roof area grows to infinity, one can ignore the fluctuations in the pressure. This is precisely analogous to the situation we refer to as the **thermodynamic limit**. Under thermodynamic limit, i.e., when N approaches infinity and V also approaches infinity, the number density (N/V), T and P acquire a fixed value.

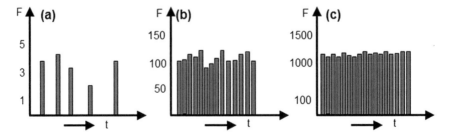

Fig. 1.1 The force versus time in different cases

1.3.1   Intensive and Extensive Variables

Consider a container of gas that has volume V. Let the temperature of the gas confined to the container is T, the pressure is P and the kinetic energy of all the gas molecules adds up to U. Now imagine dividing the container of gas in half with an imaginary plane. Now pay attention to the gas on one side of the imaginary plane. The volume of this half of the gas, let's call it V*, is by definition half that of the original container, i.e.

$$V^* = \frac{V}{2} \tag{1.1}$$

The kinetic energy of this half of the gas, let's call it U*, is clearly half that of the total kinetic energy, i.e.

$$U^* = \frac{U}{2} \tag{1.2}$$

However, the pressure P* and the temperature T* of this half of the gas are the same as for the whole container of gas, so that we can write

$$P^* = P \tag{1.3}$$

and

$$T^* = T \tag{1.4}$$

Variables which scale with dimensions of the system, for instance, V and U, are called **extensive variables**. Those which are independent of system dimensions, like P, T and number density, are known as **intensive variables**.

1.4   System Versus Surrounding

In thermodynamics, we usually confine our attention to a particular small part of the universe which we call conveniently our *system* under investigation. The rest of the universe outside our system under investigation is called *surroundings*. Note that system and surroundings collectively constitute the universe. The system and the surroundings are separated by a boundary or a wall and they may, in general, exchange energy and matter, depending on the nature of the wall. Restricting ourselves to a closed system, we shall consider only the energy exchange, i.e., no matter exchange between system and surrounding is allowed unless mentioned directly. A few other examples of thermodynamic systems are:

  (i) Combustion of fuel inside an engine
 (ii) A refrigerator
(iii) A metal rod undergoing isothermal (or adiabatic) expansion

We will discuss these thermodynamic systems later.

1.4.1   A System in Thermodynamic Equilibrium

Let us imagine a thermodynamic system whose macroscopic properties are determined by temperature (T) and pressure (P). When these parameters change either spontaneously or by virtue of outside influence (external force), the system is said to undergo a *change of state*. When such a system is not influenced by the surrounding, it is said to be an *isolated system*. In practical applications of thermodynamics, isolated systems are of little importance. We usually have to deal with a system that is influenced in some way by its surroundings or by external forces so that useful work can be obtained from such type of interaction. Let us consider the following three different situations:

Mechanical equilibrium:
When there are no unbalanced forces or torques acting on any part of the system or on the system as a whole, it is said to be in a state of *mechanical equilibrium*. When these conditions are not satisfied, either the system alone or both the system and its surroundings will undergo a change of state.

Chemical equilibrium:
When there are no chemical reactions within the system and no motion of any chemical constituent from one part of a system to another part, it is said to be in a state of *chemical equilibrium*. A system that is not in chemical equilibrium undergoes a change of state. The change of state ceases the moment the system acquires chemical equilibrium.

Thermal equilibrium:
Thermal equilibrium exists when there is no spontaneous change in the parameters of a system which is in mechanical and chemical equilibrium when it is separated from its surroundings via a diathermic wall (wall that allows heat exchange between the system and surrounding). In other words, no exchange of heat between the system and its surroundings exists. In thermal equilibrium, all parts of a system are at the same temperature, and this temperature is the same as that of the surroundings.

Whenever a system satisfies these three conditions, the system is said to be in a state of *thermodynamic equilibrium*. Under thermodynamic equilibrium, no change of state occurs. Note that states of thermodynamic equilibrium can be described in terms of macroscopic parameters (T, P, etc.) that do not involve the time, that is, in terms of thermodynamic coordinates.

1.4.2   Thermodynamic Variables or Functions of State

As we discussed in the previous section, a system is said to be in a thermodynamic equilibrium state if the corresponding macroscopic observable properties have fixed and definite values. These values are independent of how they got into the final state. These physical properties are called as *functions of state* (sometimes called *variables of state* or *thermodynamic variables* or *thermodynamic coordinates*). A function of the state is any physical quantity that has a well-defined value for each

Fig. 1.2 A gas contained in a cylinder equipped with a piston is a useful example of a system

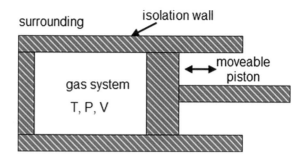

equilibrium state of the system. Thus, in thermal equilibrium, these variables of state are time independent. A few examples of the thermodynamic variable are temperature, pressure, volume and internal energy, and we will introduce a lot more in the next few chapters. Note that all these parameters are macroscopic properties of a system.

1.4.3   Equation of a State

Let us try to understand the meaning of the equation of a state. For this, we consider a system of gas (for instance in Fig. 1.2). Let a constant mass of gas is enclosed inside volume V so that the T and P may be easily determined. Now, imagine the following two situations:

(i) If we fix the volume (V) at some arbitrary value and cause the temperature to assume an arbitrarily chosen value, then we shall not be able to vary the pressure at all. That is, once V and T are fixed by us, the value of P at equilibrium is determined by nature.
(ii) Similarly, if P and T are chosen arbitrarily, then the value of V at equilibrium is fixed.

That is, of the three thermodynamic coordinates, T, P and V, only two are independent variables. This implies that there exists an equation of equilibrium which connects the thermodynamic variables in such a way that the two of them can be independently chosen but the third one is dependent on the other two. Such an equation is called an *equation of state*. In general, *equation of state* is a mathematical function relating the appropriate thermodynamic variables of a system in equilibrium. A few examples of the equation of state:

(i) For example, a system consisting of gas at very low pressure has the simple equation of state of an ideal gas

$$PV = nRT \tag{1.5}$$

where n is the number of moles and R is the molar gas constant.

(ii) At higher pressures, the equation of state is more complicated, being fairly well represented by the van der Waals equation, which takes into account particle interactions and the finite size of the particles.

$$\left(P + \frac{a}{V^2}\right)(V - b) = RT \tag{1.6}$$

where a and b are constants appropriate to the specific gas.

In general, one can say that an equation of state connects the thermodynamic variables (functions of state) of a system in equilibrium. For a gas, this takes the general form

$$f(P, V, T) = 0$$

1.4.4   Physical Meaning of dV, dP and dT

In order to explore the physical significance of dV, dP and dT, we will consider a gaseous system. Experiments show that the equilibrium state of such a system can be described by an equation of state with three thermodynamic variables, namely T, P and V. If such a system undergoes a small change of state, i.e., from initial equilibrium state to another equilibrium state very near to the initial one, then all three coordinates, in general, undergo a small change (w.r.t initial state).

If the change of, say, V is very small in comparison with V and very large in comparison with the space occupied by a few molecules, then this change of V may be written as a differential dV. Therefore, dV, in this case, refers to a change in the volume which is very small compared to the total volume V of the system, but large enough when compared to the volume occupied by a few gas molecules. If V was a geometrical quantity referring to the volume of space, then dV could be used to denote a portion of that space arbitrarily small.

Similar arguments can be applied for defining dP. If the change of P is very small in comparison with P and very large in comparison with local fluctuations of pressure caused by momentary variations in microscopic concentration, then it also may be represented by the differential dP. In general

"*Every infinitesimal change in thermodynamics must satisfy the requirement that it represents a change in a physical quantity which is small with respect to the quantity itself and large in comparison with the effect produced by the behaviour of a few molecules.*" Based upon these arguments and those we had in the previous section, we may imagine that the equation of state is solved for any coordinate in terms of the other two. Thus, we can write

$$V = \text{function of}(T, P) = V(T, P)$$

An infinitesimal change from one equilibrium state to another equilibrium state involves a dV, a dT and a dP. All these infinitesimal changes satisfy the condition discussed in the previous paragraph. Then using the fundamental theorem in partial differential calculus, we can write differential of volume V as

$$dV = \left(\frac{\partial V}{\partial T}\right)_P dT + \left(\frac{\partial V}{\partial P}\right)_T dP \qquad (1.7)$$

Here, the partial derivative $\left(\frac{\partial V}{\partial T}\right)_P$ means infinitesimal change in V for an infinitesimal change in T with pressure P held constant. Note that each of the partial derivatives can be a function of T and P. Both of these partial derivatives have an important physical meaning. We define an average coefficient of volume expansion as

$$\text{Average coefficient of vol. expansion} = \frac{\text{fractional change in volume}}{\text{change in temperature}}$$

If the temperature change is very small (infinitesimal), a corresponding change in volume also becomes infinitesimal and we get the differential coefficient of volume expansion, or, simply, the *volume expansivity*, denoted by α. If the process takes place under fixed P, we get *isobaric expansivity* α_P, and *adiabatic expansivity* α_S at fixed S. Thus

$$\alpha_P = \frac{1}{V}\left(\frac{\partial V}{\partial T}\right)_P \qquad (1.8)$$

Similarly, adiabatic expansivity is

$$\alpha_S = \frac{1}{V}\left(\frac{\partial V}{\partial T}\right)_S \qquad (1.9)$$

Another quantity of interest connected with the second differential is the *isothermal compressibility* κ_T and is defined as

$$\kappa_T = -\frac{1}{V}\left(\frac{\partial V}{\partial P}\right)_T \qquad (1.10)$$

while the *adiabatic compressibility* κ_S is defined as

$$\kappa_S = -\frac{1}{V}\left(\frac{\partial V}{\partial P}\right)_S \qquad (1.11)$$

Both quantities (κ_T and κ_S) have a minus sign so that the compressibilities are positive (this is because things get smaller when you press them, so fractional volume changes are negative when positive pressure is applied). Note that expansivity is the fractional change in volume w.r.t infinetisimal change in temperature, whereas compressibility is fractional change in volume w.r.t infinetisimal change in pressure.

If the equation of state is solved for P, then we can write

$$P = P(T, V)$$

so that

$$dP = \left(\frac{\partial P}{\partial T}\right)_V dT + \left(\frac{\partial P}{\partial V}\right)_T dV \qquad (1.12)$$

and finally, if we imagine T as a function of P and V, then

$$dT = \left(\frac{\partial T}{\partial P}\right)_V dP + \left(\frac{\partial T}{\partial V}\right)_P dV \qquad (1.13)$$

1.5   Thermodynamic Reversibility

When a system undergoes a series of changes, we say that a thermodynamic process has occurred. A process is said to be reversible if, and only if, its direction can be reversed by an infinitesimal change in the conditions. In general, for thermodynamic reversibility, two conditions need to be satisfied. First, the process must be quasi-static and second there must be no hysteresis.

For a process to be quasi-static, it must be carried out very slowly so that every state through which the system passes may be considered an equilibrium state. Strictly speaking, the process should be carried out infinitely slowly. Fast changes disturb the equilibrium between different parts of a system and such processes are said to be irreversible in nature. For instance, bursting of a tyre is an example of irreversible process.

Further, a system with hysteresis also does not retrace its previous path. In this case, system proceeds on a different path. A common example is found in the magnetization of iron. Hence, the process cannot be reversed in a system with hysteresis.

1.6   A Few Important Mathematical Results

Differential coefficients which relate the rate of change of one thermodynamic variable with another are very important in thermodynamics. They are known as thermodynamic coefficients, and since their manipulation is a vital part of thermodynamic calculation, it is important to understand their meaning and to be familiar with some basic mathematical results which are of help in handling them.

1.6.1   The Reciprocal Theorem

If we assume that three variables x, y and z are related in such a way that

$$F(x, y, z) = 0$$

Then, in principle, this equation can be rearranged to express one of the variables in terms of the other two as independent variables, i.e.

$$x = x(y, z)$$

Differentiating by parts, we obtain

$$dx = \left(\frac{\partial x}{\partial y}\right)_z dy + \left(\frac{\partial x}{\partial z}\right)_y dz \qquad (1.14)$$

here, the terms in brackets are the partial differentials of x. Equation 1.14 expresses the change in x which results from changes in both of the independent variables (y and z) on which x depends. We may write an identical equation for dz

$$dz = \left(\frac{\partial z}{\partial x}\right)_y dx + \left(\frac{\partial z}{\partial y}\right)_x dy \qquad (1.15)$$

Substituting Eqs. 1.15 in 1.14 and rearranging, we obtain

$$dx = \left(\frac{\partial x}{\partial z}\right)_y \left(\frac{\partial z}{\partial x}\right)_y dx + \left[\left(\frac{\partial x}{\partial y}\right)_z + \left(\frac{\partial x}{\partial z}\right)_y \left(\frac{\partial z}{\partial y}\right)_x\right] dy \qquad (1.16)$$

This result must hold for any pair of independent variables. If we choose x and y to be independent and take $dy = 0$ and $dx \neq 0$, then Eq. 1.16 gives

$$\left(\frac{\partial x}{\partial z}\right)_y \left(\frac{\partial z}{\partial x}\right)_y = 1$$

or

$$\left(\frac{\partial x}{\partial z}\right)_y = \frac{1}{\left(\frac{\partial z}{\partial x}\right)_y} \qquad (1.17)$$

This is the *reciprocal theorem* which allows us to replace any partial derivative by the reciprocal of the inverted derivative with the same variable(s) held constant.

1.6.2 Reciprocity Theorem

If we now substitute in Eq. 1.16 $dx = 0$ and $dy \neq 0$, the term in brackets must be identically zero, giving

$$\left(\frac{\partial x}{\partial y}\right)_z = -\left(\frac{\partial x}{\partial z}\right)_y \left(\frac{\partial z}{\partial y}\right)_x \qquad (1.18)$$

OR

$$\left(\frac{\partial x}{\partial y}\right)_z \left(\frac{\partial y}{\partial z}\right)_x \left(\frac{\partial z}{\partial x}\right)_y = -1 \qquad (1.19)$$

The above two results represent the *reciprocity theorem*. It may be written starting with any derivative and then following through the other variables in cyclic order.

1.6.3 Perfect Differential

An expression of type $F_{1(x,y)}dx + F_{2(x,y)}dy$ is said to be an exact or perfect differential if we can write it as the differential

$$df = \left(\frac{\partial f}{\partial x}\right)dx + \left(\frac{\partial f}{\partial y}\right)dy \qquad (1.20)$$

of a single-valued function f(x, y). This implies that we can define

$$F_1 = \left(\frac{\partial f}{\partial x}\right), \quad F_2 = \left(\frac{\partial f}{\partial y}\right) \qquad (1.21)$$

or in vector form, we can write it as $F = \nabla f$. Further, for an exact differential, the integral around a closed loop will be zero, i.e.

$$\oint F.dr = \oint df = \oint F_1(x, y)dx + F_2(x, y)dy = 0 \qquad (1.22)$$

By stokes theorem, which gives, $\nabla \times F = 0$, and therefore

$$\left(\frac{\partial F_1}{\partial x}\right) = \left(\frac{\partial F_2}{\partial y}\right), \quad \text{or} \left(\frac{\partial^2 f}{\partial x \partial y}\right) = \left(\frac{\partial^2 f}{\partial y \partial x}\right) \qquad (1.23)$$

Physically, it means that changing differentiation order does not influence the value of a function. In thermal physics, an important point to remember is that functions of state have exact differentials.

1.7   Solved Problems

Q.1 **The equation of state of an ideal gas is PV = nRT, where n and R are constants.**
(i) Show that the isobaric volume expansivity β is equal to 1/T.
(ii) Show that the isothermal compressibility κ_T is equal to 1/P.
(iii) Show that adiabatic compressibility κ_S is equal to $1/\gamma P$
(iv) Hence, show that $\dfrac{\kappa_T}{\kappa_S} = \gamma$

Sol: (i) The isobaric volume expansivity is defined as

$$\beta_P = \frac{1}{V}\left(\frac{\partial V}{\partial T}\right)_P$$

$$= \frac{1}{V}\left(\frac{nR}{P}\right) = \frac{1}{T}$$

(ii) Similarly, isothermal compressibility κ_T is defined as

$$\kappa_T = -\frac{1}{V}\left(\frac{\partial V}{\partial P}\right)_T$$

$$= -\frac{1}{V}\left(\frac{-nRT}{P^2}\right)$$

$$= \frac{1}{P}$$

(iii) For an adiabatic process, $PV^\gamma = C$, where γ is adiabatic index and C is constant. Now the adiabatic compressibility κ_S is defined as

$$\kappa_S = -\frac{1}{V}\left(\frac{\partial V}{\partial P}\right)_S$$

$$= -\frac{1}{V}\left[\frac{-V}{\gamma P}\right] = \frac{1}{\gamma P}$$

$\because \gamma > 1$, implies $\kappa_T > \kappa_S$

(iv) From previous results, we can write

$$\frac{\kappa_T}{\kappa_S} = \gamma = \frac{C_P}{C_V}$$

Q.2 The equilibrium states of superheated steam are represented by equation

$$(V - b) = \frac{RT}{P} - \frac{a}{T^m}$$

where b, r, a and m are constants. Calculate the isobaric volume expansivity (α_P) as a function of T and P.

Sol: We can rewrite the given equation as

$$V = b + \frac{rT}{P} - \frac{a}{T^m}$$

Therefore, the isobaric volume expansivity is

$$\alpha_P = \frac{1}{V}\left(\frac{\partial V}{\partial T}\right)_P$$
$$= \frac{1}{V}\left[\frac{r}{P} - a\left(\frac{-m}{T^{m+1}}\right)\right]$$
$$= \frac{r}{PV} + \frac{am}{VT^{m+1}}$$

Q.3 **Which of the graphs below correctly represents the variation of**

$$\kappa_T = -\frac{1}{V}\left(\frac{\partial V}{\partial P}\right)_T$$

with pressure P for an ideal gas at constant temperature? (Fig. 1.3).

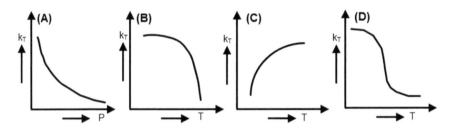

Fig. 1.3 Figures for Q.3

Sol: For an ideal gas at constant temperature, $\kappa_T = \dfrac{1}{P}$, therefore, (A) is CORRECT.

Q.4 **An ideal gas is initially at temperature T and volume V. Its volume is increased by ΔV due to an increase in temperature ΔT, pressure remaining constant. The quantity $\delta = \dfrac{1}{V}\left(\dfrac{\Delta V}{\Delta T}\right)_P$ varies with T as** (Fig. 1.4)

Sol: For an ideal gas, $PV = RT$. Therefore, at constant pressure, the quantity, δ representing isobaric expansivity is

$$\delta = \frac{1}{V}\left(\frac{\Delta V}{\Delta T}\right)_P$$
$$= \frac{1}{V}\frac{R}{P} = \frac{1}{T}$$

Therefore, (C) is CORRECT.

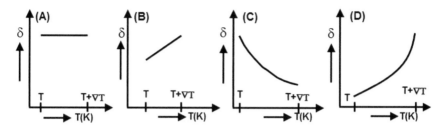

Fig. 1.4 Figures for Q.4

1.8   Multiple Choice Questions

Q.1 **Which one of the following is an extensive quantity**

    (A) Temperature   **(C)** Volume
    (B) Pressure   **(D)** Number density

Q.2 **Thermodynamic limit is valid when**

    (A) $N \to \infty$ **(C)** N/V is finite
    (B) $V \to \infty$ **(D)** All

Q.3 **Which of the following statements concerning the coefficient of volume expansion α and the isothermal compressibility κ_T of a solid is true?**
                                                 [NET-JRF-June2018]

    (A) α and κ_T are both intensive variables
    (B) α is an intensive and κ_T is an extensive variable
    (C) α is an extensive and κ_T is an intensive variable
    (D) α and κ_T are both extensive variables

Q.4 **The thermodynamics deals with**

    (A) microscopic variables   **(C)** both of these
    (B) macroscopic variable   **(D)** none of these

Q.5 **A system is said to be in thermodynamic equilibrium when it is in**

    (A) mechanical equilibrium   **(C)** chemical equilibrium
    (B) thermal equilibrium   **(D)** all three equilibrium

Q.6 **Thermodynamics deals with**

    (A) conversion of electrical energy into mechanical energy
    (B) conversion of chemical energy into electrical energy

(C)   conversion of light energy into mechanical energy
(D)   conversion of heat energy into other forms of energy

Q.7 **Isothermal compressibility is given by**

[GATE-2012, JAM-2019]

(A) $\dfrac{1}{V}\left[\dfrac{\partial V}{\partial P}\right]_T$ **(C)** $-\dfrac{1}{V}\left[\dfrac{\partial V}{\partial P}\right]_T$

(B) $\dfrac{1}{P}\left[\dfrac{\partial P}{\partial V}\right]_T$ **(D)** $-\dfrac{1}{P}\left[\dfrac{\partial P}{\partial V}\right]_T$

Q.8 **Which one of the following is a perfect differential**

(A)   work done in an isothermal process
(B)   work done in an adiabatic process
(C)   heat exchange in any process
(D)   both A and B

Q.9 **Isothermal compressibility κ_T of a substance is defined as**

$$\kappa_T = -\frac{1}{V}\left[\frac{\partial V}{\partial P}\right]_T$$

Its value for n moles of an ideal gas will be

[JAM-2009]

(A) $\dfrac{1}{P}$ **(C)** $-\dfrac{1}{P}$

(B) $\dfrac{n}{P}$ **(D)** $-\dfrac{1}{P}$

Q.10 **Which of the following quantity is path independent**

(A)   work   **(C)**   temperature
(B)   heat   **(D)**   internal energy

Q.11 **Let κ_T and κ_S represent isothermal and adiabatic compressibilities, respectively. Then the CORRECT relation giving the adiabatic index γ is**

(A) $\dfrac{\kappa_T}{\kappa_S}$ **(C)** $\dfrac{1}{\kappa_T \kappa_S}$

(B) $\dfrac{\kappa_S}{\kappa_T}$ **(D)** $\kappa_T \kappa_S$

Q.12 Expansion during heating

- **(A)**   occurs only in solids
- **(B)**   increases weight of a material
- **(C)**   decreases the density of a material
- **(D)**   occurs at the same rate for all liquids and solids

Q.13 Solids expands upon heating, because

- **(A)**   kinetic energy of atoms increases
- **(B)**   potential energy of atoms increases
- **(C)**   total energy of atoms increases
- **(D)**   the potential energy plot is asymmetric about equilibrium distance between neighbouring atoms

Q.14 When a bimetallic strip is heated, it

- **(A)**   does not bend at all
- **(B)**   gets twisted in the form of an helix
- **(C)**   bends in the form of an arc with the more expandable metal outside
- **(D)**   bends in the form of an arc with the more expandable metal inside

Q.15 A copper plate has a hole at its centre. When heated, the size of the hole

- **(A)**   will always increase
- **(B)**   will always decrease
- **(C)**   will depend on the hole and material of the plate
- **(D)**   none of these

Q.16 Adiabatic compressibility is given by

- **(A)** $\dfrac{1}{V}\left[\dfrac{\partial V}{\partial P}\right]_T$
- **(C)** $-\dfrac{1}{V}\left[\dfrac{\partial V}{\partial P}\right]_S$
- **(B)** $\dfrac{1}{V}\left[\dfrac{\partial V}{\partial P}\right]_S$
- **(D)** $-\dfrac{1}{P}\left[\dfrac{\partial P}{\partial V}\right]_S$

Q.17 Isobaric expansivity is given by

- **(A)** $\dfrac{1}{V}\left[\dfrac{\partial P}{\partial V}\right]_P$
- **(C)** $\dfrac{1}{V}\left[\dfrac{\partial V}{\partial T}\right]_P$
- **(B)** $-\dfrac{1}{V}\left[\dfrac{\partial V}{\partial P}\right]_P$
- **(D)** $-\dfrac{1}{P}\left[\dfrac{\partial P}{\partial V}\right]_P$

Q.18 **Adiabatic expansivity is given by**

(A) $\quad \dfrac{1}{V}\left[\dfrac{\partial P}{\partial V}\right]_{S}$

(C) $\quad \dfrac{1}{V}\left[\dfrac{\partial V}{\partial T}\right]_{S}$

(B) $\quad -\dfrac{1}{V}\left[\dfrac{\partial V}{\partial T}\right]_{S}$

(D) $\quad -\dfrac{1}{V}\left[\dfrac{\partial P}{\partial V}\right]_{S}$

Keys and Hints to MCQ Type Questions

Q.1 C	Q.4 B	Q.7 C	Q.10 D	Q.13 D	Q.16 C
Q.2 D	Q.5 D	Q.8 B	Q.11 A	Q.14 C	Q.17 C
Q.3 A	Q.6 D	Q.9 C	Q.12 C	Q.15 A	Q.18 C

The Laws of Thermodynamics

2

In this chapter, we are going to focus on the concept of the zeroth law of thermodynamics. Two different sign conventions are introduced and discussion is concluded with remark that both lead to same results. Some additional ideas pertaining to P-V diagrams, path dependence of work done and heat transfer will be introduced. Next, we will learn how work and heat are interconnected with each other and hence we will talk about the most important concept in this chapter, the first law of thermodynamics and its applications to various systems. Followed by this, concept of heat capacity will be introduced. At the end, we will consider various types of thermodynamic processes and learn about evaluating the work done under different types of constraints imposed on the system.

2.1   The Zeroth Law

The zeroth law of thermodynamics deals with the properties of systems in thermal equilibrium. The concept of temperature directly follows from it. Let us take an example of three different systems that are separately in thermal equilibrium. When systems A and B are brought together in thermal contact, thermal equilibrium exists in that no changes occur in the variables. Also suppose that the same is true for the systems A and C. It is an experimental observation that B and C would also be in thermal equilibrium if they were similarly brought together. This can be generalized to the statement of the zeroth law of thermodynamics (Fig. 2.1).

> If each of two systems is in thermal equilibrium with a third, they are in thermal equilibrium with one another.

This experimental observation is the basis of our concept of temperature. It follows from the zeroth law that a whole series of systems could be found that would be in

© The Author(s) 2022
S. Sharma, *Thermal and Statistical Physics*,
https://doi.org/10.1007/978-3-031-07685-5_2

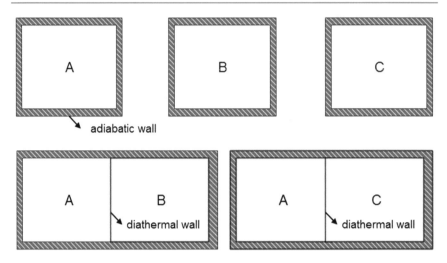

Fig. 2.1 An illustration of the zeroth law

thermal equilibrium with each other were they to be put in thermal contact-a fourth system, D, which is in thermal equilibrium with system C would also be in thermal equilibrium with A and B, and so on. All the systems possess a common property which we call the temperature, T.

> The temperature of a system is a property that determines whether or not that system is in thermal equilibrium with other systems.

2.2   Concept of Work and Sign Convention

Before we begin with the various laws of thermodynamics, we focus on understanding the concept of work. Here, we will introduce the concept of internal, external, positive and negative work, the terminology used in thermodynamics. For this purpose, we consider a system that undergoes a displacement under the action of a force. The work is said to be done (on the system) and the amount of work being equal to the product of the force and the component of the displacement along the direction of force, i.e.,

$$dW = F.dx$$

On the other hand, if a system as a whole exerts a force on its surroundings and a displacement takes place, the work that is done by the system. In both cases, the work done achieved is called ***external work***. Let us take another example of gas confined in a cylinder at uniform pressure. The gas while expanding and imparting motion to a piston does *external work* on its surroundings. However, the work done by one part of a system on another part is called ***internal work***. The interactions of molecules, atoms or electrons on one another also constitute internal work.

Note that in macroscopic thermodynamics, internal work need not be discussed. Only the work involving an interaction between a system and its surroundings has to be considered and analysed. When a system (or surrounding) does external work, the changes that take place can be described by means of macroscopic parameters (T, P or V) referring to the system as a whole. A few examples of accompanied changes in external work are

 (i) raising or lowering of a suspended weight
 (ii) the winding or unwinding of a spring
(iii) change in position of a mechanical device
(iv) extending a wire by applying an external force

One can take this as the ultimate criterion to decide whether external work is done or not. ***In rest part of the book, unless otherwise indicated, the word work will mean external work***.

2.2.1   The Sign Convention

While dealing with problems in mechanics, we are concerned with the behaviour of systems that are acted upon by external forces. As argued earlier, when the resultant force exerted on a mechanical system is in the same direction as the displacement of the system, the work of the force is positive and work is said to be done on the system, and the energy of the system increases. For thermodynamics to be consistent with mechanics, we adopt the same sign convention for work. Therefore, when work is done on the system (its internal energy content is raised), the work is taken as positive quantity. Conversely, when work is done by the system (system loses energy), the work is regarded as negative. Note that this sign convention disagrees with engineering practice, in which positive work is done by the system on an external object. On similar lines, heat supplied to the system is taken as positive whereas heat extracted from the system is taken as negative quantity. Therefore, we will ensure the use of the following rules in rest of the book.

Convention-1

 (i) Heat supplied to the system, ΔQ is positive
 (ii) Heat extracted from the system, ΔQ is negative
(iii) Work done on the system, ΔW is positive
(iv) Work done by the system on its surroundings, ΔW is negative

However, it should be kept in mind that a few books follow the conventions given below.

Convention-2

(i) Heat supplied to the system, ΔQ is positive
(ii) Heat extracted from the system, ΔQ is negative
(iii) Work done on the system, ΔW is negative
(iv) Work done by the system on its surroundings, ΔW is positive

In this case, the last two cases are different. Then the question is which one to be used? We will try to address the problem after introducing the first law of thermodynamics with some suitable examples. Note that we will stick ourselves to the first set of conventions.

2.2.2 P-V Diagrams

Let us consider a fixed mass of gas is confined to volume V as shown in Fig. 2.2. By moving the frictionless piston backward or forward the gas can be either expanded or compressed and corresponding change in pressure and volume can be monitored. One can plot the variation in pressure with volume. The resulting diagram, in which pressure is plotted along the y-axis and volume along the x-axis, is called a *P-V diagram or an indicator diagram* in engineering. A few examples of such indicator diagrams are shown in Fig. 2.3. Panel (a) represents expansion of gas from volume V_i to V_f so that $dV=V_f - V_i$ is positive and corresponding integral $-\int_{V_i}^{V_f} PdV$ is negative. Note that during expansion the pressure of the gas reduces. The integral $-\int_{V_i}^{V_f} PdV$ for the case of expansion of a gas is the shaded area under the curve labelled I. Similarly, in the case of compression, panel (b), $dV=V_f - V_i$ is negative so that the integral $-\int_{V_i}^{V_f} PdV$ becomes positive. This is in line with the adopted sign convention (1) for work. Process I represents the work done by the system and process II represents work done on the system. In panel (c) curves I and II are drawn together so that they constitute two different processes that bring the gas back to its initial state. Such a series of two or more processes representing a closed figure is called a *cycle*. In this case, the work done for process I is negative, whereas for process II it is positive. For a cycle, the net work done will be the difference between these two quantities. *The net work done in a cyclic process will be the area enclosed*

Fig. 2.2 A gaseous system confined to volume V at temperature T

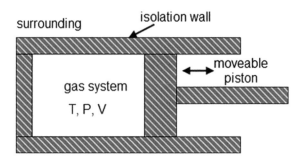

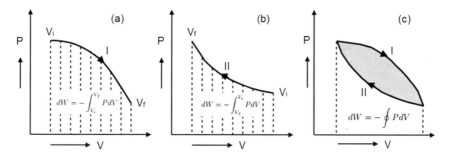

Fig. 2.3 Various PV diagrams of a gas. The shaded area represents the work done by the system or work done on the system. **a** Corresponds to expansion **b** corresponds to compression **c** processes I and II constitutes a cycle

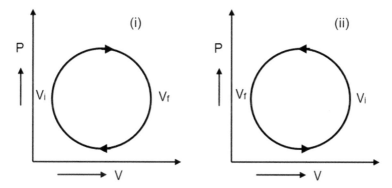

Fig. 2.4 Two cyclic processes. (i) process taken clockwise (expansion followed by compression) (ii) process traversed anticlockwise (compression followed by expansion)

by curves I and II. Note that the cycle is traversed in a direction (clockwise) such that the net work is negative, and net work is done by the system. If the cycle is taken anticlockwise (reverse the direction of processes I and II), then the net work would be positive as net work is done on the system.

Example of such processes is shown in Fig. 2.4. In case (i) net work done will be negative, whereas in case (ii) net work done will be positive.

Important Note: The discussion above is valid for sign convention-1. If we follow convention-2, then in Fig. 2.3, panel (a), work is positive, panel (b), work is negative, panel (c), net work is positive. Similarly, one can draw conclusion for cyclic processes shown in Fig. 2.4.

2.2.3   Path Dependence of Work Done

Let us take an example of PV diagram shown in Fig. 2.5. Here, initial and final equilibrium states labelled as i and f are characterized respectively by coordinates

Fig. 2.5 Diagram illustrating path dependence of work done in a thermodynamic system

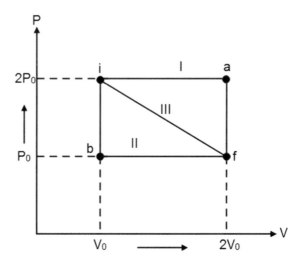

(P_i, V_i, T_i) and (P_f, V_f, T_f). As we see, the system can move from state i to f through various routes (I, II and III) shown in figure. Let us calculate the work done in moving from state i to f through path I (i to a and then a to f). The work done is

$$W_I = W_{i-a-f}$$
$$= -2P_0 (2V_0 - V_0) + 0 = -2P_0 V_0 \qquad (2.1)$$

Similarly for path II (i to b and then b to f), the work done is

$$W_{II} = W_{i-b-f}$$
$$= 0 - P_0 (2V_0 - V_0) = -P_0 V_0 \qquad (2.2)$$

For path III (i to f), the work done is

$$W_{III} = W_{i-f}$$
$$= -\frac{1}{2} (2V_0 - V_0) (2P_0 - P_0) - P_0 V_0 = -\frac{3}{2} P_0 V_0 \qquad (2.3)$$

These results clearly indicate that work done by a system depends not only on the initial and final states but also on the intermediate states. Therefore, the work done is path dependent.

2.2.4   Path Dependence of Heat-Transfer

The amount of heat entering into the system (or moving out of the system), when it passes from state i (at temp T_i) to state f (at temp T_f) also depends on how the system is heated (i.e., depends on constraints). For instance, one can heat the gas at initial pressure P_i (= $2P_0$) until temperature increases to T_f and then lower the pressure to P_f (= P_0) by keeping the temperature fixed at T_f. In an alternate way, we can first lower the pressure to P_f, and then heat it to the temperature T_f, by keeping the pressure fixed at $P_f=P_0$. In each process, a different value of $\int dQ$ will be obtained. Thus, we notice that the heat flowing into (or out of) a system depends not only on the initial and final states but also on the path of the process, i.e., intermediate states. Hence, both work and heat do depend on the path of the process and both are referred to as *imperfect differential*. It should be noted that in a special case of adiabatic process (dQ = 0, so that dU = dW), *the work done becomes independent of the path of a process and behave like a perfect differential*.

2.3   Work Done in Various Systems

Here we will take example of various systems and processes to evaluate the work done. It is always possible to express the work done on a system in terms of macroscopic parameters if the changes in which the work is performed are *thermodynamically reversible*. In such a case, the macroscopic parameters also describe the action of external forces. Let us take an example to understand this point. For instance, assume that a fluid is enclosed in a cylinder that has a piston which can move back and forth. If frictional forces (contact between piston and cylinder) are not negligible, then the force that has to be applied to the piston to overcome the friction and compress the fluid is greater than the force exerted on the fluid by the piston. Therefore, the external work done on the whole system (fluid and its container) is greater than the work done on the fluid alone and cannot be expressed in terms of the fluid's parameters of state. However, if the friction is negligible, however, the work done on the fluid becomes equal to that done by the external forces and both can be expressed in terms of the system parameters. When changes occur reversibly, the work done on a system can be written as

$$dW = \sum_i X_i x_i$$

where X_i and x_i are generalized force and displacement, respectively. We shall show that work can be described by this relation in many cases. Here we shall take a few examples to demonstrate the utility of this equation.

2.3.1   Work Done in Changing the Area of a Surface Film

We take an example of a soap film, which is stretched across a wire framework and consists of two surfaces enclosing water. As we see in Fig. 2.6a, the right side of the wire framework is movable. If the movable wire has a length L and the surface tension in one surface is γ then the external force F exerted on both surfaces is equal

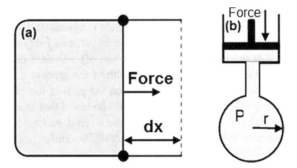

Fig. 2.6   a A soap film stretched across a rigid wire framework, having a movable wire on the right. An external force F displaces the movable wire at an infinitesimal distance dx. **b** A spherical droplet of liquid of radius r is suspended from a thin pipe connected to a piston which maintains the pressure P of the liquid

in magnitude to $2\gamma L$. For an infinitesimal displacement dx of the wire, the work done is

$$dW = 2\gamma L dx$$

Now

$$2Ldx = dA$$

where A is the total surface area. Hence, we can write

$$dW = \gamma dA$$

This result is also valid for single surface, for instance boundary between the bulk of a liquid and the gaseous environment. It is found experimentally that γ is normally independent of area and depends only on temperature, so that for a finite increase of area from A_i to A_f we can write

$$W = \int_{A_i}^{A_f} \gamma dA \qquad (2.4)$$

2.3.2   Work Done in Changing the Area of a Liquid Drop

Let us take an example of a liquid surface with surface area A (see Fig. 2.6b). Let an external force $\mathbb{F}$ is applied so that the piston moves downward by dx and surface area of the drop changes by dA. Here, the expression for the work is given by

$$dW = \mathbb{F}.dx = PdV \qquad (2.5)$$

This amount of work will be done on the liquid which is incompressible in nature. The droplet radius will therefore increase by an amount dr such that

$$dV = 4\pi r^2 dr$$

and the change in surface area of the drop is

$$dA = 4\pi (r + dr)^2 - 4\pi r^2 \approx 8\pi r dr \qquad (2.6)$$

Therefore, in terms of surface tension γ, the work done is

$$dW = \gamma dA = 8\pi \gamma r dr$$

where *gamma* is the surface tension. Equating the above expression with

$$dW = PdV = 4\pi P r^2 dr$$

yields

$$P = \frac{2\gamma}{r} \qquad (2.7)$$

The pressure P in this expression is the pressure difference between the pressure in the liquid and the atmospheric pressure against which the surface of the drop pushes.

2.4   Historical Background Related to First Law of Thermodynamics

The first law of thermodynamics is essentially an extension of the principle of the conservation of energy to include systems in which there is flow of heat. Historically, it marks the recognition of heat as a form of energy. It is interesting to approach the first law through historical developments that led to present understanding.

In the beginning of the nineteenth century, the heat was considered as an indestructible substance (caloric) which flowed from a hot body, rich in caloric, to a cold body which had less caloric. At that time, the heat was quantified by the temperature rise it produced in a unit mass of water, taken as a standard reference substance.

The experiments of Black at the end of the eighteenth century had also shown that, when two bodies were put in thermal contact, the heat lost by one in this 'method of mixtures' was equal to the heat gained by the other. This seemed to confirm that heat was a conserved entity.

Later, Benjamin Thompson while supervising the boring of cannon, noticed that large amount of heat was produced, as measured by the temperature rise in the cooling water. Further, when he made use of a blunt boring tool, he noticed that he could even boil the water, with the supply of heat being apparently inexhaustible. With these

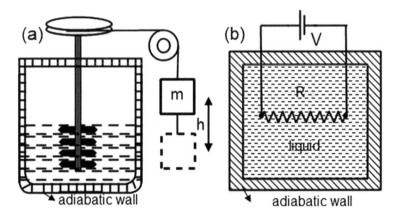

Fig. 2.7 a A schematic illustration of Joule's apparatus. **b** Another method to take away the system between two equilibrium states by performing adiabatic work

observations, he concluded that heat could not be a finite substance such as caloric and that there was a direct relation between the work done and the heat produced.

The precise relation between heat and mechanical work was established by Joule some fifty years later in a careful series of experiments between 1840 and 1849. Joule made use of a tub containing a paddle wheel which could be rotated by the action of weights falling outside the tub (See Fig. 2.7a). Water inside the tub could thus be stirred (irreversibly because of turbulence, friction between water and paddle make the process irreversible), raising its temperature between two equilibrium states. The walls of the tub were insulating, so the work was performed under adiabatic conditions. We call such work *adiabatic work*. Joule's observations are as below

(i) Joule noted that it required 4.2 kJ of work to raise the temperature of one kg of water through one degree kelvin. This is known as the mechanical equivalent of heat J.

(ii) It does not matter how the adiabatic work was performed, it always required the same amount of work to take the water system between the same two equilibrium states. He also varied the adiabatic work by changing the weights and the number of drops.

(iii) He also performed the same amount of adiabatic work electrically by allowing the current produced by an electrical generator to be dissipated in a known resistance immersed in the water.

Figure 2.7b represents another way to change the system between two equilibrium states by performing adiabatic work. Note that, it does not matter how the change is brought, system can attain same initial and final equilibrium state, irrespective of the way change is brought in.

2.5   First Law of Thermodynamics

Above results may be summarized as below *If a system which is thermally isolated from surrounding, is changed from one equilibrium state to another, then the work required to achieve this change is independent of the process used.*

This means that the adiabatic work $\Delta W_{adiabatic}$ necessary to bring the system from one equilibrium state to another is path independent. In other words, $\Delta W_{adiabatic}$ depends only on end equilibrium points. Therefore, there must exist a state function whose difference between these two end points is equal to $\Delta W_{adiabatic}$. We call this state function as internal energy U and we write as

$$\Delta W_{adiabatic} = U_f - U_i$$

Now suppose that the system we are talking about is not thermally isolated. In this case, it is found that the work done (W) in taking the system between a pair of equilibrium points is function of the path. Initial and final equilibrium points are fixed, i.e., for a given change; $\Delta U = U_f - U_i$ is fixed, but now W is not simply equal to $U_f - U_i$. In other words there exist a difference between the adiabatic work and non-adiabatic work required to bring about a same change between two equilibrium states. Latter can have an infinite number of possible values to attain same equilibrium states for which $\Delta U = U_f - U_i$.

Let us denote this difference between ΔU and ΔW as heat ΔQ and in place of the above equation we write as

$$\Delta U = \Delta Q + \Delta W \qquad (2.8)$$

Equation 2.8 gives the mathematical statement of the first law of thermodynamics. It implies that the internal energy of a system can be enhanced either by supplying heat to the system or by doing work on the system. In this form, it is true for reversible as well as irreversible processes. For a differential change, we write the above equation as

$$dU = dQ + dW \qquad (2.9)$$

Important note on sign convention: It should be noted that, Eq. 2.9 uses sign convention-1, i.e., $dW = -$ PdV, is positive (negative) when work is done on (by) the system. However, if one prefers to use second sign convention, i.e., $dW = $ PdV, is positive (negative) when work is done by (on) the system, we simply need to make a minor change in fundamental definition of the first law of thermodynamics. It has to be defined as below

$$dU = dQ - dW \qquad (2.10)$$

It is already discussed that Eq. 2.9 ensures that whenever work is done on (dW is positive, convention-1) the system, it is accompanied with increase in internal energy of the system. Let us check whether, second definition [Eq. 2.10] of first law follows this logic or not. According to **convention-2, work done on the system is taken as negative, so that the internal energy increases [dU = dQ-(-dW) = dQ + dW]**. Therefore, both definitions of the law follow the same logic and we need not worry, which one we adopt. We will take one example to clear the confusion.

Ex 2.1   **A system gains 1500 J of heat from its surroundings, and 2200 J of work is done by the system on the surroundings. In another situation, the system also gains 1500 J of heat, but now 2200 J of work is done on the system by the surroundings. In each case, determine the change in the internal energy of the system.**

Sol: We will solve this problem by making use of two sign conventions.

Sign Convention-1 Here $dQ = 1500$ J (system gains energy), $dW = -2200$ J (work done by the system, taken negative). The change in internal energy, according to the first law of thermodynamics is

$$dU = dQ + dW = 1500\,\text{J} - 2200\,\text{J} = -700\,\text{J}$$

The negative sign for dU indicates that the internal energy has decreased, as expected.

For next situation, $dQ = 1500$ J (system gains energy), $dW = 2200$ J (work done on the system, taken positive). The change in internal energy according to the first law of thermodynamics is

$$dU = dQ + dW = 1500\,\text{J} + 2200\,\text{J} = 3700\,\text{J}$$

The positive sign for dU indicates that the internal energy has increased, as expected.

Sign Convention-2 Here $dQ = 1500$ J (system gains energy), $dW = +2200$ J (work done by the system, taken positive). The change in internal energy, according to the first law of thermodynamics is

$$dU = dQ - dW = 1500\,\text{J} - (+2200\,\text{J}) = -700\,\text{J}$$

The negative sign for dU indicates that the internal energy has decreased, as expected.

For next situation, $dQ = 1500$ J (system gains energy), $dW = -2200$ J (work done on the system, taken negative). The change in internal energy, according to the first law of thermodynamics is

$$dU = dQ - dW = 1500\,\text{J} - (-2200\,\text{J}) = 3700\,\text{J}$$

The positive sign for dU indicates that the internal energy has increased, as expected. Thus, irrespective of the convention used, final answer is same. The only thing that we need to keep in mind is right use of Eq. 2.9 and Eq. 2.10 with Convention-1 and Convention-2, respectively.

2.5.1   Applying First Law to Various Systems

Here we will take two examples; stretching of a thermally isolated wire and compression of a gas in isolated cylinder. We want to apply the first law of thermodynamics in these cases. Let us continue with the first one.

Stretching a Thermally Isolated Wire

Let us consider a thermally isolated system (a wire shown in Fig. 2.8a), that is a system which cannot exchange heat ($dQ = 0$) with its surroundings. In this case, we note that Eq. 2.9 becomes

$$dU = dW$$

Because no heat can enter or leave a thermally isolated system. Let a force $\mathbb{F}$ is applied to the wire and it causes a displacement dx in it. Then the work done on the system (wire) which is thermally isolated is

$$dW = \mathbb{F}.dx$$

Therefore, the first law becomes

$$dU = dW = \mathbb{F}.dx \qquad (2.11)$$

Compression of Gas in an Isolated Container

Similar approach can be adopted to calculate the work done by compressing a gas (pressure P, volume V) by a piston (see Fig. 2.8b). In this case the force is $\mathbb{F}=PA$,

Fig. 2.8 a Stretching a thermally isolated wire. **b** A piston moving inward causes a negative change in volume (V_f-V_i=-dV)

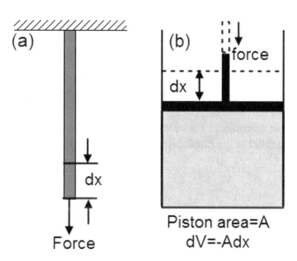

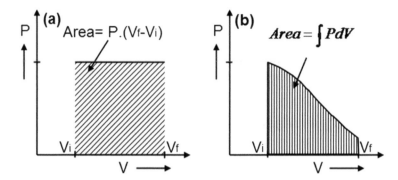

Fig. 2.9 a Evaluating work done when volume changes but pressure stays constant. **b** work done evaluation when pressure changes along with volume

where A is the area of the piston, and $Adx = -dV$, so that the work done on the gas (system) becomes

$$dW = -P(V_f - V_i) = -PAdx = PdV \qquad (2.12)$$

The negative sign in this equation ensures that the work dW done on the gas is positive when dV is negative (for compression $dV = V_f - V_i$ is negative), i.e. when the gas is being compressed. Therefore, the first law becomes

$$dU = dW = -PdV \qquad (2.13)$$

In this case also dQ=0, because we considered the system to be isolated. If it is not an isolated system, then the first law will read

$$dU = dQ + dW$$

$$or$$

$$dU = dQ - PdV \qquad (2.14)$$

2.5.2   Compression Work and Utility of Eq. 2.12

Let us consider the cylinder piston system shown in Fig. 2.8b. The applied force is parallel to the displacement dx, so that we can write the work done on the gas as

$$dW = \mathbb{F}.dx \qquad (2.15)$$

In the next step we will replace, $\mathbb{F}$ with PA, where P is the pressure exerted by the gas inside the container and A is the area of the piston. But in order to do this, we have to ensure that as the gas is compressed it is always in internal equilibrium, so that pressure is uniform from place to place. This is possible when piston moves slowly so that gas has sufficient time to continually equilibrate to changing conditions. Or in other words the process is quasi-static. Therefore, for quasi-static compression, the force exerted on the gas equals pressure times area of the piston. Therefore, (following sign convention)

$$dW = -PAdx = PdV \; (quasi\text{-}static \; compression) \qquad (2.16)$$

Here, $Adx = A(x_f - x_i) = (V_f - V_i) = -dV$, is minus (absolute) change in volume of the gas (because the volume reduces during compression). The same equation holds for expansion also. In this case, $Adx = A(x_f - x_i) = dV$, positive, so that $dW = -PdV$ as required for expansion (work done by the gas).

It should be noted that there is one flaw in using Eq. 2.16. In general, the pressure P will change during compression (or expansion). In that case, which pressure value should you use, initial, final, average or what else? When the volume change is very small (infinitesimal), then corresponding change in pressure will be negligible. But, we can always replace a large change as a bunch of small changes, one after another. Hence, if pressure changes significantly during compression, we need to mentally divide the process into many tiny steps, apply Eq. 2.16 to each small step and add all little works to get total work during a typical process. This procedure can be understood easily with the help of Fig. 2.9. If the pressure is constant, the work done (compression) is simply minus the area under a P-V graph (-PdV). If pressure is not constant, we divide the process into a bunch of small steps, compute the area under P-V graph for each step, then add up all tiny areas to get total work. In other words, the work is still minus the total area under the P verses V graph. If we know pressure as a function of volume, P(V), then one can easily evaluate the total work as an integral:

$$W = \int_{V_i}^{V_f} dW = - \int_{V_i}^{V_f} P(V) dV \; (quasi\text{-}static) \qquad (2.17)$$

This result is valid whether the pressure changes or not during the process. This result is general and is applicable for all kinds of processes involving gases.

Importance of First Law

Before we move further, let us recall the significance of the first law of thermodynamics. The following points reflect the significance of the law.

(i) This law enables us to define internal energy 'U' as a function of state of a system.
(ii) The law also emphasizes that heat is energy in transit rather than a fluid (Calorie) as believed earlier.

(iii) The law also excludes the possibility of constructing a machine that can work on its own without any external stimulus. Machine of this kind is called as perpetual motion machine of first kind.

Limitations of First Law

Some of the serious limitations of the first law of thermodynamics are mentioned below

 (i) The law does not provide any information about unidirectional flow of heat from hot to cold body. In simple words, it does not rule out the possibility of heat flow from cold body to hot body. It simply focuses on energy conservation.
(ii) The law also does not rule out the feasibility of a machine that can take heat from ambient and transforms it fully into work. Such machine is known as perpetual machine of second kind, which is physically not allowed.

2.6   Heat Capacities

The first law of thermodynamics allows us to express the internal energy change during a reversible change as

$$dU = dQ + dW \qquad (2.18)$$

or for a gaseous system, one can rewrite the above equation (recall that $dW = -PdV$)

$$dQ = dU + PdV \qquad (2.19)$$

The quantity $dQ/d\alpha$ is called a general heat capacity of the system and expresses the rate at which heat is absorbed when the variable α is changed, α may be any function of state (T, V, P, etc). As usual, the heat absorbed in a given change is not defined unless the path of the change is defined. Hence $dQ/d\alpha$ remains undefined in the absence of constraints. To define the path of a change in an n-parameter system, we need (n -1) constraints. The heat capacity is then defined and it may be written as

$$\frac{dQ_{\beta,\gamma,\ldots}}{d\alpha} = C^{\alpha}_{\beta,\gamma,\ldots} \qquad (2.20)$$

For instance, heat capacity at constant volume is defined as

$$C^T_V = \frac{dQ_V}{dT} = \left(\frac{dQ}{dT}\right)_V \qquad (2.21)$$

It is the rate at which heat is absorbed as temperature is changed by keeping the volume constant. Notice that the right side of Eq. 2.21 is not a derivative of a function, but, rather, the ratio of two small experimental quantities dQ and dT.

In the case of thermal capacities for increase of temperature, the superscript is usually omitted, as with the principal heat capacities, C_P and C_V. Clearly, the heat capacities are extensive quantities which change when mass or volume of a system is changed. Standardization occurs when an extensive quantity is divided by the mass of an arbitrary sample. The quantity when divided by mass is called as *specific quantity*. For instance, heat capacity per unit mass is called as *specific heat capacity*. This is also referred to as *specific heat*. The specific heat capacities are denoted by small letter symbols, for instance c_p and c_v are the principal specific heat capacities. It should be noted that heat capacity is an extensive variable, whereas specific heat is intensive one.

We can understand these general remarks by taking an example of a simple system subject to work by hydrostatic pressure. Using definition of work from Eq. 2.12, the first law becomes

$$dQ = dU + PdV \tag{2.22}$$

Suppose we wish to use T and V as independent variables. That is the internal energy U can be expressed as U=U(T,V). Hence a small change in U can be related to change in T and V by

$$dU = \left(\frac{\partial U}{\partial T}\right)_V dT + \left(\frac{\partial U}{\partial V}\right)_T dV \tag{2.23}$$

Rearranging the above two equations, we have

$$dQ = \left(\frac{\partial U}{\partial T}\right)_V dT + \left[P + \left(\frac{\partial U}{\partial V}\right)_T\right] dV \tag{2.24}$$

Now, if we apply the constraint that V is constant (i.e. dV=0), we obtain the usual expression for heat capacity at constant volume,

$$C_V = \left(\frac{dQ}{dT}\right)_V = \left(\frac{\partial U}{\partial T}\right)_V \tag{2.25}$$

This is an obvious result since if there is no volume change no work can be done and any change in internal energy must simply be equal to the heat entering the system. The corresponding specific heat capacity (heat capacity per unit mass) is

$$c_v = \frac{1}{m}\left(\frac{\partial U}{\partial T}\right)_V = \left(\frac{\partial u}{\partial T}\right)_v \tag{2.26}$$

However, if we keep T constant, we have a new kind of heat capacity:

$$C_T = \left(\frac{dQ}{dV}\right)_T = P + \left(\frac{\partial U}{\partial V}\right)_T \tag{2.27}$$

This is the amount of heat absorbed per unit volume increase as the system moves along an isotherm: a sort of latent heat, since temperature does not change. Further, the heat capacity at constant pressure is defined using Eq. 2.24 as

$$C_P = \left(\frac{\partial Q}{\partial T}\right)P$$
$$= \left(\frac{\partial U}{\partial T}\right)_V + \left[P + \left(\frac{\partial U}{\partial V}\right)_T\right]\left(\frac{\partial V}{\partial T}\right)_P \tag{2.28}$$

Rearranging the above equation, we get

$$C_P - C_V = \left[P + \left(\frac{\partial U}{\partial V}\right)_T\right]\left(\frac{\partial V}{\partial T}\right)_P \tag{2.29}$$

Further, we may obtain similar expressions if we choose P and T as our independent variables. That is considering U=U(T,P) and V=V(T,P), we first obtain dU and dV in terms of dP and dT:

$$dU = \left(\frac{\partial U}{\partial P}\right)_T dP + \left(\frac{\partial U}{\partial T}\right)_P dT \tag{2.30}$$

and

$$dV = \left(\frac{\partial V}{\partial P}\right)_T dP + \left(\frac{\partial V}{\partial T}\right)_P dT \tag{2.31}$$

Putting these equations in Eq. 2.22 and rearranging we get

$$dQ = \left[\left[\frac{\partial U}{\partial P}\right]_T + P\left[\frac{\partial V}{\partial P}\right]_T\right]dP + \left[\left[\frac{\partial U}{\partial T}\right]_P + P\left[\frac{\partial V}{\partial T}\right]_P\right]dT \tag{2.32}$$

In this case, the two heat capacities are given by

$$C_P = \left(\frac{\partial Q}{\partial T}\right)_P = \left[\left(\frac{\partial U}{\partial T}\right)_P + P\left(\frac{\partial V}{\partial T}\right)_P\right] \tag{2.33}$$

and

$$C_T = \left(\frac{\partial Q}{\partial P}\right)_T = \left[\left(\frac{\partial U}{\partial P}\right)_T + P\left(\frac{\partial V}{\partial P}\right)_T\right] \tag{2.34}$$

Substituting Eqs. 2.33 and 2.34 back in Eq. 2.32 we can write

$$dQ = C_T dP + C_P dT \tag{2.35}$$

2.6.1   Simpler form of C_P, Definition of Enthalpy

From the preceding discussion, we learned that C_V took a particularly simple form in terms of the internal energy:

$$C_V = \left(\frac{\partial U}{\partial T}\right)_V \tag{2.36}$$

but the other principal heat capacity C_P had a more complicated form:

$$C_P = \left(\frac{\partial U}{\partial T}\right)_P + P\left(\frac{\partial V}{\partial T}\right)_P \tag{2.37}$$

It would be more convenient to construct an energy function H, which would give C_P a form similar to Eq. 2.25. That is,

$$C_P = \left(\frac{\partial H}{\partial T}\right)_P \tag{2.38}$$

Equating Eqs. 2.37 and 2.38, we get

$$\left(\frac{\partial H}{\partial T}\right)_P = \left(\frac{\partial U}{\partial T}\right)_P + P\left(\frac{\partial V}{\partial T}\right)_P = \left[\frac{\partial}{\partial T}(U + PV)\right] \tag{2.39}$$

A suitable function is therefore

$$H = U + PV \tag{2.40}$$

H is called *enthalpy*. As all the terms on the right-hand side of Eq. 2.40 are functions of state, H must also be a function of state. Its differential form follows immediately from (2.40):

$$dH = dU + PdV + VdP$$
$$= dQ + VdP \tag{2.41}$$

Equations 2.38 and 2.41 imply that when a system undergoes a reversible isobaric change, the change in H is equal to the heat absorbed, i.e.,

$$dH = dQ_P \tag{2.42}$$

2.6.2   Molar Heat Capacity

In previous subsection, we talked about principal heat capacities (C_P and C_V) followed by a discussion on *specific heat capacities*. We also emphasized that C_P and C_V are extensive variables whereas c_v and c_p are intensive ones. Note that specific heat capacity is obtained when heat capacity is normalized (i.e., C_V/m or C_P/m) with mass of substance. There is another way of standardization of these physical quantities. This is obtained when the heat capacities are standardized to the *same amount of substance* (a different mass for each different substance) called a *mole*.

As we know, a mole (abbreviated "mol") is defined as the amount of substance that contains as many elementary entities (atoms, molecules, ions, electrons or other particles) as there are atoms in 0.012 kg of ^{12}C. This number of atoms of ^{12}C is called Avogadro's number N_A and is equal to 6.022×10^{23} particles per mole. If we assume that the mass of an atom is m, then the mass of a mole of atoms is mN_A. This quantity, the *molar mass*, is also called the "*molecular weight*". If we designate the molar mass by M, we have

$$M = mN_A$$

This gives the number of moles as

$$n = \frac{\text{total mass}}{M}$$

Further, if C is the heat capacity of n moles, then one can write the molar heat capacity c as

$$c = \frac{C}{n} = \frac{1}{n}\frac{dQ}{dT}$$

It is expressed in terms of $JK^{-1}mol^{-1}$. Note that the specific heat and the molar heat capacity are expressed as lower-case c, whereas the heat capacity of an arbitrary sample is expressed as capital C. All three quantities are state functions, because they are a measure of the change of internal energy in an isochoric process.

2.6.3   Specific Heats of Gas

After giving a general definition of heat capacity via Eq. 2.20, we note that gases being compressible in nature, their heat capacity or specific heat (C/m) may vary from 0 to ∞. For instance, if a gas is compressed, its temperature rises without supply of heat, i.e., dQ=0 and $C = \frac{dQ}{dT} = 0$. On the other hand if a gas expands freely, without any change in temperature, i.e., dT=0, then $C \longrightarrow \infty$. For this reason, heat capacity of a gas is defined by considering either T or P constant. Thus gases have two heat capacities, C_V and C_P.

C_P is Greater than C_V

Suppose heat is supplied to a gas which is expanding at constant pressure. The heat supplied is used in doing two tasks: (i) first it raises the temperature of the gas (increase in internal energy of the gas) and (ii) it also does work in expanding the gas against external pressure (as $dQ = dU + PdV$ for expansion). On the other hand when gas is heated at constant volume, no work is done by the gas and entire heat supplied is used to raise the temperature of the gas and hence its internal energy. Hence, more heat is required when process is carried out at constant pressure rather than at constant volume.

2.7 Principle of Calorimetry

Let us consider two different systems A and B in contact with an adiabatic wall which surrounds both systems. For both systems, application of the first law gives

$$U_f^A - U_i^A = Q^A + W^A$$
$$U_f^B - U_i^B = Q^B + W^B$$

here, A and B refer to two different systems and i and f refer to initial and final states of system. When both systems form a composite system, we can add the above equations to obtain

$$\left(U_f^B + U_f^A\right) - \left(U_i^B + U_i^A\right) = \left(Q^B + Q^A\right) + \left(W^A + W^B\right)$$

where $\left(U_f^B + U_f^A\right) - \left(U_i^B + U_i^A\right)$ is the net change in internal energy of the composite system and $\left(W^A + W^B\right)$ is the work done on it. As the composite system is isolated, net heat flow is zero, i.e., $\left(Q^B + Q^A\right)=0$, i.e.,

$$Q^A = -Q^B \tag{2.43}$$

That is heat lost by system B is equal to heat gained by system A. This is the principle of calorimetry, which says that total energy is conserved.

2.8 Various Thermodynamic Processes, Revisiting the First Law

2.8.1 Isothermal Process

A process which is carried out at constant temperature is called as an *isothermal* process. In an isothermal process,

$$dT = 0$$

For an ideal gas, we have the relation $dU = C_V dT$, and hence this implies that for an isothermal process

$$dU = C_V dT = 0$$

Because, internal energy U is a function of T only, hence, dU=0. Therefore, for this case, the first law becomes

$$dU = dQ + dW = 0$$
$$OR$$
$$dW = -dQ \tag{2.44}$$

Work Done During Isothermal Process

Let us consider a special case where an ideal gas undergoes isothermal change from volume V_i to V_f. The work done for this case can be calculated as below

$$W = \int_{V_i}^{V_f} dW = -\int_{V_i}^{V_f} P dV = -Nk_B T \int_{V_i}^{V_f} \frac{1}{V} dV$$
$$= -Nk_B T \ln \frac{V_f}{V_i} = Nk_B T \ln \frac{V_i}{V_f} \tag{2.45}$$

Equation 2.45 gives general expression for work done during a isothermal process. When ideal gas is compressed at constant T, $V_f < V_i$. In that case work done on the gas is positive ($\ln \frac{V_i}{V_f}$ is positive). On the other hand for expansion $V_f > V_i$, so that work done by the gas is negative ($\ln \frac{V_i}{V_f}$ is negative) as required by sign convention adopted.

Heat Changes During Isothermal Process

Let us now evaluate the heat exchange between system (ideal gas here) and surrounding. Using Eqs. 2.44 and 2.45

$$W = \int dW = -\int dQ$$
$$or$$
$$Q = Nk_B T \ln \frac{V_f}{V_i} \tag{2.46}$$

For isothermal expansion, $V_f > V_i$, so that Q>0, therefore an ideal gas absorbs heat $Q = Nk_B T \ln \frac{V_f}{V_i}$ during expansion. On the other hand during isothermal compression of the gas the heat $Q = Nk_B T \ln \frac{V_f}{V_i} < 0$ flow out into the environment.

Therefore, we conclude that for compression, Q is negative as heat leaves the gas; for isothermal expansion, heat must enter the gas as Q is positive.

Sign convention-2

In case if we restrict our self to sign convention-2, then Eq. 2.45 reduces to

$$W = \int_{V_i}^{V_f} dW = \int_{V_i}^{V_f} P dV$$

$$= Nk_B T \int_{V_i}^{V_f} \frac{1}{V} dV = Nk_B T \ln \frac{V_f}{V_i}$$

Similarly, heat changes during isothermal expansion (work done by system, +ve in convention-2) process become [here we will use dU=dQ-dW=0 and dW=PdV].

$$Q = \int dQ = \int dW = Nk_B T \ln \frac{V_f}{V_i} \tag{2.47}$$

Therefore, final result is same in both cases.

2.8.2 Adiabatic Process

A process that involves no heat exchange between system and surrounding is known as an adiabatic process. Let us take an example of adiabatic expansion of an ideal gas. In an adiabatic, there is no flow of heat and we have

$$dQ = 0$$

The first law therefore implies that

$$dU = dW \tag{2.48}$$

Now for an ideal gas, $dU = C_V dT$ and $dW = -PdV$. Thus for one mole of an ideal gas,

$$C_V dT = -PdV = -\frac{RT}{V} dV$$

If we assume that during adiabatic expansion the volume changes from V_1 to V_2 and corresponding temperature change is from T_1 to T_2, then after rearranging above equation we get

$$\ln \frac{T_2}{T_1} = -\frac{R}{C_V} \ln \frac{V_2}{V_1} \tag{2.49}$$

For an ideal gas, $C_P = C_V + R$, after dividing by C_V we obtain

$$\gamma = \frac{C_P}{C_V} = 1 + \frac{R}{C_V}$$

Above equations lead to

$$\ln \frac{T_2}{T_1} = (1 - \gamma) \ln \frac{V_2}{V_1}$$
$$= \ln \left(\frac{V_2}{V_1} \right)^{1-\gamma} = \ln \left(\frac{V_1}{V_2} \right)^{\gamma-1} \qquad (2.50)$$

After rearranging we get

$$T_2 V_2^{\gamma-1} = T_1 V_1^{\gamma-1}$$

or

$$T V^{\gamma-1} = constant \qquad (2.51)$$

For an ideal gas, $PV \propto T$, thus

$$PV V^{\gamma-1} = constant$$
$$or$$
$$PV^{\gamma} = constant \qquad (2.52)$$

Or equivalently, in terms of T and P we can write $V \propto T/P$, therefore Eq. 2.51 yields

$$T \left(\frac{T}{P} \right)^{\gamma-1} = T^{\gamma} P^{1-\gamma} = constant \qquad (2.53)$$

Adiabatic Verses Isothermal Processes on Same P-V Diagram

An ideal gas that undergoes isothermal process will satisfy the equation PV=RT=constant. That is

$$P = \frac{RT}{V}$$

or

$$\left[\frac{dP}{dV} \right]_{iso} = -\frac{1}{V^2} RT = -\frac{P}{V} \qquad (2.54)$$

Fig. 2.10 a Indicator diagram for an isothermal and adiabatic process. **b** Indicator diagram for two adiabatice processes corresponding to monoatomic [2] and diatomic [1] gases

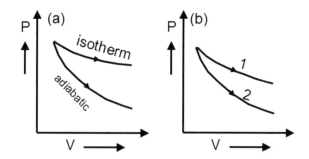

Similarly, if gas undergoes an adiabatic change, then it is bound to satisfy Eq. 2.52. One can rewrite this equation as $P = \dfrac{C}{V^\gamma}$, where C is a constant. Therefore, for a small change

$$\left[\frac{dP}{dV}\right]_{adia} = -\gamma \frac{C}{V^{\gamma+1}} = -\frac{\gamma}{V}\frac{C}{V^\gamma} = -\gamma\left(\frac{P}{V}\right) \qquad (2.55)$$

Or we can rewrite it as

$$\left[\frac{dP}{dV}\right]_{adia} = \gamma\left[\frac{dP}{dV}\right]_{iso} \qquad (2.56)$$

Since γ is larger than one, adiabatic curve is more steeper than an isothermal curve. This is shown in Fig. 2.10 panel a.

P-V Diagram for Monoatomic and Diatomic Gases Undergoing an Adiabatic Change

For an adiabatic process, we just noted that

$$\left[\frac{dP}{dV}\right]_{adia} = \gamma\left[\frac{dP}{dV}\right]_{iso} \qquad (2.57)$$

For monoatomic gas (He say), $\gamma = 5/3 = 1.6$ and for a diatomic gas (O_2 say), $\gamma = 7/5 = 1.4$. Therefore, slope of P-V curve for a monoatomic gas will be steeper than that for a diatomic gas. Hence, in Fig. 2.10 panel b, curve 2 corresponds to He and curve 1 corresponds to O_2.

Work Done During Adiabatic Expansion

When an ideal gas undergoes adiabatic change, P and V obey the relationship

$$PV^\gamma = constant$$

$$or$$

$$P = \frac{C}{V^\gamma} \tag{2.58}$$

The adiabatic work done can be evaluated with the help of Eq. 2.17.

$$
\begin{aligned}
W &= -\int_{V_1}^{V_2} P dV = -\int_{V_1}^{V_2} \frac{C}{V^\gamma} dV \\
&= -C \left. \frac{V^{-\gamma+1}}{-\gamma+1} \right|_{V_1}^{V_2} = -\frac{C}{1-\gamma} \left[V_2^{1-\gamma} - V_1^{1-\gamma} \right] \\
&= -\frac{C}{1-\gamma} \left[\frac{1}{V_2^{\gamma-1}} - \frac{1}{V_1^{\gamma-1}} \right] = -\frac{C}{1-\gamma} \left[\frac{P_2}{P_2 V_2^{\gamma-1}} - \frac{P_1}{P_1 V_1^{\gamma-1}} \right] \\
&= \frac{1}{\gamma-1} [P_2 V_2 - P_1 V_1] = \frac{nR}{\gamma-1} [T_2 - T_1]
\end{aligned} \tag{2.59}
$$

Note that while writing the last expressions we have made use of Eq. 2.52 and fact that an ideal gas obeys PV = nRT equation. Above equation gives the work done during an adiabatic process. Note that if we follow the convention-2, the final expression for work done will be

$$
\begin{aligned}
W &= \int_{V_1}^{V_2} P dV = \int_{V_1}^{V_2} \frac{C}{V^\gamma} dV \\
&= \frac{1}{\gamma-1} [P_1 V_1 - P_2 V_2] = \frac{nR}{\gamma-1} [T_1 - T_2]
\end{aligned} \tag{2.60}
$$

The readers can reproduce this result themselves following the same steps as above.

2.8.3 Free Expansion

If a closed system, say a gas, expands in such a way that no heat enters or leaves the system (adiabatic process) and also no work is done by or on the system, then the expansion is called the '*free expansion*'.

Consider a thermally isolated container divided into two parts, one containing a gas and the other evacuated. When the partition is suddenly removed, the gas molecules rush into the vacuum and expand freely. Let U_f and U_i be the final and initial internal energies of the gas. Applying the first law of thermodynamics we obtain

$$U_f - U_i = Q + W$$

Since the container is thermally isolated, no heat enters or leaves the system, i.e., Q=0. Further, the walls of the container are rigid and the gas expands into the vacuum, there is no external work done by the gas, i.e., W=0. Hence, we can rewrite the first law as

$$U_f - U_i = 0$$

$$or$$

$$U_f = U_i$$

Hence, for such an expansion in vacuum, the initial and the final internal energies are equal. We call such an expansion as *free expansion*.

2.8.4   Cyclic Process

Consider a closed system which is taken from an initial to a final equilibrium state by one or more processes, and then brought back to the initial state by some other one or more processes. During this cyclic process, system comes back to same initial equilibrium state and the net change in internal energy of the system is zero (i.e., $\Delta U = 0$). This is because the internal energy of the system depends only on the initial and final state. Hence, the first law of thermodynamics allows us to write

$$0 = \Delta Q + \Delta W$$

Or in differential form one can write

$$0 = dQ + dW$$

So that net work done during a cyclic process becomes

$$W = \oint dW = - \oint dQ \tag{2.61}$$

Therefore, for a closed system which is undergoing through a cyclic process, the cyclic (closed) integral of heat is equal to the cyclic (closed) integral of work.

2.8.5   Isochoric Process

A process which takes place at constant volume (i.e., $\Delta V = 0$) is called an 'isochoric' process. In such a process, the work done on or by the system is always zero (W = 0). Therefore, applying the first law of thermodynamics, we get

$$dU = dQ + dW = dQ$$

$$or$$

$$\Delta U = \int dQ = Q$$

Therefore, we notice that in an isochoric process, the heat added to (or taken from) the system increases (or decreases) the internal energy of the system.

2.8.6 Isobaric Process

A few natural processes take place at constant pressure. For instance, the conversion of boiling water to steam or the freezing of water to ice takes place at a constant pressure (and also at a constant temperature). Such processes, which take place at constant pressure are called as 'isobaric processes'.

Work Done and Internal Energy Change in an Isobaric Process

Boiling of Water

As an example of isobaric process, we consider boiling of water that takes place at $100\,^{\circ}$C and at atmospheric pressure. Let m be the mass of water and assume that in liquid state the volume be V_i and in vapour state it is V_f. Let L be the latent heat of vaporization. The heat absorbed by the mass m during the change of state (liquid to vapour) is

$$Q = mL$$

When water changes into vapours (at constant background pressure), its volume increases (V_l to V_{vap}). Therefore, the work done by the mass m in expanding from volume V_l to volume V_{vap} against the constant external pressure P is

$$W = -\int_{V_l}^{V_{vap}} P\,dV = -P\int_{V_l}^{V_{vap}} dV = -P\left[V_{vap} - V_l\right]$$

Let ΔU be the change in internal energy, then from the first law of thermodynamics we follow

$$\Delta U = Q + W = mL - P\left[V_{vap} - V_l\right] \qquad (2.62)$$

This is the required expression for the change in the internal energy of the system.

Freezing of Water

As an another example, we will consider freezing of water into ice at $0\,^{\circ}$C and atmospheric pressure. When this process takes place, a definite amount of heat (Latent) is given out by water. It is known that ice occupies a larger volume than an equal mass of water (i.e., water expands on freezing). Therefore, during freezing process water does work (loose heat) against the atmospheric pressure. Let L be the latent heat of fusion, then for mass m of water, the heat released is

$$Q = -mL$$

Note the negative sign here, because water loses heat during freezing. The external work done (by water in changing its volume from V_l to V_{ice}) is

$$W = -P \int_{V_l}^{V_{ice}} = -P\,[V_{ice} - V_l] \qquad (2.63)$$

Therefore, using the first law, the internal energy change is

$$\Delta U = Q + W = -mL - P\,[V_{ice} - V_l] \qquad (2.64)$$

2.8.7   Quasi-static Process

Consider an ideal gas confined to a cylinder so that at any instant of time, T, P and V characterize it. That is the gas is in equilibrium and has well-defined values for the variables. Now, suppose a slight change (δP) in P takes place, so that $\delta P <<< P$ and one can assume the system to be in equilibrium. Such a process during which a system is never far away from an equilibrium state is a 'quasi-static' process. For instance, one can obtain a quasi-static expansion of gas by an infinitesimal change in pressure or some other variable. In fact, a quasi-static process is an ideal process which can never be realized practically. But it can be approximated under some circumstances.

2.8.8   Reversible Process

A process when reversed in such a way that all the changes taking place in the forward direction are exactly repeated when process is operated in reverse direction. For instance, suppose a direct process takes place by supplying an amount of heat to a system and certain amount of work is obtained from it. When process is reversed, same amount of heat should be obtainable by performing the same amount of work on the system.

Conditions for Reversibility

A reversible process is subjected to satisfy the following conditions:

(i) *Absence of dissipative forces such as electrical resistance, friction, viscosity, inelasticity and hysteresis, etc.* Let us take an example to illustrate this point in detail. Consider a gas contained in a chamber fitted with a piston is placed in contact with a constant-temperature source. If we assume that the pressure exerted by the piston on the gas is exactly balanced by the pressure of the gas on the piston, so that it does not move. Now, if we reduce the load on the piston, the gas will expand by pushing up the piston and hence doing an external work against the frictional forces between piston and wall of the chamber. The gas acquired necessary heat for performing external work from the constant temperature source. Now, for a time being assume that frictional forces are removed (absent) and piston is moving inward and has compressed the gas. The work used in (during expansion) pushing the piston outward is now recovered. Note that if frictional force is present, more work need to be done. The expansion is therefore irreversible in this case. Similarly, other dissipative forces cause the processes to be irreversible. (ii) *Process must be quasi-static.* Suppose during expansion, gas does W amount of work. A part of this work imparts kinetic energy to the piston, and this work cannot be recovered during the reverse process. Contrary to this, more work is required to be done to give kinetic energy to the piston. Therefore, to make this process a reversible the pressure difference, i.e., pressure exerted by gas on piston and that exerted by the piston on the gas should differ infinitesimally. If that is the case, the expansion or compression will take place infinitely slowly and no kinetic energy will be produced. Such conditions can never be realized in practice. Therefore, a reversible process is only an ideal concept.

2.8.9   Irreversible Process

In simple words, any process which is not reversible is an irreversible process. All practical processes for instance free expansion, Joule–Thomson expansion, electrical heating of wire, boiling of an egg, diffusion of liquids or gases, etc. are irreversible. All natural processes such as flow of water, bursting of a tyre, conduction, radiation, radioactive decay, etc. are also irreversible.

Diffusion of Two Different Gases
When two or more gases diffuse into one another, there is a change in chemical composition which makes the process irreversible. Similarly when ink drop is thrown into clean water, it diffuses and spread, again composition changes occur making this process irreversible.

Heat Transfer from Hot to Cold Body
This is another example of an irreversible process. Because heat cannot be transferred back from the cold body to the hot body without producing any change somewhere else.

Heating a Wire Electrically
This is another example of an irreversible process. As the electrical energy used for producing heat energy cannot be converted back into electrical energy.

Heat Transfer by Radiation
Thermal radiation from a hot body cannot be radiated back to the hot body without producing some change elsewhere in the universe, hence it is also an example of irreversible process.

Free Expansion or Joule Expansion of a Perfect Gas

In free expansion of an ideal gas, the volume of the gas changes from some initial volume V_i to final volume V_f without any change in temperature of the gas. To bring the gas back to its initial equilibrium state, it needs to be compressed by some external mean. The external work performed on the gas would heat the gas (because process of expansion took place in isolation). To bring the gas into initial equilibrium state, this heat has to be converted completely into work without altering the surrounding. The last step is not feasible, hence the process of free expansion is irreversible.

Joule–Thomson Expansion

This is another example of irreversible change and will be discussed in detail in Chap. 8 of this book.

2.9   Solved Problems

Q.1 **A gas expands from state A to state B by three different paths shown in Fig. 2.11 a. Let W_1, W_2 and W_3 be the work done by the gas along three different paths, respectively, then**

(A) $W_1 > W_2 > W_3$ (C) $W_1 = W_2 = W_3$

(B) $W_1 < W_2 < W_3$ (D) $W_1 < W_2, W_1 = W_3$

Sol: On an indicator diagram, work done (irrespective of sign) is equal to area under a curve. Therefore, $W_1 < W_2 < W_3$. B is CORRECT.

Q.2 **Starting with same initial conditions, an ideal gas expands from volume V_1 to V_2 in three different ways. The work done by the gas is W_1 if the process is purely isothermal, W_2 if purely isobaric, W_3 if purely adiabatic. Then**

(A) $W_2 > W_1 > W_3$ (C) $W_1 > W_2 > W_3$

(B) $W_2 > W_3 > W_1$ (D) $W_1 > W_3 > W_2$

Sol: The required conditions are illustrated in Fig. 2.11b, c. Therefore, we note that $W_2 > W_1 > W_3$. Option A is CORRECT.

Q.3 **In a given process for an ideal gas, $dW = 0$ and $dQ < 0$. Then for the gas**

(A)   temperature will decrease   (C)   pressure will remain constant

(B)   volume will increase   (D)   temperature will increase

Sol: Given, $dW = 0$ and $dQ < 0$, therefore $Q_f - Q_i < 0$. Or $Q_f < Q_i$. That is heat is flowing out from system. Therefore, $T_f < T_i$. A is CORRECT.

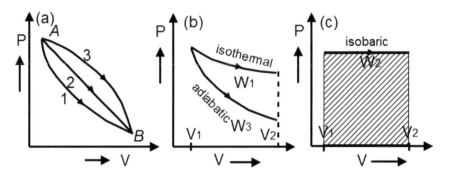

Fig. 2.11 Diagram **a** for Q.1 and **b**, **c** for Q.2

Q.4 **One mole of a monoatomic perfect gas initially at temperature T_0 expands from volume V_0 to $2V_0$, (a) at constant temperature (b) at constant pressure. Calculate the work of expansion and heat absorbed in each case.**

Sol: (a) At constant temperature T_0, the work done is

$$W = -\int_{V_0}^{2V_0} P\,dV = -RT_0 \int_{V_0}^{2V_0} \frac{dV}{V} = -RT_0 \ln 2$$

At constant temperature, the change in internal energy is zero. Therefore, the first law allows us to write

$$Q = -W = RT_0 \ln 2$$

(b) At constant pressure, the work done is

$$W = -\int_{V_0}^{2V_0} P\,dV = -PV_0 = -RT_0$$

The change in internal energy will be

$$\Delta U = \frac{3}{2}R\Delta T = \frac{3}{2}P\Delta V = \frac{3}{2}PV_0 = \frac{3}{2}RT_0$$

Using First law

$$\Delta U = \Delta W + \Delta Q$$

$$\Delta Q = \Delta U - \Delta W = \frac{3}{2}RT_0 - (-RT_0) = \frac{5}{2}RT_0$$

Q.5 **For a diatomic ideal gas at room-temperature, what fraction of the heat supplied is available for external work if the gas is expanded at constant pressure? At constant temperature?**

Sol: For expansion at constant pressure, the work done by the gas is

$$W = -P\Delta V = -P\,(V_2 - V_1) = -nR\Delta T$$

The increase in internal energy of the system is

$$\Delta U = C_v \Delta T$$

Therefore, fraction of heat supplied available for external work is

$$\frac{W}{Q} = \frac{W}{\Delta U - W} = \frac{nR\Delta T}{C_v \Delta T + nR\Delta T} = \frac{nR}{C_v + nR} = \frac{2}{7}$$

In second case when expansion takes place at constant temperature, the internal energy does not change. Therefore,

$$\frac{W}{Q} = \frac{W}{W} = 1$$

Q.6 **(a) An ideal gas undergoes quasi-static expansion under isothermal conditions. Show that the work done by an ideal gas during such expansion from an initial pressure P_i to a final pressure P_f is given by**

$$W = nRT \ln \frac{P_f}{P_i} \qquad (2.65)$$

(b) Calculate the work done when the pressure of 1 mol of an ideal gas is decreased quasi-statically from 20 to 4 atm under isothermal condition (20°C), (R = 8.31J mol^{-1}deg^{-1}).

Sol: (a) The work done during isothermal expansion of an ideal gas is given by Eq. 2.45 as below

$$W = Nk_B T \ln \frac{V_i}{V_f}$$

Here, N is the total number of ideal gas molecules. In terms of number of mole (n) and Avogadro's number (N_A), we can rewrite n=N/N_A, so that the above equation reduces to

$$W = nN_A k_B T \ln \frac{V_i}{V_f} = nRT \ln \frac{V_i}{V_f}$$

Since ideal gas undergoes quasi-static isothermal expansion, it must satisfy the equation of state. This implies $P_i V_i = P_f V_f$. Therefore,

$$\frac{V_i}{V_f} = \frac{P_f}{P_i}$$

Hence, work done is

$$W = nN_A kT \ln \frac{P_f}{P_i}$$

(b) here, n = 1, R = 8.31 J mol^{-1}deg^{-1}, T = 273 + 20 = 293 K, P_i = 20 atm and P_f = 4 atm. Therefore, the work done is

$$W = 1 \times 8.31 \times 293 \ln \frac{1}{4} = -3.37 \times 10^3 \text{ J}$$

Q.7 A real gas (n =1 mol) obeying the van der Walls equation of state

$$\left(P + \frac{a}{V^2}\right)(V - b) = RT$$

undergoes isothermal expansion quasi-statically. Calculate the work done upon expansion from volume V$_i$ to a volume V$_f$.

Sol: The work done for a process undergoing quasi-static change is given by

$$W = -\int P dV = -RT \int_{V_i}^{V_f} \frac{dV}{V - b} + a \int_{V_i}^{V_f} \frac{1}{V^2} dV$$
$$= -RT \ln \left[\frac{V_f - b}{V_i - b}\right] - a \left[\frac{1}{V_f} - \frac{1}{V_i}\right]$$

NOTE: In case if we follow convention 2, the work done will be

$$W = -\int P dV = RT \ln \left[\frac{V_f - b}{V_i - b}\right] + a \left[\frac{1}{V_f} - \frac{1}{V_i}\right]$$

Q.8 An ideal gas is made to undergo a cyclic process as shown in Fig. 2.12 a. For each of the steps A, B and C determine whether each of the following is positive, negative or zero.
(i) the work done on the gas
(ii) the change in energy content of the gas
(iii) the heat added to the gas
(iv) then determine the sign of each quantity for whole cycle

Sol: **Step-A**
(i) Let volume changes from V_1 to V_2, so that the work done on the gas is

$$\Delta W = -\int_{V_1}^{V_2} P dV = -P(V_2 - V_1) < 0 \qquad (2.66)$$

(ii) The change in internal energy while step A is performed. Since initial

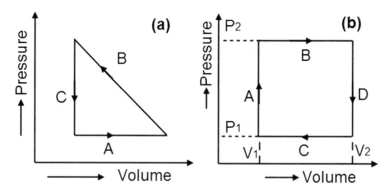

Fig. 2.12 P-V diagram **a** for Q.8 and **b** for Q.9

and final states are different, and process is performed at constant pressure. The change in internal energy will occur. From equipartition theorem,

$$U = N f \frac{1}{2} k_B T$$

N is total number of gas molecules, f are degrees of freedom and k_B is Boltzmann constant. The change in internal energy of an Ideal gas is

$$\Delta U = N f \frac{1}{2} k_B \Delta T$$

Now,

$$T = \frac{PV}{N k_B}$$

$$\Delta T = \frac{1}{Nk} \Delta (PV)$$

Therefore,

$$\Delta U = \frac{f}{2} \Delta (PV) = \frac{f P}{2} (V_2 - V_1) > 0$$

(iii) Heat added to the system during step A

$$\Delta U = \Delta Q + \Delta W$$
$$\Delta Q = \Delta U - \Delta W$$
$$= \frac{f P}{2} (V_2 - V_1) + P (V_2 - V_1) = P_1 (V_2 - V_1) \left[1 + \frac{f}{2} \right] > 0$$

STEP-B

It occur at constant volume, dV=0

(i) ΔW=0, because $\Delta V = 0$

(ii)The internal energy change is

$$\Delta U = \frac{f}{2}\Delta(PV) = \frac{f}{2}V_2(P_2 - P_1) > 0$$

(iii) Heat energy change is

$$\Delta U = \Delta Q + \Delta W = \Delta Q + 0 = \frac{f}{2}V_2(P_2 - P_1) > 0$$

STEP-C

For path C, the pressure can be written as

$$P(V) = P_1 + \alpha(V - V_1)$$

where $\alpha = \dfrac{P_2 - P_1}{V_2 - V_1}$

Q.9   **An ideal diatomic gas, in a cylinder with movable piston undergoes a rectangular cyclic process shown in Fig. 2.12 b. Assume the temperature is such that rotational degrees of freedom are active but vibrational degrees are frozen. For each of the four steps A, B, C and D find the work done on gas, heat added and change in energy content of the gas.**

Sol: **STEP-A**

(i) $W = 0$, because $\Delta V = 0$

(ii) Internal energy change is

$$\Delta U = \frac{f}{2}\Delta(PV) = \frac{f}{2}V_1(P_2 - P_1) > 0$$

(iii) Now $\Delta Q = \Delta U = \frac{f}{2}V_1(P_2 - P_1) > 0$, where f=5 for diatomic gas.

STEP-B

(i) W=$-P_2(V_2 - V_1) < 0$

(ii) $\Delta U = \frac{f}{2}P_2(V_2 - V_1) > 0$

(iii) $\Delta Q = \Delta U - W = \left(\frac{f}{2} + 1\right)P_2(V_2 - V_1) > 0$

STEP-C

(i) W=0, because $\Delta V = 0$

(ii) $\Delta U = \frac{f}{2}V_2(P_1 - P_2) < 0$

(iii) $\Delta Q = \Delta U - W = \frac{f}{2}V_2(P_1 - P_2) < 0$

STEP-D

(i) $W = -P_1(V_1 - V_2) > 0$

(ii) $\Delta U = \frac{f}{2}P_1(V_1 - V_2) < 0$

(iii) $\Delta Q = \Delta U - W = \dfrac{f}{2} P_1 (V_1 - V_2) + P_1(V_1 - V_2) = \left(\dfrac{f}{2} + 1 \right)$

$P_1(V_1 - V_2) < 0$

Q.10 An ideal gas is compressed adiabatically from an initial volume V to a final volume αV and a work W is done on the system in doing so. The final pressure of the gas will be $\left[\gamma = \dfrac{C_p}{C_v} \right]$

[JEST-2015]

(A) $\dfrac{W}{V^\gamma} \dfrac{1 - \gamma}{\alpha - \alpha^\gamma}$
(B) $\dfrac{W}{V^\gamma} \dfrac{\gamma - 1}{\alpha - \alpha^\gamma}$

(C) $\dfrac{W}{V} \dfrac{1 - \gamma}{\alpha - \alpha^\gamma}$
(D) $\dfrac{W}{V} \dfrac{\gamma - 1}{\alpha - \alpha^\gamma}$

Sol: The work done during adiabatic process is

$$W = - \int_{V_1}^{V_2} P dV = \dfrac{1}{\gamma - 1} [P_2 V_2 - P_1 V_1]$$

Because, $P_2 V_2^\gamma = P_1 V_1^\gamma$, this implies

$$P_1 = P_2 \left(\dfrac{V_2}{V_1} \right)^\gamma = P_2 \alpha^\gamma$$

Therefore,

$$W = \dfrac{P_2 \alpha V - P_2 \alpha^\gamma V}{\gamma - 1}$$

$$P_2 = \dfrac{W}{V} \left[\dfrac{\gamma - 1}{\alpha - \alpha^\gamma} \right] \qquad (2.67)$$

NOTE:
In case we follow the other sign convention for evaluating the work done

$$W = \int_{V_1}^{V_2} P dV = \dfrac{1}{1 - \gamma} [P_2 V_2 - P_1 V_1]$$

Then the final pressure will be

$$P_2 = \dfrac{W}{V} \left[\dfrac{1 - \gamma}{\alpha - \alpha^\gamma} \right]$$

Q.11 **The equation of state for one mole of a non-ideal gas is given by**

$$PV = A\left(1 + \frac{B}{V}\right)$$

where the coefficient A and B are temperature dependent. If the volume changes from V_1 to V_2 in an isothermal process, the work done by the gas is

[JAM-2018]

(A) $AB\left(\dfrac{1}{V_1} - \dfrac{1}{V_2}\right)$　　　　**(C)** $A\ln\left(\dfrac{V_2}{V_1}\right) + AB\left(\dfrac{1}{V_1} - \dfrac{1}{V_2}\right)$

(B) $AB\ln\dfrac{V_2}{V_1}$　　　　　　　**(D)** $A\ln\left(\dfrac{V_2 - V_1}{V_1}\right) + B$

Sol: From given equation, we obtain P as

$$P = \left[\frac{A}{V} + \frac{AB}{V^2}\right]$$

Therefore work done by gas during isothermal expansion is

$$W = -\int_{V_1}^{V_2} P\,dV$$

$$= -\int_{V_1}^{V_2} \frac{A}{V}\,dV - \int_{V_1}^{V_2} \frac{AB}{V^2}\,dV = -A\ln\left[\frac{V_2}{V_1}\right] + \left[\frac{AB}{V}\right]_{V_1}^{V_2}$$

$$= -A\ln\left[\frac{V_2}{V_1}\right] + AB\left[\frac{1}{V_2} - \frac{1}{V_1}\right]$$

This is the required work done. However, we note that none of the results matches with this. This is because we have obtained result using sign convention-1. If we use convention-2, then final result will be

$$W = \int_{V_1}^{V_2} P\,dV = A\ln\frac{V_2}{V_1} + AB\left[\frac{1}{V_1} - \frac{1}{V_2}\right]$$

Q.12 **In the thermodynamic cycle shown in the Fig. 2.13 (a), one mole of a monoatomic ideal gas is taken through a cycle. AB is a reversible isothermal expansion at a temperature of 800 K in which the volume of the gas is doubled. BC is an isobaric contraction to the original volume in which the temperature is reduced to 300 K. CA is a constant volume process in**

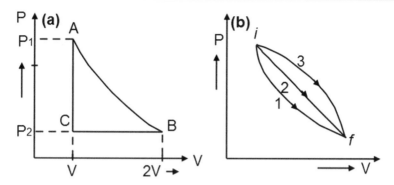

Fig. 2.13 P-V diagram **a** for Q.11 and **b** for Q.12

which the pressure and temperature return to their initial values. The net amount of heat (in Joules) absorbed by the gas in one complete cycle is ....

[JAM-2015]

Sol: For isothermal expansion A-B, $T_A = T_B = 800$ K. At A let pressure is P_A and $V_A = $ V, So that $P_B = P_A/2$ and $V_B = 2$V given. Since process is isothermal expansion, internal energy will not change. Therefore, heat absorbed during isothermal expansion will be

$$Q_1 = nRT_A \ln\left[\frac{V_B}{V_A}\right] = 4608 \text{ J}$$

For isobaric (contraction) process, B-C, $P_B = P_C = P_A/2$, $V_C = V_A$ and $T_C = 300$ K. The heat rejected is

$$Q_2 = nC_P\Delta T = n\frac{\gamma R}{\gamma - 1}\Delta T = \left[\frac{\gamma R}{\gamma - 1}\right] R[300 - 800] = -10344 J$$

Note, here n=1, $\gamma = 5/3$. Similarly for isochoric process C-A, $T_C=300$ K, $T_A=800$ K.

$$Q_3 = nC_V\Delta T = n\frac{R}{\gamma - 1}\Delta T = \left[\frac{R}{\gamma - 1}\right] R[800 - 300] = 6194 J$$

Therefore, total heat exchanged is

$$Q = Q_1 + Q_2 + Q_3 = 458 \text{ J}$$

Q.13   **The P-V diagram in Fig. 2.13 b shows three possible paths for an ideal gas to reach the final state f from an initial state i. Which among the following statement(s) is (are) correct?**

[JAM-2016]

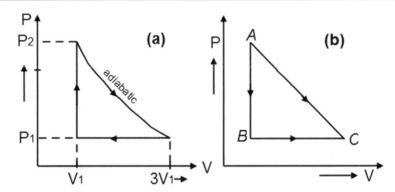

Fig. 2.14 P-V diagram **a** for Q.14 and **b** for Q.15

(A)   The work done by the gas is maximum along path-3.
(B)   Minimum change in the internal energy occurs along path-2.
(C)   Maximum heat transfer is for path-1.
(D)   Heat transfer is path independent.

Sol: We learned that area under the curve in a P-V diagram gives the work done, so it is maximum in path 3. Note that change in internal energy is same for all paths Therefore, to maintain same internal energy change, heat exchange will be maximum in path 3.

$$\Delta U = Q_3 + W_3 = Q_2 + W_2 + Q_1 + W_1$$

Because, $W_3 > W_2 > W_1$, and W is negative, implies $Q_3 > Q_2 > Q_1$. Therefore, A and C are CORRECT.

Q.14   **Consider a monoatomic ideal gas operating in a closed cycle as shown in the P-V diagram given below in Fig.2.14a. The ratio P_1/P_2 is (Specific your answer up to two digits after the decimal point)**

[JAM-2018]

Sol: For a monoatomic gas, $\gamma = 5/3$. Further, for an adiabatic process between two points, $P_1 V_1^{\gamma} = P_2 V_2^{\gamma}$, this gives

$$\frac{P_1}{P_2} = \left[\frac{V_1}{3V_1}\right]^{5/3} = \left[\frac{1}{3}\right]^{5/3} = 0.16$$

Q.15   **A given quantity of gas is taken from the state A→C reversibly, by two paths, A→C directly and A→B→C as shown in the Fig. 2.14b. During the process A→C the work done by the gas is 100 J and the heat absorbed is 150 J. If during the process A→B→C the work done by the gas is 30 J, the heat absorbed is**

[NET-JRF-2012]

(A)   20 J   (C)   220 J
(B)   80 J   (D)   280 J

Sol: for path AC, $\Delta W = -100\,J$, $\Delta Q = 150$ J, therefore using first law

$$\Delta U = \Delta Q + \Delta W = 150\,J - 100\,J = 50\,J$$

Because, internal energy is a state function and perfect differential, therefore for path A-B-C, internal energy change will also be the same. For this case $\Delta W = -30$ J and heat absorbed is

$$\Delta Q = \Delta U - \Delta W = 50\,J - (-30\,J) = 80\,J$$

Therefore, B is CORRECT.

Q.16 **The pressure P of a fluid is related to its number density $\rho = n/V$ by the equation of state**

$$P = a\rho + b\rho^2$$

where a and b are constants. If the initial volume of the fluid is V_0, the work done on the system when it is compressed, so as to increase the number density from an initial value of ρ_0 to $2\rho_0$ is

[NET-JRF-2014]

(A) $a\rho_0 V_0$
(B) $(a + b\rho_0)\,\rho_0 V_0$

(C) $\left(\dfrac{3a}{2} + \dfrac{7\rho_0 b}{3}\right)\rho_0 V_0$
(D) $(a \ln 2 + b\rho_0)\,\rho_0 V_0$

Sol: Given pressure as

$$P = a\rho + b\rho^2 = a\frac{n}{V} + b\frac{n^2}{V^2} \tag{2.68}$$

The work done on the system (compression)

$$W = \int_{V_1}^{V_2} P\,dV = an \int_{V_1}^{V_2} \frac{dV}{V} + bn^2 \int_{V_1}^{V_2} \frac{dV}{V^2}$$

$$= an \ln\left(\frac{V_2}{V_1}\right) + bn^2(-1)\left(\frac{1}{V_2} - \frac{1}{V_1}\right)$$

$$= -an \ln 2 - bn\rho_0 = -n\,[a \ln 2 + b\rho_0] = -\rho_0 V_0\,[a \ln 2 + b\rho_0]$$

Note that additional negative sign is due to sign convention-1. Using convention-2 the final answer will be

$$W = -\int_{V_1}^{V_2} P\,dV = \rho_0 V_0 \,[a \ln 2 + b\rho_0]$$

Here, we have used, $V_1 = V_0 = n/\rho_0$ and $V_2 = n/2\rho_0$.

Q.17 Show that for an ideal monoatomic gas

    (i) $C_P - C_V = R$

    (ii) $\gamma = \dfrac{C_P}{C_V} = \dfrac{5}{3}$

Sol. (i) As we discussed in the previous article, the relation connectin C_P and C_V is

$$C_P - C_V = \left[P + \left(\frac{\partial U}{\partial V}\right)_T\right]\left(\frac{\partial V}{\partial T}\right)_P$$

$$(2.69)$$

For an ideal monatomic gas, the internal energy U is entirely due to the kinetic energy, and hence $U = \frac{3}{2}RT$ per mole. This means that U is simply function of T only. Therefore,

$$\left(\frac{\partial U}{\partial V}\right)_T = 0 \qquad\qquad (2.70)$$

For one mole of an ideal gas, PV=RT, so that V=RT/P, and hence

$$\left(\frac{\partial V}{\partial T}\right)_P = \frac{R}{P} \qquad\qquad (2.71)$$

Substituting the value of these derivatives in Eq. 2.69, we get

$$C_P - C_V = R \qquad\qquad (2.72)$$

(ii) For an ideal gas, $U = \frac{3}{2}RT$, therefore,

$$C_V = \left(\frac{\partial U}{\partial T}\right)_V = \frac{3}{2}R$$

$$and$$

$$C_P = C_V + R = \frac{5}{2}R$$

The adiabatic index is defined as

$$\gamma = \frac{C_P}{C_V} = \frac{5}{3}$$

Q.18 Show that for a real gas

$$C_P - C_V = R\left[1 + \frac{2a}{RTV^3}\right] \tag{2.73}$$

where terms have their usual meaning.

Sol. For a real gas, the internal energy change with volume, because work has to be done against intermolecular forces. For a real gas, the equation of state is

$$\left(P + \frac{a}{V^2}\right)(V - b) = RT$$

We will show later that for a real gas

$$\left(\frac{\partial U}{\partial V}\right)_T = \frac{a}{V^2}$$

Now using these results and the general equation between C_P and C_V

$$C_P - C_V = \left[P + \left(\frac{\partial U}{\partial V}\right)_T\right]\left(\frac{\partial V}{\partial T}\right)_P \tag{2.74}$$

we get

$$C_P - C_V = \frac{RT}{(V - b)}\left(\frac{\partial V}{\partial T}\right)_P \tag{2.75}$$

Differentiating van der Waals equation of state, we get

$$\left(P + \frac{a}{V^2} - (V - b)\frac{2a}{V^3}\right)\left(\frac{\partial V}{\partial T}\right)_P = R$$

so that

$$\left(\frac{\partial V}{\partial T}\right)_P = \frac{R}{\left(P + \frac{a}{V^2} - (V - b)\frac{2a}{V^3}\right)}$$

After multiplying the numerator and denominator by (V-b), this equation reduces to

$$\left(\frac{\partial V}{\partial T}\right)_P = \frac{R(V - b)}{RT - \frac{2a}{V^3}(V - b)^2} \tag{2.76}$$

such that

$$\frac{1}{(V-b)} \left(\frac{\partial V}{\partial T}\right)_P = \frac{1}{T \left[1 - \frac{2a}{RTV^3}(V-b)^2\right]} \tag{2.77}$$

For small a, under Binomial expansion we get

$$\frac{1}{(V-b)} \left(\frac{\partial V}{\partial T}\right)_P = \frac{1}{T}\left[1 + \frac{2a}{RTV^3}(V-b)^2\right] \tag{2.78}$$

This gives finally

$$C_P - C_V = R\left[1 + \frac{2a}{RTV^3}(V-b)^2\right] \tag{2.79}$$

Q.19 **A vessel has two compartments of volume V_1 and V_2, containing an ideal gas at pressures P_1 and P_2, and temperatures T_1 and T_2, respectively. If the wall separating the compartments is removed, the resulting equilibrium temperature will be**

[NET-JRF-2013]

(A) $\dfrac{P_1 T_1 + P_2 T_2}{P_1 + P_2}$ (C) $\dfrac{P_1 V_1 + P_2 V_2}{\left(\dfrac{P_1 V_1}{T_1}\right) + \left(\dfrac{P_2 V_2}{T_2}\right)}$

(B) $\dfrac{V_1 T_1 + V_2 T_2}{V_1 + V_2}$ (D) $[T_1 T_2]^{1/2}$

Sol: The ideal gas is at different pressures P_1 and P_2, so that there number of moles are n_1 and n_2, respectively.

$$n_1 = \frac{P_1 V_1}{RT_1}, n_2 = \frac{P_2 V_2}{RT_2}, n = n_1 + n_2 = \frac{1}{R}\left[\frac{P_1 V_1}{T_1} + \frac{P_2 V_2}{T_2}\right]$$

Let T be the final temperature of the mixture, therefore, we can write $U_{total} = nC_V T = (n_1 + n_2)C_V T$

$$U_{total} = U_1 + U_2$$
$$(n_1 + n_2)\, C_V T = n_1 C_V T_1 + n_2 C_V T_2$$

This allows us to write

$$T = \frac{n_1 T_1 + n_2 T_2}{n_1 + n_2} = \frac{\left(\frac{P_1 V_1}{RT_1}\right) T_1 + \left(\frac{P_2 V_2}{RT_2}\right) T_2}{\frac{1}{R}\left(\frac{P_1 V_1}{T_1} + \frac{P_2 V_2}{T_2}\right)}$$

$$= \frac{P_1 V_1 + P_2 V_2}{\left(\frac{P_1 V_1}{T_1}\right) + \left(\frac{P_2 V_2}{T_2}\right)}$$

Therefore C is CORRECT.

Q.20 **Two different thermodynamic systems are described by the following equations of state:**

$$\frac{1}{T_1} = \frac{3RN_1}{2U_1}, \frac{1}{T_2} = \frac{5RN_2}{2U_2}$$

where $T_{1,2}$, $N_{1,2}$ and $U_{1,2}$ are, respectively, the temperatures, the mole numbers and the internal energies of the two systems, and R is the gas constant. Let U_{tot} denotes the total energy when these two systems are put in contact and attain thermal equilibrium. The ratio U_1/U_{tot} is

                                                                                    [NET-JRF-2013]

(A) $\dfrac{5N_2}{3N_1 + 5N_2}$ (C) $\dfrac{N_1}{N_1 + N_2}$

(B) $\dfrac{3N_1}{3N_1 + 5N_2}$ (D) $\dfrac{N_2}{N_1 + N_2}$

Sol: Given

$$\frac{1}{T_1} = \frac{3RN_1}{2U_1}, \frac{1}{T_2} = \frac{5RN_2}{2U_2}$$

These equations allow us to write the total energy for the mixture as

$$U_{total} = U_1 + U_2 = \frac{3}{2}RN_1 T_1 + \frac{5}{2}RN_2 T_2 \qquad (2.80)$$

Therefore

$$\frac{U_1}{U_{total}} = \frac{\frac{3}{2}RN_1 T_1}{\frac{3}{2}RN_1 T_1 + \frac{5}{2}RN_2 T_2} = \frac{3N_1 T_1}{3N_1 T_1 + 5N_2 T_2}$$

When, equilibrium is attained, $T_1 = T_2$ and hence,

$$\frac{U_1}{U_{total}} = \frac{3N_1}{3N_1 + 5N_2}$$

Therefore, B is CORRECT.

Q.21 **When an ideal monoatomic gas is expanded adiabatically from an initial volume V_0 to $3V_0$, its temperature changes from T_0 to T. Then the ratio T/T_0 is**

[NET-JRF-2016]

(A) $1/3$ (C) $(1/3)^{1/3}$
(B) $(1/3)^{2/3}$ (D) 3

Sol: For an adiabatic process, $TV^{\gamma-1}=C$. Also $\gamma=1/3$ (for monoatomic gas). Therefore,

$$T = T_0 \left[\frac{V_0}{3V_0}\right]^{\gamma-1} = T_0 \left[\frac{1}{3}\right]^{2/3}$$

Hence, B is CORRECT.

Q.22 **A box containing 2 moles of a diatomic ideal gas at temperature T_0 is connected to another identical box containing 2 moles of a monoatomic ideal gas at temperature $5T_0$. There are no thermal losses and the heat capacity of the boxes is negligible. Find the final temperature of the mixture of gases (ignore the vibrational degrees of freedom for the diatomic molecules).**

[IIT-JAM-2009]

(A) T_0 (C) $2.5T_0$
(B) $1.5T_0$ (D) $3T_0$

Sol: The total internal energy of the system will remain conserved after mixing. That is $U_{total} = U_{monoatomic} + U_{diatomic}$. Let $C_{V1} = 3R/2, C_{V2} = 5R/2$ be the heat capacities of monoatomic and diatomic gases, respectively. So that

$$U_{monoatomic} = n_1 C_{V1} T_1, U_{diatomic} = n_2 C_{V2} T_2$$

Further, $n_1 = n_2 = 2, T_1 = 5T_0, T_2 = T_0$. Assume that C_V be the common heat capacity and T be the final temperature of the mixture, so that we can write

$$C_V = \frac{n_1 C_{V1} + n_2 C_{V2}}{n_1 + n_2}.$$

$$n_1 C_{V1} T_1 + n_2 C_{V2} T_2 = (n_1 + n_2) C_V T = [n_1 C_{V1} + n_2 C_{V2}] T$$

Rearranging the terms, we obtain

$$T = \frac{n_1 C_{V1} T_1 + n_2 C_{V2} T_2}{n_1 C_{V1} + n_2 C_{V2}} = \frac{2 \times \frac{3}{2} R \times 5 T_0 + 2 \times \frac{5}{2} R \times T_0}{2 \times \frac{3}{2} R + 2 \times \frac{5}{2} R} = 2.5 T_0$$

Q.23   A trapped air bubble of volume V_0 is released from a depth h measured from the water surface in a large water tank. The volume of the bubble grows to $2V_0$ as it reaches just below the surface. The temperature of the water and the pressure above the surface of water (10^5Nm^{-2}) remain constant throughout the process. If the density of water is (1000 kg m^{-3}) and the acceleration due to gravity is 10ms^{-2}, then the depth h is.

[IIT-JAM-2010]

(A)   1m   (C)   50 m
(B)   10 m   (D)   100 m

Sol: Let pressure at surface be P_0. Then at a depth h, the pressure is $P_1 = P_0 + \rho g h$, and volume is $V_1 = V_0$. At the surface given pressure $P_2 = P_0$ and volume $V_2 = 2V_0$. Since the temperature of the water does not change (hence air bubble), we can write $P_1 V_1 = P_2 V_2$. This gives

$$(P_0 + \rho g h) V_0 = 2 P_0 V_0$$

$$h = \frac{P_0 V_0}{\rho g V_0} = \frac{P_0}{\rho g} = \frac{10^5}{10^3 \times 10} = 10m$$

Q.24   A cylinder contains 16g of O_2. The work done when the gas is compressed to 75% of the original volume at constant temperature of 27° C is........(Universal gas constant R=8.31 $JK^{-1}mol^{-1}$)

[JAM-2016]

Sol: The number of moles n=16/32 = 0.5. T=27+273=300 K. The work done on gas when compressed is

$$W = \int_{V_i}^{V_f} P dV = nRT \ln \left[\frac{V_f}{V_i} \right] = \frac{1}{2} \times 8.31 \times 300 \ln \left[\frac{75}{100} \right] = 358 J$$

Q.25   A di-atomic gas undergoes adiabatic expansion against the piston of a cylinder. As a result, the temperature of the gas drops from 1150 K to 400 K. The number of moles of the gas required to obtain 2300 J of work from the expansion is........(The gas constant R=8.314 J mol^{-1} K^{-1}.

[JAM-2019]

Sol: The work done during adiabatic expansion (convention-1) [γ=1.4]

$$W = \frac{nR}{\gamma - 1}[T_2 - T_1], 2300J = \frac{n \times 8.314}{1.4 - 1}[400 - 1150], n = -0.1475$$

According to sign convention-2,

$$W = \frac{nR}{\gamma - 1}[T_1 - T_2], 2300J = \frac{n \times 8.314}{1.4 - 1}[1150 - 400], n = 0.1475$$

Therefore, convention merely introduces a negative sign, else n=0.1475.

Q.26 **An ideal gas has a specific heat ratio $C_P/C_V = 2$. Starting at a temperature T_1 the gas undergoes an isothermal compression to increase its density by a factor of two. After this, an adiabatic compression increases its pressure by a factor of two. The temperature of the gas at the end of the second process would be**

[JEST-2016]

(A) $T_1/2$ (C) $2T_1$

(B) $\sqrt{2}T_1$ (D) $T_1/\sqrt{2}$

Sol: During first isothermal step, $T=T_1$ stays constant. Let at the end of this process the thermodynamic coordinates are $A(P_1, T_1)$. Adiabatic process starts at this point and ends at $B(P_2, T_2)$. For an adiabatic process, $T_1^\gamma P_1^{\gamma-1} = T_2^\gamma P_2^{\gamma-1}$. This gives

$$T_2 = \left[\frac{P_1}{P_2}\right]^{\frac{1-\gamma}{\gamma}} = \left[\frac{P_1}{2P_1}\right]^{\frac{1-2}{2}} T_1 = \sqrt{2}T_1$$

Q.27 **In a thermodynamic process the volume of one mole of an ideal varies with T as $V = aT^{-1}$ where a is constant. The adiabatic exponent of the gas is γ. What is the amount of heat received by the gas if the temperature of the gas increases by ΔT in the process?**

[JEST-2018]

(A) $R\Delta T$ (C) $\dfrac{R\Delta T}{2 - \gamma}$

(B) $\dfrac{R\Delta T}{1 - \gamma}$ (D) $R\Delta T\left[\dfrac{2 - \gamma}{1 - \gamma}\right]$

Sol: Given $V = \dfrac{a}{T}$ so that $dV = -\dfrac{a}{T^2} dT$. The work done is given by (as T increases, it is expansion, hence by convention-1, W is negative)

$$\Delta W = -\int P dV = -\int \frac{RT}{V} dV$$

$$= -\int \frac{RT^2}{a} \left(-\frac{a}{T^2}\right) dT = \int R dT = R\Delta T$$

Now change in internal energy of an ideal gas in terms of temperature is

$$\Delta U = C_V \Delta T = \frac{R}{\gamma - 1} \Delta T$$

By using the first law of thermodynamics,

$$\Delta Q = \Delta U - \Delta W = \frac{R}{\gamma - 1} \Delta T - R\Delta T = R\Delta T \left[\frac{2 - \gamma}{1 - \gamma}\right]$$

NOTE: The final answer doesn't change if we use convention-2. Let us check, how? In this case

$$\Delta W = \int P dV = -R\Delta T \quad \text{and} \quad \Delta U = \frac{R}{\gamma - 1} \Delta T$$

Applying the First law, i.e.,

$$\Delta Q = \Delta U + \Delta W = \frac{R}{\gamma - 1} \Delta T - R\Delta T = R\Delta T \left[\frac{2 - \gamma}{1 - \gamma}\right]$$

Reader must recall here that two sign conventions have different ways of defining the First law. However, both statements follow the same logic, i.e., when system does work, its internal energy decrease, on the other hand internal energy decrease when work is done on the system.

2.10   Multiple Choice Questions

Q.1 **The internal energy of a perfect gas is**

    (A)   entirely potential energy of its molecules
    (B)   entirely kinetic energy of its molecules
    (C)   both kinetic and potential energy of its molecules
    (D)   Neither kinetic nor potential energy of molecules

Q.2 Internal energy of a perfect gas depends on

(A)	Temperature	**(C)**	Volume
(B)	Pressure	**(D)**	All

Q.3 Internal energy of a real gas depends on

(A)	Temperature	**(C)**	Volume
(B)	Pressure	**(D)**	All

Q.4 Internal energy (U) is a unique function of any state because change in U

(A) depends on path
(B) independent of path
(C) corresponds to adiabatic process
(D) corresponds to isothermal process

Q.5 The first law of thermodynamics is a special case of

(A)	Newton's law	**(C)**	conservation of energy
(B)	Charles law	**(D)**	law of heat exchange

Q.6 The specific heat of a gas

(A) has only two values C_P and C_V
(B) has unique value at given temperature
(C) can have any value between 0 and ∞
(D) depends on mass of gas

Q.7 The temperature of an ideal gas is kept constant as it expands. the gas does external work. During this process, the internal energy of gas

(A)	decrease	**(C)**	remain constant
(B)	increase	**(D)**	cannot be predicted

Q.8 A system performs $\triangle W$ amount of work when heat $\triangle Q$ is added to the system and internal energy change is $\triangle U$. A unique function of initial and final state is

(A)	$\triangle Q$	**(C)**	$\triangle U$
(B)	$\triangle W$	**(D)**	$\triangle Q$ and $\triangle W$

Q.9 **Free expansion of a gas means expansion**

   (A) at 0 K
   (B) against P=0 when heat is not supplied
   (C) at 1 atm
   (D) at constant T

Q.10 **An ideal gas expands freely in a rigid insulated container against vacuum. The gas undergoes**

   (A) decrease in temperature   **(C)** decrease in T and P
   (B) no change in internal energy   **(D)** None of these

Q.11 **Let ΔW be the work done in a quasi-static reversible thermodynamic process. Which of the following statements about ΔW is correct ?**

                                                      [NET-JRF-2012]

   (A) ΔW is a perfect differential if the process is isothermal
   (B) ΔW is a perfect differential if the process is adiabatic
   (C) ΔW is always a perfect differential
   (D) ΔW cannot be a perfect differential

Q.12 **When a gas expands adiabatically from volume V_1 to V_2 by a quasi-static reversible process, it cools from temperature T_1 to T_2. If now the same process is carried out adiabatically and irreversibly, and T_2' is the temperature of the gas when it has equilibrated, then**

                                                      [NET-JRF-2014]

   (A) $T_2' = T_2$
   (B) $T_2' > T_2$
   (C) $T_2' = T_2 \left[\dfrac{V_2 - V_1}{V_2} \right]$
   (D) $T_2' = \left[\dfrac{T_2 V_1}{V_2} \right]$

Q.13 **An isolated box is divided into two equal parts by a partition. One of the compartments contains a van der Waals gas while the other compartment is evacuated. If the partition between two compartments is removed, and the gas has filled the entire box and equilibrium has been achieved, which of the following statement(s) is (are) correct**

                                                      [JAM-2017]

(A) Internal energy of the gas has not changed
(B) Internal energy of the gas has decreased
(C) Temperature of the gas has increased
(D) Temperature of the gas has decreased

Q.14 **During free expansion of an ideal gas under adiabatic condition, the internal energy of the gas.**

[JAM-2019]

(A) decreases
(B) increases
(C) initially decreases and then increases
(D) remains constant

Keys and Hints to MCQ Type Questions

Q.1 B	Q.3 D	Q.5 C	Q.7 C	Q.9 B	Q.11 B	Q.13 A&D
Q.2 A	Q.4 B	Q.6 C	Q.8 C	Q.10 B	Q.12 B	Q.14 D

Hint.13 This is an example of free expansion of a real gas. Internal energy of the gas will not change. For a real gas

$$dU = C_V dT + \frac{a}{V^2} dV$$

In order to keep internal energy fixed, if dV increases then dT must decrease.

Exercises

1. State and explain Zeroth law of thermodynamics

2. What is the first law of thermodynamics?. Explain that it is general law of conservation of energy.

3. Show that the slope of an adiabatic curve through a point P in a P-V diagram of an ideal gas is γ times the slope of an isothermal curve through the same point.

4. Show that for an ideal gas undergoing an adiabatic process, PV^γ=constant.

5. Show that, for an ideal has undergoing an isothermal expansion, work done is given by

$$W = nRT \ln \frac{V_2}{V_1}$$

6. Obtain adiabatic equation of a perfect gas. Also find out work done during an adiabatic process.

7. Starting with the definition of work done (W) by a system as $W = P \int_{V_1}^{V_2} dV$, show that the work done in adiabatic expansion of an ideal gas from a state (P_1, V_1) to state (P_2, V_2) is given by

$$W = \frac{1}{\gamma - 1} [P_1 V_1 - P_2 V_2]$$

8. If n mole of real gas (obeying the van der Waals equation of state) undergo an isothermal expansion. Show that the work done ($W = \int P dV$) is given by

$$W = nRT \ln \left[\frac{V_2 - b}{V_1 - b} \right] + n^2 a \left[\frac{1}{V_2} - \frac{1}{V_1} \right]$$

8. Explain following in brief
(i) Free expansion?.
(ii) Indicator diagram
(iii) function of a state
(iv) Is work done a perfect differential? Give example.

9. A real gas contained in a cylinder of volume V and has a pressure P. If the gas undergoes isothermal change, then show that the work done is given by

$$W = nRT \ln \left(\frac{V_2 - b}{V_1 - b} \right) + n^2 a \left(\frac{1}{V_2} - \frac{1}{V_1} \right)$$

10. What is thermodynamic equilibrium? What conditions must a system satisfy to be in thermodynamic equilibrium?.

11. What is an indicator diagram? State its importance.

12. Discuss how the first law leads to the concept of internal energy.

13. Distinguish between isothermal and adiabatic processes. Show that for an adiabatic change in a perfect gas

$$PV^{\gamma} = \text{constant}$$

14. For a quasi-static adiabatic process of an ideal gas, show that

$$\frac{T}{(P)^{(\gamma-1)/\gamma}} = \text{constant}$$

15. Show that equation of state for an adiabatic change is $V^{\gamma-1} T = \text{constant}$.

16. Obtain an expression for work done and change in internal energy for isothermal and adiabatic processes.

17. Explain why gases have two heat capacities? Why C_P is greater than C_V ?.

18. What is meant by isothermal and adiabatic curves? Show that the slope of an adiabatic curve through a point on indicator diagram of an ideal gas is γ times the slope of an isothermal curve through same point.

19. State and explain Zeroth law of thermodynamics. What is its importance? How does it lead to the concept of temperature?

20. What do you understand by internal and external work? What kind of work is important in thermodynamics?

21. Give a general definition for heat capacity and specific heat. Show that

$$C_P = C_V = \left[P + \left(\frac{\partial U}{\partial V} \right)_T \right] \left(\frac{\partial V}{\partial T} \right)_P$$

Second Law of Thermodynamics

3

In this chapter, we introduce the second law of thermodynamics. The first law empha-sizes the importance of the conservation of energy in a particular process. It can be converted from one form into another but the total energy content remains unchanged. However, the first law is silent over the issue of heat flow between different systems. Indeed, the second law of thermodynamics focuses on the conditions which allow heat flow between two different systems, precisely speaking a system and its sur-rounding. It also imposes limits on processes involving the conversion of heat into work. The latter also leads to the development of the thermodynamic temperature scale and to define the concept of entropy. Before we begin with the second law of thermodynamics, we shall first discuss a few examples which are helpful to under-stand its physical meaning.

3.1 Converting Work into Heat

Let us take an example of a water tub fixed with a movable paddle. When it is rotated by doing some mechanical work on it, the moving paddle imparts kinetic energy to the water molecules, and as a result, water becomes warmer, see Fig. 3.1a. In another example, suppose we are rubbing two stones inside the same water tub (reservoir), see Fig. 3.1b. During the rubbing process, the force of friction comes into play and the work done against this force appears as a rise in the internal energy of the stones. As a result, the temperature of stones rises and this increased energy of stones is transferred to the surrounding water. In the third example shown in Fig. 3.1c, we consider the heating of water by means of an electrical resistor through which an electric current is passed. Due to resistive heating, the temperature of surrounding water increases. In this case, the electrical work is converted completely into heat. In this case, there is no change in the thermodynamic coordinates of the resistive

© The Author(s) 2022
S. Sharma, *Thermal and Statistical Physics*,
https://doi.org/10.1007/978-3-031-07685-5_3

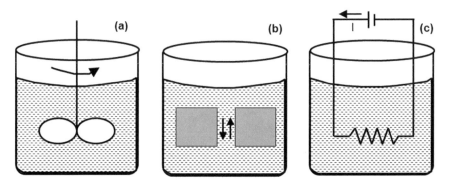

Fig. 3.1 Schematic illustration of converting work into heat. **a** A paddle moving inside water, **b** rubbing two stones inside water, frictional forces causes water to heat and **c** heating of water via Joule heating

wire. In all three cases discussed above, we have a complete conversion of work (mechanical or electrical) into heat. There is no change in the state of the system (paddle, stones and resistor in the third case) during the entire process. Therefore, during these processes, system merely acts as a mediator whose state is unaffected by the process of energy transfer and this process can continue indefinitely. Next immediate question is whether complete conversion of heat into work is feasible or not? *And whether this conversion can continue indefinitely without any change in the state of the system?*

Various examples of processes can be thought of where a complete conversion of heat into work occurs. ***But in all such processes, a net change in the state of the system takes place***. For instance, it appears that during *isothermal expansion* of n mole of an ideal gas, the work done by the gas is

$$W = -\int_{V_1}^{V_2} P dV = -nRT \int_{V_1}^{V_2} \frac{dV}{V} = -nRT \ln\left(\frac{V_2}{V_1}\right) \quad (3.1)$$

For an isothermal expansion, $dT = 0$ and so $dU = 0$. The first law then becomes

$$Q = -W > 0 \quad (3.2)$$

where Q is the net amount of heat supplied. Equation 3.2 implies that in an isothermal expansion complete conversion of heat into work occurs. This process involves a change in the state of the system and hence doesn't violate the second law of thermodynamics. During this process, the volume increases from V_1 to V_2 and pressure drops until atmospheric pressure is reached and ultimately process will stop. Therefore, we conclude that process of isothermal expansion cannot be used indefinitely for converting heat into work. Note that while writing the expression for work in Eq. 3.1, we have taken into account the sign convention that the work done by the system is negative (that's why a negative sign appears in equation).

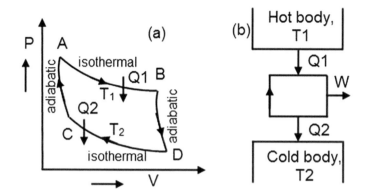

Fig. 3.2 a A cyclic process on P-V diagram and **b** Schematic representation of a heat engine working in a cycle

For a continuous supply of work, we need a cyclic process in which the system is brought back to its initial state. Each such cyclic process involves heat transfer between bodies (system and reservoir) of nearly equal temperature. Such a cycle consists of a reservoir at a higher temperature (T_1) than the system and a reservoir at a lower temperature (T_2) than the system.

For illustration purpose, a cyclic process with ideal gas as a working substance is shown in Fig. 3.2a.

For one complete cycle, let

Q_1 represent the heat exchanged between the reservoir at temperature T_1 and the system.

Q_2 represent the heat exchanged between the reservoir at temperature T_2 and the system.

W represent the work exchanges between the system and its surrounding.

Note that the quantities Q_1, Q_2 and W are all absolute quantities. These can take positive as well as negative values. When Q_1 is larger than Q_2 and W is done by the system on surrounding, then the machine that forces the system to undergo a cycle is called a *heat engine*. On the other hand, an engine that extracts heat from a cold body and delivers heat to a hot body when work is performed on the engine is termed as a *refrigerator* and we shall denote it by R.

A general heat engine E is represented as shown below in Fig. 3.2b. Heat Q_1 is supplied from a reservoir at temperature T_1 to the working substance. Heat Q_2 is given by the system to reservoir at temperature T_2 lower than T_1. W is the work done by the engine in one complete cycle. As a measure of the efficiency (η) of the engine, we define η as the work delivered divided by the heat input.

$$\eta = \frac{W}{Q_1} \tag{3.3}$$

Application of first law gives

$$\Delta U = Q_1 - Q_2 - W \tag{3.4}$$

In writing this equation, we have taken into account the sign convention that heat into the system is taken as positive and work done by the system as negative. During the cycle, the working substance does not change; therefore, $\Delta U = 0$, and this implies that $Q_1 - Q_2 = W$. Therefore, the efficiency of the engine becomes

$$\eta = 1 - \frac{Q_2}{Q_1} \tag{3.5}$$

It is clear from this equation that η will be unity (efficiency 100%) if Q_2 is zero. In other words, if an engine could be built to operate in a cycle in which no heat flows away from the working substance to low-temperature reservoir, then 100% conversion of heat (from a high-temperature source) into work can be achieved. However, we shall see later that there must always be an outflow of heat from an engine, so the efficiency of a heat engine is always less than 100%.

3.2   Second Law of Thermodynamics

It is important to understand that the second law of thermodynamics deals with a different aspect of nature than does the first law of thermodynamics. The second law basically deals with the natural tendency of heat to flow from hot to cold, whereas the first law deals with energy conservation and focuses on both heat and work. The operation of various devices depends on heat and work, and to really understand such devices, both laws are needed. For instance, an automobile engine is a type of heat engine that uses heat to produce work.

There are two well-known classical statements of the second law of thermodynamics. The first statement places the emphasis on the efficiency of conversion of heat into work and the second emphasizes the irreversibility (spontaneous flow of heat) of nature. Let us continue with the first one. It is an experimental fact that there exists no device (engine) that extracts heat from a reservoir at high temperature and converts it completely into an equivalent amount of work without rejecting some heat to the reservoir at a relatively lower temperature. This statement is derived from everyday experience and has been formulated in different ways by different persons. Below we give two such widely accepted statements. The first one is a joint statement by Kelvin and Planck and another one is given by Clausius. Later, we will also demonstrate that these statements are equivalent.

Fig. 3.3 Schematic representation of the Kelvin–Planck statement of the second law of thermodynamics

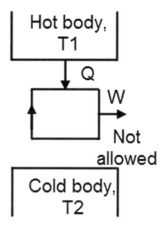

3.2.1 Kelvin–Planck Statement

It is impossible to construct a device (engine) that, operating in a cycle, will produce no effect other than extraction of heat from a single reservoir at fixed temperature and the performance of an equivalent amount of work

OR precisely

No process (cyclic) is possible whose sole result is the complete conversion of heat into work.

Schematic diagram correlating this statement is given in Fig. 3.3. The second law ensures that some amount of heat must be rejected by the heat engine to a reservoir at a low temperature. As shown in the previous section, the efficiency of a heat engine operating in the cycle is

$$\eta = 1 - \frac{Q_2}{Q_1} \tag{3.6}$$

If Q_2 is zero, one could have an engine with 100% efficiency. Was this statement untrue, one could drive a ship across the sea by extracting the heat from the sea and converting it into an equivalent amount of work.

3.2.2 Clausius Statement

So for discussion was centred on a heat engine that operates in a cyclic manner and takes the working substance through a sequence of processes where some heat(Q_1) is absorbed by the working substance from a heat reservoir at a higher temperature, relatively a smaller amount of heat (Q_2) is rejected to a reservoir at a lower temperature and a fixed amount of work (W) is done by the system(working substance)

Fig. 3.4 Schematic diagram for **a** refrigerator and **b** the Clausius statement of second law of thermodynamics

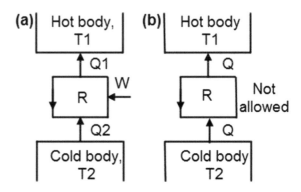

on surroundings. Let us imagine a cycle performed in a direction opposite to that of an engine. In such a cycle, a small amount of heat (Q_2) is absorbed by the working substance from a heat reservoir at a lower temperature, and a LARGER amount of heat (Q_1) is rejected to a heat reservoir at high temperature when a fixed amount of work is done on the system (working substance) by the surrounding. A machine that works in such a manner is called a *refrigerator*. The working substance is the also called a system or refrigerant. Let us consider Fig. 3.4a. Let us assume that amount of heat rejected by the working substance to high-temperature reservoir is Q_1, the amount of heat absorbed by the working substance from a low-temperature reservoir is Q_2 and amount of work done on the working substance by the surrounding is W. As the working substance goes through a cycle, the net change in internal energy, i.e., $\Delta U = 0$. Under these conditions, the first law will give $\Delta U = Q_2 - Q_1 + W$. In writing this the relation we haven't ignored the sign convention where heat into the system and work done on the system are both taken as positive. Therefore,

$$Q_1 - Q_2 = W$$
$$Q_1 = Q_2 + W \tag{3.7}$$

Equation 3.7 says that heat rejected by the system (working substance) to the reservoir at a high temperature is larger than the heat extracted from the reservoir at low temperature by the amount of work done on the system.

A refrigerator extracts heat Q_1 from a low-temperature reservoir at the expenditure of electrical work W (done on refrigerant or system). Work is always required to transfer heat from a low-temperature reservoir to a high-temperature reservoir. This is in fact a natural tendency that "*heat flows spontaneously from a substance at a higher temperature to a substance at a lower temperature and does not flow spontaneously in the reverse direction*". This negative statement leads us to the Clausius statement of the second law of thermodynamics.

No process is possible whose sole result is the transfer of heat from a colder to a hotter body.

At first sight, it appears that the Kelvin–Planck and the Clausius statements are unconnected, but below we will shall prove that they are equivalent.

3.2.3 Equivalence of the Kelvin–Planck and the Clausius Statements

Now we will prove that the two statements of the second law of thermodynamics are equivalent to each other. This can be proved by showing that the falsity of each implies the falsity of the other. Let us first assume that the Kelvin–Planck statement is untrue. This means that we can have an engine E that takes Q_1 from a hot reservoir and delivers an equivalent amount of work $W = Q_1$ in one cycle. Further, we assume that this engine is driving a refrigerator R as shown in Fig. 3.5a. If we adjust the size of the working cycles so that W is sufficient to drive the refrigerator around one cycle. Assume that the refrigerator extracts heat Q_2 from the cold body. Then the heat delivered by it to the hot body is

$$Q_2 + W = Q_2 + Q_1$$

Now, we may regard the engine E and the refrigerator R as a composite refrigerator shown by the dotted line in Fig. 3.5b. This composite refrigerator extracts Q_2 from the cold body and delivers a net amount of heat

$$Q_1 + Q_2 - Q_1 = Q_2 \tag{3.8}$$

to the hot body at temperature T_2, but no work is done. Therefore, we have a violation of the Clausius statement, which prohibits the transfer of heat from a cold reservoir to a hot reservoir without additional work done on the system.

In the second case, we assume that the Clausius statement is untrue. This means that we can have a refrigerator that extracts a heat Q_2 from a reservoir and delivers the same heat Q_2 to a body at high temperature in one cycle without performing any work. This is shown in Fig. 3.6a. Let us now imagine an engine operating between the same two bodies and somehow if we adjust the size of its working cycle so that, in one cycle, it extracts heat Q_1 from the hot body and gives up the same amount of heat Q_2 to the cold body as was extracted by the refrigerator so that it delivers the work which is equal to

$$W = Q_1 - Q_2$$

Fig. 3.5 Schematic diagram illustrating that if the Kelvin–Planck statement of the second law is false, then Clausius statement is also false

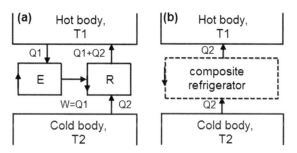

Fig. 3.6 Schematic diagram
illustrating that if the
Clausius statement of the
second law is false, then the
Kelvin–Planck statement is
also false

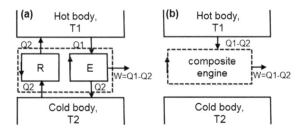

Now we can imagine the engine and the refrigerator as the composite engine enclosed
by the dotted line, as shown in Fig. 3.6b. This composite engine takes $Q_1 - Q_2$
amount of heat from the hot body and delivers the amount of work which is equal to

$$W = Q_1 - Q_2$$

Hence, we have a violation of the Kelvin–Planck statement, which prohibits the
complete conversion of heat into work during a cyclic process. This proves the
equivalence of the two different statements of the second law of thermodynamics.

3.3   Carnot Heat Engine

The Carnot engine is based upon a process called as the Carnot cycle, illustrated
in Fig. 3.7a. A Carnot cycle consists of two reversible adiabatic and two reversible
isothermal processes. The engine operates between two heat reservoirs. One of the
reservoirs at higher temperature T_1 and the other at lower temperature T_2. Note
that during adiabatic processes, no heat exchange between the system (gas) and
surrounding occurs. Heat can enter and leave only during the reversible isothermal
processes. Let us assume that heat Q_1 enters during the isothermal expansion from
A to B and heat Q_2 leaves during the isothermal compression from C to D. Because
the process is cyclic, the internal energy (a state function) change going around
the cycle (ABCDA) is zero. Therefore, the work output by the engine during the
complete cycle is

$$W = Q_1 - Q_2 \tag{3.9}$$

Fig. 3.7 Carnot cycle on **a**
P-V diagram and **b** on T-S
diagram

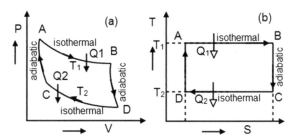

Let us consider the cycle shown in Fig. 3.7, performed reversibly. We assume that the working substance is an ideal gas. The four stages are characterized as follows.

Stage A to B: *isothermal expansion*:
From A to B, the gas expands from volume V_A to V_B at a constant high temperature T_1. During isothermal expansion, the pressure drops as the volume increases [ideal gas law, $P = nKT/V$]. The gas does work as it expands and its temperature will reduce if heat is not supplied externally. Then the application of the first law gives

$$\Delta U = dQ + dW$$
$$0 = dQ - PdV$$
$$dQ = PdV$$

Because during an isothermal process $\Delta T = 0$ and hence $\Delta U = 0$. Further, we make use of the definition of Work as $dW = -PdV$, the work done in a reversible expansion of gas (1 mole) from volume V_A to V_B. Let Q_1 be the total amount of heat taken by the gas from stage A to B, then from the above equation

$$Q_1 = \int_{V_A}^{V_B} dQ = -\int_{V_A}^{V_B} dW$$
$$= \int_{V_A}^{V_B} PdV = \int_{V_A}^{V_B} RT_1 \frac{dV}{V} = RT_1 \ln \frac{V_B}{V_A} \qquad (3.10)$$

Because, for an expansion $V_B > V_A$, and therefore $Q_1 > 0$.

Stage B to C: *adiabatic expansion*:
During this step, the gas undergoes adiabatic expansion and no energy transfer by heating takes place, i.e., $dQ = 0$ for each small step in pressure and volume. Note that the temperature of the gas drops as gas loses energy by doing work (because gas is no longer heated to compensate it for the work); this results in a faster drop in pressure as the gas expands. For this step, we make use of an equation that is valid for an adiabatic process. For stage B to C, we have

$$T_1 V_B^{\gamma-1} = T_2 V_C^{\gamma-1}$$
$$\left(\frac{T_1}{T_2}\right) = \left(\frac{V_C}{V_B}\right)^{\gamma-1} \qquad (3.11)$$

Stage C to D: *isothermal compression*:
During this stage, the gas is compressed at constant temperature T_2. The heat energy leaves the gas and goes into the cold reservoir at temperature T_2. During this stage, let Q_2 be the total amount of heat exiting from the gas. Then similar to step 1 (A to B), the heat entering the gas along isotherm CD is $RT_2 \ln \frac{V_D}{V_C}$. This is negative as $V_D < V_C$. This implies that heat flows out of the gas. But, we have defined positive Q_2 as the amount of heat flowing out from the gas. Therefore,

$$Q_2 = -RT_2 \ln \frac{V_D}{V_C} = RT_2 \ln \frac{V_C}{V_D} \qquad (3.12)$$

Stage D to A: *adiabatic compression*:
This step takes back the gas to its initial equilibrium state. And we have

$$\left(\frac{T_2}{T_1}\right) = \left(\frac{V_A}{V_D}\right)^{\gamma-1} \tag{3.13}$$

Equations 3.11 and 3.13 lead to

$$\left(\frac{V_B}{V_A}\right) = \left(\frac{V_C}{V_D}\right) \tag{3.14}$$

Dividing Eq. 3.10 by Eq. 3.12 and using Eq. 3.14

$$\frac{Q_2}{Q_1} = \frac{T_2}{T_1} \tag{3.15}$$

or, equivalently,

$$\frac{Q_2}{T_2} = \frac{Q_1}{T_1} \tag{3.16}$$

The schematic presentation of the Carnot engine is shown in Fig. 3.2b. It draws a heat Q_1 from the hot reservoir at temperature T_2 and rejects Q_2 amount of heat to reservoir at low temperature T_2. W is the amount of work obtained during one cycle from the heat engine. The efficiency is defined as the the ratio of work out to the heat in. Therefore,

$$\eta = \frac{W}{Q_1}$$
$$= \frac{Q_1 - Q_2}{Q_1} = 1 - \frac{Q_2}{Q_1} = 1 - \frac{T_2}{T_1} \tag{3.17}$$

In writing the above result, we have used Eq. 3.16. We conclude with the following points:

Equation 3.17 signifies that the efficiency of Carnot's reversible engine is independent of the nature of the working substance and depends only on the absolute temperatures of two reservoirs (source and sink).

Since work out (W) can never be larger than heat in (i.e., W< Q_1), $\eta < 1$. The efficiency must be below 100% for all practical real engines.

For the engine to be 100% efficient ($\eta = 1$), the sink temperature T_2 must be zero. Since one cannot have a sink at absolute zero, it is practically impossible to have an engine that is 100% efficient.

Back to the Carnot cycle, when the gas expands, it does work on the surrounding. When gas is compressed, an external agency (reservoir) must do work on it. What is the net effect of doing this? The expansion occurs at a higher pressure than the compression. The work done by the gas during expansion is larger than during compression. The net result is that non-zero (positive in magnitude) work is done by the gas on the surrounding. According to sign convention, this work done (dW = − PdV, $V_f > V_i$) will be negative.

3.3.1 Efficiency of Heat Engine in Terms of Adiabatic Expansion Ratio (p)

An alternative expression for the efficiency of Carnot's engine can be obtained as follows. The efficiency of the heat engine in terms of temperature T_1 and T_2 of source and sink is

$$\eta = 1 - \frac{T_2}{T_1} = 1 - \left(\frac{V_B}{V_C}\right)^{\gamma-1} \tag{3.18}$$

where we have made use of Eq. 3.11. If we know the volume expansion ratio ($p = V_f/V_i$) and γ, the corresponding value of η can be evaluated. In terms of p, the above equation reads

$$\eta = 1 - \left(\frac{1}{p}\right)^{\gamma-1} \tag{3.19}$$

3.3.2 Increasing Efficiency of Heat Engine

The efficiency of a Carnot engine working between temperatures T_1 and T_2 is given by

$$\eta = 1 - \frac{T_2}{T_1}$$

To increase η, we must reduce T_2/T_1. This can be achieved either by decreasing T_2 or by increasing T_1. As $T_2 < T_1$, a decrease in T_2 will be more effective than an equal increase in T_1.

3.3.3 Entropy Change During Carnot Cycle

In Carnot cycle, the gas returns to its initial state. The change in internal energy $\Delta U = 0$. There is no net change in entropy of the gas during such a cycle. As both internal energy and entropy are perfect differentials. Let us check what is the net change in entropy of gas, hot and cold reservoirs. During the first step (A to B), gas acquires Q_1 amount of energy by heating at temperature T_1. Hence, the entropy of gas increases by Q_1/T_1. At the same time, hot reservoir loses an equal amount

Table 3.1 Entropy change in working substance (gas) during various stages of Carnot cycles. As the gas returns to its initial state, the sum of entropy changes for the gas must be equal to zero. Here ΔS_1 and ΔS_2 are the entropy changes corresponding to hot and cold reservoirs, respectively

Stage	ΔS_g	ΔS_1	ΔS_2
A to B	$\dfrac{Q_1}{T_1}$	$-\dfrac{Q_1}{T_1}$	–
B to C	Zero	–	–
C to D	$-\dfrac{Q_2}{T_2}$	–	$\dfrac{Q_2}{T_2}$
D to A	Zero	–	–
Net change	Zero	–	–

of energy (Q_1) and its entropy reduces by $-Q_1/T_1$. Steps B–C and D–A involve no heat exchange (adiabatic processes), hence no entropy change occurs during these processes. From C–D, the gas gives $-Q_2$ units to the cold reservoir at T_2. Corresponding entropy change for gas is $-Q_2/T_2$, and for reservoir, it is Q_2/T_2. All these changes are tabulated in Table 3.1. The entropy of each reservoir changes during each cycle. The sum of these entropy changes for reservoirs is zero. Because

$$\frac{Q_2}{T_2} = \frac{Q_1}{T_1}$$

Similarly, net entropy change for the gas during one complete cycle is

$$\Delta S_g = \frac{Q_1}{T_1} - \frac{Q_2}{T_2} = \frac{Q_1}{T_1} - \frac{Q_1}{T_1} = 0$$

Hence, the net entropy change for the gas during the Carnot cycle is zero.

3.3.4   Two Carnot Engines Connected in Series

Let us consider two Carnot engines [Fig. 3.8a] which are operating in series. The first engine absorbs a quantity of heat Q_1 at a temperature T_1 and after doing work W_{12} rejects the remaining heat Q_2 at a lower temperature T_2. The second engine absorbs the heat Q_2 at temperature T_2 (rejected by first) and after doing work W_{23} rejects the remaining heat Q_3 at a still lower temperature T_3. This is shown in the diagram below. We want to evaluate the efficiency of the combination. The total work by the combination in one complete cycle is

$$W_{13} = W_{12} + W_{23}$$
$$= [Q_1 - Q_2] + [Q_2 - Q_3] = Q_1 - Q_3$$

Fig. 3.8 Schematic diagram for **a** two Carnot engines connected in series and **b** Composite diagram for (**a**)

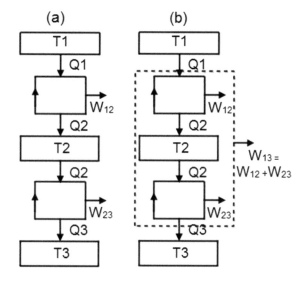

As clear from Fig. 3.8b, the heat absorbed by the combination is Q_1 and heat rejected is Q_3. Therefore, the efficiency of the combination is

$$\eta = \frac{\text{work done}}{\text{heat absorbed}}$$
$$= \frac{W_{13}}{Q_1} = \frac{Q_1 - Q_3}{Q_1} = 1 - \frac{Q_3}{Q_1} = 1 - \frac{T_3}{T_1} \qquad (3.20)$$

This is the same as the efficiency of a single engine operating between temperatures T_1 and T_3. Here, in last step, we have used $\dfrac{T_3}{T_1} = \dfrac{Q_3}{Q_1}$.

3.3.5   Carnot Theorem

Now we will prove Carnot's theorem which states that *no engine operating between two reservoirs can be more efficient than a Carnot engine operating between those same two reservoirs.*

To prove this, let us assume that such a hypothetical engine E' does exist with an efficiency η'. As shown in Fig. 3.9a, this engine extracts heat Q_1' from the hot reservoir and performs the work W', and heat delivered to the cold reservoir is

$$Q_2' = Q_1' - W'$$

Let us also consider a Carnot engine E with efficiency η working between two reservoirs so that it extracts heat Q_1 from the hot reservoir, performs work W and delivers Q_2 heat to the cold reservoir which is equal to

$$Q_2 = Q_1 - W$$

Fig. 3.9 Schematic diagram for **a** A hypothetical engine E', which is more efficient than a Carnot engine, is connected to a Carnot engine E and (**b**) The composite refrigerator obtained from (**a**)

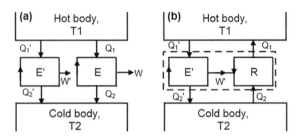

Let us also adjust the size of the cycle to make this engine perform the same amount of work as the hypothetical engine E' does. Since the hypothetical engine is assumed to be more efficient than the Carnot engine

$$\frac{W'}{Q'_1} > \frac{W}{Q_1} \quad (W' = W)$$

Therefore,

$$Q_1 > Q'_1$$

As the Carnot engine is a reversible engine, it can run backwards as a refrigerator as shown in Fig. 3.9b. One can imagine the hypothetical engine E' and the Carnot refrigerator R together act as a composite refrigerator. The heat extracted by this composite device (shown by the dotted line) from the cold reservoir is

$$Q_2 - Q'_2 = (Q_1 - W) - (Q'_1 - W'_1)$$
$$= Q_1 - Q'_1$$

As clear from Fig. 3.9b, the composite device also delivers the same amount of heat to the hot reservoir with no external work being required. But reservoirs are just large bodies with very large heat capacity where the temperature is unchanged upon the addition of heat. This clearly indicates the violation of the Clausius statement of the second law. Therefore, the assumed engine E' cannot exist and our starting assumption that $\eta' > \eta$ is incorrect.

Further, what happens when $\eta' = \eta$?. In this case, $Q_1 = Q'_1$, i.e., the composite device transfers no heat for no work, which is allowed. Hence, we conclude that, for any real engine

$$\eta \le \eta_{Carnot}$$

3.3.6   Two Carnot Engines Operating Between Similar Reservoirs

We imagine two Carnot engines E and E' which are operating between two identical reservoirs (at T_1 and T_2). Let the size of the working cycles is adjusted so that each

Fig. 3.10 The arrangement
of engine E and E' to prove
that all Carnot engines
operating between identical
reservoirs have the same
efficiency

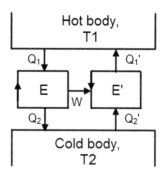

engine delivers the same amount of work W. Let engine E runs E' backwards as
shown in Fig. 3.10. Following the arguments we had in the previous section,

$$\eta_E \leq \eta'_E$$

If E' runs E backwards, then we can write

$$\eta'_E \leq \eta_E$$

Therefore, from these two results, we conclude $\eta_E = \eta'_E$. That is, "*All Carnot engines
operating between the same two reservoirs have the same efficiency*".

3.4   Carnot Engine Running Backwards

If heat engines run in reverse by putting in work in to move heat, two distinct
applications can be achieved. The first example is that of a refrigerator. The second
case belongs to a heat pump. Let us continue with the former.

3.4.1   Carnot Refrigerator

A refrigerator is a heat engine that operates in a reverse manner so that you put work
in (by external power supply) and cause a heat flow from a cold body to a hot body.
In the present case, the food inside the refrigerator serves as a cold reservoir and the
kitchen serves as a hot reservoir. The heat is withdrawn from the food (cold reservoir)
and given to the kitchen (hot reservoir). In the refrigerator, we want the heat to be
sucked out of the contents of the cold reservoir by doing electrical work. The first
law allows us to write

$$Q_1 = Q_2 + W$$

Fig. 3.11 A Carnot heat
engine running backwards
known as refrigerator

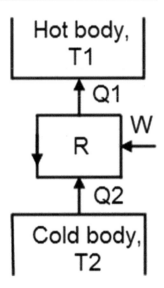

Therefore, for refrigerator, the efficiency is defined as

$$\eta_R^C = \frac{\text{heat extracted}}{\text{work done}}$$

$$= \frac{Q_2}{W}$$

$$= \frac{Q_2}{Q_1 - Q_2} = \frac{1}{\frac{Q_1}{Q_2} - 1}$$

To avoid confusion with the efficiency of the Carnot heat engine, the quantity η_R^C is also called as **coefficient of performance** (COP). The first and second laws can be used to understand the limits on COP in terms of T_1 and T_2. Note that there is no upper limit of COP. If we take the realistic values of T_1 and T_2 equal to 293 and 273 K, COP $= 13.6$. For real refrigerators, COP varies between 4 and 5 (Fig. 3.11).

3.4.2 Heat Pump

A heat pump transfers heat from a reservoir to a place where it is desired to add heat. A heat pump does it so at the expense of energy supplied externally by a motor. Thus, it does not violate any law of thermodynamics. Therefore, a heat pump is essentially a refrigerator, which is utilized in a different way. The inner part of the earth is usually at a higher temperature and one can use a heat pump to pump heat from there into a house that needs heating. Suppose we want to add heat Q_1 to the house and this requires W amount of work (electric) to be done. Therefore, the efficiency of the heat pump is defined as

$$\eta = \frac{Q_1}{W} \tag{3.21}$$

Since $Q_1 > W$, η is always greater than unity and hence the efficiency of a heat pump is always larger than 100%. This illustrates why heat pumps are attractive for heating purposes.

3.5   The Thermodynamic Temperature Scale

In Sect. 3.3, it was proven that the efficiency of a Carnot engine is independent of the nature of the working substance and depends only on the temperature of the reservoirs. Taking this idea, Lord Kelvin defined a temperature scale that does not depend upon the properties of any particular substance. This is known as "Kelvin's absolute thermodynamic scale of temperature". Therefore, a Carnot engine serves as a basis for the thermodynamic scale of temperature.

Let us consider the operation of a reversible engine E_{12} working between reservoirs at temperatures T_1 and T_2 as shown in Fig. 3.8 The efficiency of this reversible engine is

$$\eta = 1 - \frac{Q_2}{Q_1}$$

Comparing this with Eq. 3.15, we can write

$$\frac{T_1}{T_2} = \frac{Q_1}{Q_2} \tag{3.22}$$

Suppose we have a second Carnot engine E_{23} operating between the reservoir at T_2 and a third reservoir at T_3. If we assume that E_{23} absorbs the same amount of heat Q_2 from the reservoir at T_2 as was rejected by engine E_{12} to the reservoir at T_2. When the two engines run together, the reservoir at T_2 is therefore unchanged. Equation 3.22 can therefore be written as

$$\frac{T_1}{T_3} = \frac{Q_1}{Q_3} \tag{3.23}$$

Note that this equation does not involve the intermediate temperature T_2. Since the reservoir at T_2 is unchanged, we may consider the two engines E_{12} and E_{23}, acting together, to be a composite Carnot engine E_{13} that operates between the two reservoirs at T_1 and T_3. This composite engine is denoted by the dotted line in Fig. 3.8b. Therefore, the application of Eq. 3.22 shows that relation Eq. 3.23 is precisely the one that holds for the composite engine.

In general, we can rewrite Eq. 3.22 as

$$\frac{Q_1}{Q_2} = \frac{f(T_1)}{f(T_2)} = f(T_1, T_2) \tag{3.24}$$

Since $Q_1 > Q_2$, hence the function $f(T_1) > f(T_2)$ when $T_1 > T_2$. It means that the function f(T) increases as the temperature rises. Hence, it can be used to measure temperatures. This equation defines the absolute thermodynamic scale of temperature. On this scale, the ratio of any two temperatures equals the ratio of heat absorbed and rejected by a Carnot engine operating between these two reservoirs.

For a Carnot engine that is operating between two reservoirs, one at temperature T and another one at triple point of water T_{TP} (273.16 K), one can write

$$\frac{Q}{Q_{TP}} = \frac{T}{T_{TP}}$$

equivalently,

$$T = 273.16K \frac{Q}{Q_{TP}} \tag{3.25}$$

The ratio of heats $\frac{Q}{Q_{TP}}$ that are transferred during two isothermal processes bounded by two adiabatic curves can be measured with precision. As a result, this approach is used to measure temperatures below 1 K.

3.5.1   Absolute Zero and Size of a Degree on Absolute Scale of Temperature

Let us consider a Carnot engine operating between two reservoirs, one at the steam point and another one at the ice point. Corresponding indicator diagram for the engine is shown in Fig. 3.12. Here, AB and CD represent the isothermals for steam point and ice point, respectively. Then the efficiency of such an engine is given by

$$\eta = 1 - \frac{T_{ice}}{T_{steam}}$$

The work done by the engine during one complete cycle is numerically equal to the area bounded by the curve ABCD. One can divide this area into hundred equal parts by simply drawing isotherms parallel to AB or CD. In such a case, each isotherm will be at a temperature that is one degree lower than the upper isotherm and the area of each part will correspond to one degree on the absolute scale of temperature. Therefore, one degree on the absolute scale of temperature may be defined as the temperature difference, (between subsequent values) such that a Carnot engine working between that range absorbs (or rejects) energy equal to one-hundredth

Fig. 3.12 P-V diagram for a
Carnot engine working
between different
temperatures

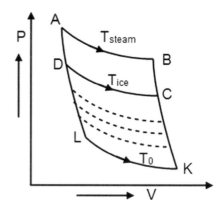

of the area of the indicator diagram corresponding to Carnot's cycle obtained when the
engine operates between steam point and ice point. In other words, the temperature
difference between the steam point and ice point on this scale is one hundred.

If one draws isotherms below the ice point (CD), an isotherm representing zero
of absolute scale can be drawn. The isotherm KL represents this zero and is called
as *absolute zero* of the thermodynamic scale.

If we consider an engine operating between steam point and absolute zero, the
efficiency in that case is given by

$$\eta = 1 - \frac{T_0}{T_{steam}} = 1$$

Because $T_0 = 0\,\mathrm{K}$ and also the efficiency can be written as

$$\eta = 1 - \frac{Q_2}{Q_1} = 1$$

Therefore,

$$\frac{Q_2}{Q_1} = 0 \ \text{ or } \ Q_2 = 0$$

Therefore, zero on the absolute scale represents the temperature of the sink at which
no heat can be rejected to it. For a sink at absolute zero, the efficiency of the heat
engine becomes unity, i.e., when sink temperature corresponds to absolute zero,
a Carnot engine converts entire heat (extracted from the hot reservoir) into work.
*Thus, the absolute temperature may be defined as the temperature at which a system
undergoes a reversible isothermal process without any transfer of heat to a cold
reservoir.* In other words, one can say that at absolute zero, isotherm and adiabatic
processes are identical.

3.5.2   Absolute Zero and Efficiency of a Carnot Engine

It follows that the efficiency of Carnot's engine in terms of temperatures of source and sink is given by

$$\eta = 1 - \frac{T_2}{T_1}$$

For a Carnot engine to be hundred percent efficient, T_2 must be equal to zero. It is only when the sink is at absolute zero that the entire heat will be converted into work. Note that nature does not provide us a reservoir at absolute zero, and a hundred percent efficient heat engine is practically impossible.

3.5.3   Feasibility of Negative Temperature on Absolute Scale

Let us try to understand if the negative temperature on absolute scale of temperature is feasible or not. For this, we assume that $T_2 = -TK$. Then, for an engine working between T_1 and $-TK$, the efficiency is

$$\eta = 1 - \frac{T_2}{T_1} = 1 - \frac{-T}{T_1} = 1 + \frac{T}{T_1} > 1$$

Thus, in such a case, the efficiency of the heat engine is greater than unity, a fact contradicting the second law of thermodynamics. Hence, the negative temperature on an absolute scale is not feasible.

3.6   The Equivalence Between Thermodynamic and the Ideal Gas Scales

Until this point, we have used the symbol T for absolute temperature as defined on the ideal gas scale. In this section, we will use another symbol T_{g1} and T_{g2} for the gas scale temperatures and use T for absolute scale temperatures. In moments we shall prove them to be identical. The efficiency of Carnot's engine when ideal gas is used as a working substance is given by

$$\eta = 1 - \frac{Q_2}{Q_1} = 1 - \frac{T_{g2}}{T_{g1}} \qquad (3.26)$$

Here, Q_1 and Q_2 are heats taken and rejected, at temperatures T_{g1} and T_{g2} of source and sink, respectively. Note that these temperatures are measured on the perfect gas scale and not on the absolute scale of temperature.

On the absolute scale of temperature, the efficiency of the heat engine is given by

$$\eta = 1 - \frac{Q_2}{Q_1} = 1 - \frac{T_2}{T_1} \qquad (3.27)$$

Comparing Eqs. 3.26 and 3.27, we obtain

$$\frac{T_{g2}}{T_{g1}} = \frac{T_2}{T_1} \tag{3.28}$$

The above equation indicates that the ratio of any two temperatures on the ideal gas scale is the same as on the perfect gas scale. let us try to understand this point from the following discussion.

Suppose that sink is at absolute zero temperature, i.e., $T_2 = 0$ K, we must have $T_{g2} = 0$ so that zero of absolute scale is identical to zero of ideal gas scale. Further, if T_{steam} and T_{ice} correspond to steam and ice temperatures, respectively, then on the absolute scale,

$$T_{steam} - T_{ice} = 100 \tag{3.29}$$

on the other hand, the difference between these two points measured with an ideal gas scale will be

$$T_{gsteam} - T_{gice} = 100 \tag{3.30}$$

For a Carnot engine operating between these two temperatures, the efficiency (when T is measured on the absolute scale) is

$$\eta = 1 - \frac{T_{ice}}{T_{steam}} = \frac{T_{steam} - T_{ice}}{T_{steam}} = \frac{100}{T_{steam}}$$

When the ideal gas scale is used

$$\eta = 1 - \frac{T_{gice}}{T_{gsteam}} = \frac{T_{gsteam} - T_{gice}}{T_{gsteam}} = \frac{100}{T_{gsteam}}$$

In both cases, the efficiency must be the same; therefore, we obtain

$$\frac{100}{T_{steam}} = \frac{100}{T_{gsteam}}$$

or

$$T_{gsteam} = T_{steam}$$

The above equation indicates that irrespective of temperature scale (absolute scale or ideal gas scale), the steam point has the same value. Further, using above result and Eqs. 3.29 and 3.30, we can show that

$$T_{gice} = T_{ice} \tag{3.31}$$

Therefore, the ice point has the same numerical value on both scales. The above results indicate that temperature on the ideal gas scale T_g agrees with the corresponding temperature T shown on the absolute scale. Hence, the absolute scale of temperature is exactly identical to the ideal gas scale.

3.7   Solved Problems

Q.1 **An engine has an efficiency of 22% and produces 2510 J of work. How much heat is rejected by the engine?**

Sol: The efficiency of the heat engine is

$$\eta = \frac{W}{Q_1} = \frac{2510 \text{ J}}{Q_1} = 0.22$$

this gives $Q_1 = 11409$ J. Further, for an engine, the amount of heat rejected is

$$Q_2 = Q_1 - W = 11409 - 2510 = 8899 \text{ J}$$

Q.2 **A steam turbine is operated with an intake temperature of 400 °C and an exhaust temperature of 150 °C. What is the maximum amount of work the turbine can do for a given heat input Q? Under what conditions is the maximum achieved?**

Sol: Here, $Q_1 = Q$, $T_1 = 673$ K and $T_2 = 423$ K. The maximum work achieved in heat engine is

$$W = Q_1 - Q_2 = Q_1 \left(1 - \frac{Q_2}{Q_1} \right) = Q \left(1 - \frac{T_2}{T_1} \right) = 0.37Q$$

Note that this maximum work is achievable only when steam turbine can work reversibly.

Q.3 **A Carnot engine has a cycle given below.**
(a) What thermodynamic processes are involved at boundaries AD and BC; AB and CD?
(b) Where is the work put in and where is it extracted?
(c) If the above is a steam engine with $T_{in} = 450$ K, operating at room temperature, calculate the efficiency.

Sol: (a) DA and BC are adiabatic processes whereas AB and CD are isothermal processes.
(b) Work is extracted in processes AB and BC whereas work is put in during processes CD and DA.
(c) The efficiency of the steam engine is

$$\eta = 1 - \frac{T_2}{T_1} = 1 - \frac{300}{450} = \frac{1}{3}$$

Q.4 **What is the maximum possible efficiency of an engine operating between two thermal reservoirs, one at $100\,^\circ$C and the other at $0\,^\circ$C?**

Sol: The efficiency is

$$\eta = 1 - \frac{T_2}{T_1} = 1 - \frac{273}{373} = 0.27$$

Q.5 **For a Carnot engine using an ideal gas, the adiabatic expansion ratio is 5 and the value of $\gamma = 1.4$. Calculate the efficiency of the engine.**

Sol: The efficiency of the engine is given by

$$\eta = 1 - \frac{T_2}{T_1} = 1 - \left(\frac{V_i}{V_f}\right)^{\gamma - 1} = 1 - \left(\frac{1}{5}\right)^{1.4 - 1} = 1 - 0.52 = 0.48$$

Therefore, the efficiency of the engine is 48%.

Q.6 **A monatomic ideal gas at $170\,^\circ$C is adiabatically compressed to 1/8 of its original volume. The temperature after compression is**

[JEST-2012]

(A)	2.1°	(C)	-200.5°
(B)	17°	(D)	1499°

Sol: For an adiabatic process

$$TV^{\gamma - 1} = constant, \quad T_1 V_1^{\gamma - 1} = T_2 V_2^{\gamma - 1}$$

$$T_2 = T_1 \left(\frac{V_1}{V_2}\right)^{\gamma - 1} = 443 \times 8^{5/3 - 1} = 1772K = 1499\,^\circ C$$

Q.7 **Efficiency of a perfectly reversible (Carnot) heat engine operating between absolute temperature T and zero is equal to**

[JEST-2012]

(A)	0	(C)	0.75
(B)	0.5	(D)	1

Sol: The efficiency is

$$\eta = 1 - \frac{T_2}{T_1} = 1 - \frac{0}{T} = 1$$

The correct option is D.

Q.8 **For a diatomic ideal gas near room temperature, what fraction of the heat supplied is available for external work if the gas is expanded at constant pressure?**

[JEST-2013]

(A) $\dfrac{1}{7}$ (C) $\dfrac{3}{4}$

(B) $\dfrac{5}{7}$ (D) $\dfrac{2}{7}$

Sol: During expansion at constant pressure, the volume changes from V_1 to V_2 and temperature changes from T_1 to T_2. As a result, we can write

$$PV_1 = nRT_1, \; PV_2 = nRT_2$$

During this expansion, the work done by the system on the outside world is

$$W = -P(V_2 - V_1) = -nR\Delta T$$

The corresponding increase in internal energy is

$$\Delta U = C_v \Delta T$$

The fraction of heat available for work is

$$\frac{W}{Q} = \frac{W}{\Delta U + W} = \frac{nR}{C_v + nR} = \frac{2}{7}$$

Q.9 **The entropy–temperature diagram of two Carnot engines, A and B, are shown in Fig. 3.14a. The efficiencies of the engines are η_A and η_B, respectively. Which one of the following equalities is correct?**

[JEST-2015]

(A) $\eta_A = \dfrac{\eta_B}{2}$ (C) $\eta_A = 3\eta_B$

(B) $\eta_A = \eta_B$ (D) $\eta_A = 2\eta_B$

Sol: The efficiency is defined as

$$\eta = \frac{W}{Q_1}$$

where W is the area bounded by the T-S curve and Q_1 is the area under high-temperature curve.

$$\eta_A = \frac{(2T - T)\,(4S - 1S)}{2T\,(4S - 1S)} = \frac{1}{2}$$

Similarly, η_B is

$$\eta_B = \frac{(4T - 3T)\,(4S - 1S)}{4T\,(4S - 1S)} = \frac{1}{4}$$

Therefore, $\eta_A = 2\eta_B$.

Q.10 As shown in Fig. 3.14b, the P-V diagram AB and CD are two isotherms at temperatures T_1 and T_2, respectively ($T_1 > T_2$). AC and BD are two reversible adiabats. In this Carnot cycle, which of the following statements are true?

[JAM-2015]

(A) $\dfrac{Q_1}{T_1} = \dfrac{Q_2}{T_2}$

(B) The entropy of the source decreases.

(C) The entropy of the system increases.

(D) Work done by the system $W = Q_1 - Q_2$

Sol: Options A, B and D are correct

Q.11 An ideal gas has a specific heat ratio $C_P/C_V = 2$. Starting at a temperature T_1 the gas undergoes an isothermal compression to increase its density by a factor of two. After this, an adiabatic compression increases its pressure by a factor of two. The temperature of the gas at the end of the second process would be

[JEST-2015]

(A) $\dfrac{T_1}{2}$

(B) $\sqrt{2}T_1$

(C) $2T_1$

(D) $\dfrac{T_1}{\sqrt{2}}$

Sol: Isothermal process takes place at constant temperature $T = T_1$. Let us assume the adiabatic process starts at point A(P_1, T_1) and finishes as B(P_2, T_2). For an adiabatic process between two equilibrium states

$$P_1^{1-\gamma}T_1^{\gamma} = P_2^{1-\gamma}T_2^{\gamma}$$

This gives

$$T_2 = T_1 \left(\frac{P_1}{P_2}\right)^{\frac{1-\gamma}{\gamma}} = T_1 \left(\frac{P_1}{2P_1}\right)^{\frac{1-2}{2}} = \sqrt{2}T_1$$

Fig. 3.13 CarnotCycle
diagram for Problem 21

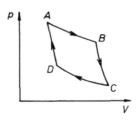

Q.12 **After the detonation of an atom bomb, the spherical ball of gas was found to be of 15 m radius at a temperature of 3×10^5 K. Given the adiabatic expansion coefficient $\gamma = 5/3$, what will be the radius of the ball when its temperature reduces to 3^3 K?**

[JEST-2017]

(A)   156 m   (C)   150 m
(B)   50 m   (D)   100 m

Sol: For an adiabatic process

$$TV^{\gamma-1} = constant$$

Therefore,

$$T_1 V_1^{\gamma-1} = T_2 V_2^{\gamma-1}$$

$$V_2 = V_1 \left(\frac{T_1}{T_2}\right)^{\frac{1}{1-\gamma}} = V_1 \left(\frac{T_1}{T_2}\right)^{3/2}$$

Hence, the radius of the ball after temperature reduces to 3×10^3 K is

$$R_2 = R_1 \left(\frac{T_1}{T_2}\right)^{1/2} = 15 \left(\frac{3 \times 10^5}{3 \times 10^3}\right)^{1/2} = 150 \text{ m}$$

Q.13 **Consider a Carnot engine operating between temperatures of 600 and 400 K. The engine performs 1000 J of work per cycle. The heat (in Joules) extracted per cycle from the high-temperature reservoir is.......(Specify your answer to two digits after the decimal point).**

[JAM-2017]

Sol: The efficiency of the heat engine is

$$\eta = \frac{W}{Q_1} = 1 - \frac{T_2}{T_1} = 1 - \frac{400}{600} = \frac{1}{3}$$

Therefore, $\dfrac{W}{Q_1} = \dfrac{1000}{Q_1} = \dfrac{1}{3}$ implies that $Q_1 = 3000$ J (Fig. 3.13).

Fig. 3.14 a T-S diagram for Q.9 and **b** P-V diagram for Q.10

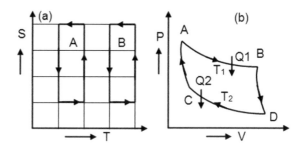

3.8 Multiple Choice Questions

Q.1 **First law of thermodynamics is the consequence of conservation of**

(**A**) work
(**B**) energy

(**C**) heat
(**D**) all

Q.2 **Which of the following is not a thermodynamical function**

(**A**) Enthalpy
(**B**) Work done

(**C**) Gibbs energy
(**D**) internal energy

Q.3 **The internal energy change in a system that has absorbed 2 kcals of heat and done 500 J of work is**

(**A**) 6400 J
(**B**) 5400 J

(**C**) 7900 J
(**D**) 8900 J

Q.4 **110 J of heat is added to a gaseous system whose internal energy is 40 J. Then the amount of external work done is**

(**A**) 150 J
(**B**) 70 J

(**C**) 110 J
(**D**) 40 J

Q.5 **An ideal gas goes from state A to state B via three different processes as indicated in the P-V diagram left panel in Fig. 3.15: If Q_1, Q_2, Q_3 indicate**

Fig. 3.15 Left panel for Q.5 and right panel for Q.6

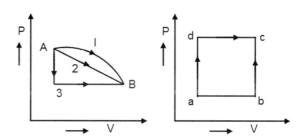

the heat absorbed by the gas along the three processes and ΔU_1, ΔU_2, ΔU_3 indicate the change in internal energy along the three processes, respectively, then

(A) $Q_1 > Q_2 > Q_3$ and $\Delta U_1 = \Delta U_2 = \Delta U_3$
(B) $Q_3 > Q_2 > Q_1$ and $\Delta U_1 = \Delta U_2 = \Delta U_3$
(C) $Q_1 = Q_2 = Q_3$ and $\Delta U_1 > \Delta U_2 > \Delta U_3$
(D) $Q_3 > Q_2 > Q_1$ and $\Delta U_1 > \Delta U_2 > \Delta U_3$

Q.6 A system is taken from state a to state c by two paths adc and abc as shown in the right panel in Fig. 3.15. The internal energy at a is $U_a = 10$ J. Along the path adc the amount of heat absorbed $\delta Q_1 = 50$ J and the work done $\delta W_1 = 20$ J whereas along the path abc the heat absorbed $\delta Q_2 = 36$ J. The amount of work done along the path abc is

(A) 6 J   (C) 12 J
(B) 10 J   (D) 36 J

Q.7 A gas is compressed isothermally to half its initial volume. The same gas is compressed separately through an adiabatic process until its volume is again reduced to half. Then

(A) Compressing the gas isothermally will require more work to be done.
(B) Compressing the gas through an adiabatic process will require more work to be done.
(C) Compressing the gas isothermally or adiabatically will require the same amount of work.
(D) Which of the case (whether compression through isothermal or through the adiabatic process) requires more work will depend upon the atomicity of the gas.

Q.8 An ideal gas is compressed to half its initial volume by means of several processes. Which of the process results in the maximum work done on the gas?

(A) isobaric   (C) isothermal
(B) isochoric   (D) adiabatic

Q.9 A monoatomic gas at a pressure P, having a volume V expands isothermally to a volume 2 V and then adiabatically to a volume 16 V. The final pressure of the gas is take $\gamma = \dfrac{5}{3}$

(A) 64P   (C) $\dfrac{P}{64}$
(B) 32P   (D) 16P

Fig. 3.16 Diagram for Q.10

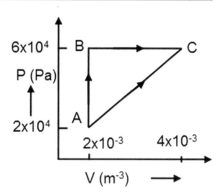

Q.10 **Figure 3.16 below shows two paths that may be taken by a gas to go from a state A to a state C. In process AB, 400 J of heat is added to the system and in process BC, 100 J of heat is added to the system. The heat absorbed by the system in the process AC will be**

(A) 500 J   (C) 300 J
(B) 460 J   (D) 380 J

Q.11 **Which of the following relations does not give the equation of an adiabatic process, where terms have their usual meaning?**

(A) $P^\gamma T^{1-\gamma} = \text{constant}$ (C) $PV^\gamma = \text{constant}$
(B) $P^{1-\gamma}T^\gamma = \text{constant}$ (D) $TV^{\gamma-1} = \text{constant}$

Q.12 **A monatomic ideal gas at $170\,^\circ$C is adiabatically compressed to 1/8 of its original volume. The temperature after compression is**
                                                                    [JEST-2012]

(A) 2.1° (C) -200.5°
(B) 17° (D) 1499°

Q.13 **Isothermal compressibility κ_T of a substance is defined as**

$$\kappa_T = \frac{1}{V}\left(\frac{\partial V}{\partial T}\right)_T$$

Its value for n moles of an ideal gas will be
                                                                    [JAM-2009]

(A) $\dfrac{1}{P}$ (C) $-\dfrac{1}{P}$
(B) $\dfrac{n}{P}$ (D) $-\dfrac{n}{P}$

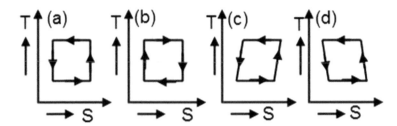

Fig. 3.17 Figure for Q.15

Q.14 **In a heat engine based on the Carnot cycle, heat is added to the working substance at constant**

[JAM-2019]

(A) Entropy   (C) Temperature
(B) Pressure   (D) Volume

Q.15 **Which one of the figures correctly represents the T-S diagram of a Carnot engine (Fig. 3.17)?**

[JAM-2018]

Keys and Hints to MCQ Type Questions

Q.1	B	Q.4	B	Q.7	B	Q.10	B	Q.13	C
Q.2	B	Q.5	B	Q.8	D	Q.11	A	Q.14	C
Q.3	C	Q.6	A	Q.9	C	Q.12	C	Q.15	B

Hint.3

$$dU = dQ + dW = -500 \text{ J} + (2 \times 4.2 \times 1000) \text{ J} = 7900 \text{ J}$$

Hint.4   dQ = +110 J (Heat added), dU = 40 J. By first law

$$dU = dQ + dW$$
$$40 \text{ J} = 110 \text{ J} + dW, \quad dW = -70 \text{ J}$$

Therefore, work is done by the system. The negative sign appears just by the choice of sign convention.

Hint.5   Internal energy is a perfect differential. Therefore,

$$\Delta U_1 = \Delta U_2 = \Delta U_3$$

Applying the First Law to three different processes

$$\Delta U_1 = Q_1 + W_1$$
$$\Delta U_2 = Q_2 + W_2$$
$$\Delta U_3 = Q_3 + W_3$$

Since $W_1 > W_2 > W_3$, this implies $Q_3 > Q_2 > Q_1$.

Hint.6 $U_a = 16$, for path adc: $\delta Q_1 = 50$ J, $\delta W_1 = -20$ J

$$U_c - U_a = 50 \text{ J} - 20 \text{ J} = 30 \text{ J}$$
$$U_c = 30 \text{ J} + 16 \text{ J} = 46 \text{ J}$$

Now for path abc:

$$U_c - U_a = 46 \text{ J} - 16 \text{ J} = 30 \text{ J}$$
$$= \delta Q_2 + \delta W_2 = 36 \text{ J} + \delta W_2$$
$$\delta W_2 = 30 \text{ J} - 36 \text{ J} = -6 \text{ J}$$

Where a negative sign merely indicates that work is done by the system.

Hint.9   For isothermal process $P_1 V_1 = P_2 V_2$, implying $P_2 = P/2$. For an adiabatic process,

$$P_2 V_2^\gamma = P_3 V_3^\gamma$$
$$\frac{P}{2}(2V)^\gamma = P_3(16V)^\gamma, \quad P_3 = \frac{P}{64}$$

Hint.10   While following path ABC:

$$\Delta U_{ABC} = \delta Q_{AB} + \delta Q_{BC} + \delta W = 400 \text{ J} + 100 \text{ J} + 120 \text{ J} = 380 \text{ J}$$

For path AC:

$$\Delta U_{AC} = \delta Q_{AC} - 80 \text{ J}$$
$$\delta Q_{AC} = \Delta U_{AC} + 80 \text{ J} = 380 \text{ J} + 80 \text{ J} = 460 \text{ J}$$

Hint.13   For an ideal gas $PV = nRT$. The isothermal compressibility is

$$\kappa_T = \frac{1}{V}\left(\frac{\partial V}{\partial P}\right)_T = -\frac{1}{P}$$

3.9   Exercises

1. Why must you put air conditioner in the window of a building and not in the middle of a room?

2. Is it possible to cool off your kitchen by leaving the refrigerator door open?

3. Is it possible to completely convert heat into work? Give reason to support your answer.

4. Explain the need of a cyclic process.

5. An isothermal process along with first law of thermodynamics implies $dT = 0$ and $dQ = -dW$. Does it mean complete conversion of heat into work is feasible?. Explain how it doesn't violate the second law of thermodynamics?

6. Show that efficiency (η) of a Carnot engine is given by $\eta = 1 - \dfrac{T_2}{T_1}$, where T_1 and T_2 are source and sink temperatures. Explain under what conditions the 100% efficiency can be achieved.

7. What are Kelvin–Planck and Clausius statements of second law of thermodynamics? Prove that they are equivalent.

8. Explain the principle of absolute scale of temperature. Compare the ideal gas scale with absolute scale of temperature. Why is the zero on this scale considered as lowest possible temperature on this scale?.

9. Deduce Kelvin's thermodynamic scale of temperature. Prove that
   (i) This scale of temperature is equivalent to that of ideal gas
   (ii) This scale of temperature does not depend on nature of working substance.

10. Show that net entropy change of working substance in a Carnot cycle is zero.

11. Obtain an expression for efficiency of a composite Carnot engine consisting of two Carnot engines connected in series.

12. State and prove Carnot theorem. Show that net entropy change of a working substance during Carnot cycle is zero.

13. Explain the working of Carnot refrigerator. Define coefficient of performance of a refrigerator and obtain an expression for it.

14. Show that no engine between two given temperatures can be more efficient than a reversible Carnot engine working between same temperatures.

15. Prove the equivalence of Clausius and Kelvin–Planck statement of second law of thermodynamics.

16. Explain the working principle of a refrigerator. Define coefficient of performance.

17. Obtain the efficiency of Carnot engine working between steam point and ice point.   (26.8%)

18. A Carnot engine whose low-temperature reservoir is at 280 K, has an efficiency of 50%. It is desired to increase the efficiency to 70%. By how many degrees should the temperature of high-temperature reservoir must be increased to achieve it?
                                                                        (280 K)

19. A Carnot engine absorbs 10^4 Calories of heat from a reservoir at 627 °C and rejects heat to a sink at 27 °C. What is the efficiency of engine? How much work does it perform? (given 1 Cal = 4.2 J)
                                                          (66.7%, W = 2800 J)

Entropy

4

This chapter begins with the mathematical definition of the second law of thermo-dynamics. This is followed by the introduction to the concept of entropy. We will define entropy change as a reversible process and learn how it can be evaluated for an irreversible process. We will also discuss various traditional examples of entropy change calculation including two non traditional cases of mass-spring system and a capacitor.

4.1   Clausius Theorem

Here we will discuss the Clausius theorem that led to the concept of entropy. It has two parts, one valid for a reversible process and another one for an irreversible process. Let us continue with the first case.

4.1.1   Reversible Case

Let us take an example of a reversible process i-f as shown in Fig. 4.1. Let us draw steps i-x, x-y and y-f such that the former represents an adiabatic process, second one represents an isothermal process and the latter again an adiabatic process. All three processes are reversible in nature. If somehow we can adjust these intermediate processes in such a way that work done (area under the curve) during process i-f is equal to work done during process ixyf, i.e.

$$W_{if} = W_{ixyf} \tag{4.1}$$

© The Author(s) 2022
S. Sharma, *Thermal and Statistical Physics*,
https://doi.org/10.1007/978-3-031-07685-5_4

Fig. 4.1 i-f is a reversible process, i-x a reversible adiabatic process, x-y a reversible isothermal process and y-f a reversible adiabatic process

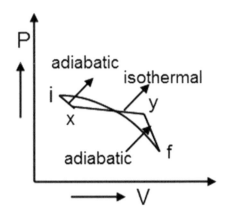

Application of first law to reversible process i-f gives

$$U_f - U_i = Q_{if} + W_{if}$$
$$Q_{if} = U_f - U_i - W_{if} \qquad (4.2)$$

For process i-x-y-f, we can write

$$Q_{ixyf} = U_f - U_i - W_{ixyf} \qquad (4.3)$$

Then, Eqs. 4.1, 4.2 and 4.3 imply

$$Q_{if} = Q_{ixyf}$$

Since processes ix and yf involve no heat exchange (adiabatic), the only heat exchange involved is during isothermal process xy. Therefore, we can write

$$Q_{if} = Q_{xy} \qquad (4.4)$$

The above equation implies that if a reversible process can be replaced between same equilibrium states by a zigzag process involving alternate reversible adiabatic and isothermal processes, then the heat change involved during an isothermal segment of a new process is equivalent to net heat change in the original process.

Now we will utilize the above result for a cyclic process. An example of such a process is shown in Fig. 4.2. It displays a closed curve on a generalized work diagram. This entire cyclic process is divided into a large number of zigzag closed paths, each consisting of alternate isothermal and adiabatic processes such that the area bounded by the closed curve is equal to the area enclosed by all zigzag closed paths. Each such zigzag close path represents a Carnot cycle (two reversible isotherms and two reversible adiabatic processes). Let us pick the first zigzag closed path ABCD. It consists of two isothermal processes at temperature T_1 during which heat Q_1 is absorbed, and CD at temperature T_2, during which heat Q_2 is rejected. These two

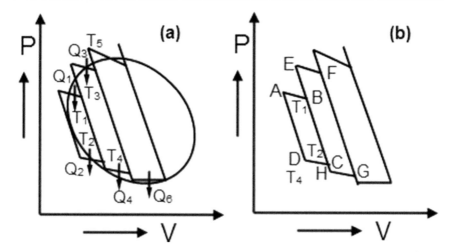

Fig. 4.2 a A closed reversible cycle and zigzag closed path constituting alternate reversible adiabatic and reversible isothermal processes **b** enhanced view of reversible cycles, ABCD and EFGH. Isothermal process AB occurs at temperature T_1, followed by an adiabatic process BC, an isothermal process CD at T_2 and at the end another adiabatic process DA

isotherms are bounded by two adiabatic curves BC and AD, and therefore, cycle ABCD constitutes a Carnot cycle. Therefore, we can write

$$\frac{Q_1}{T_1} = \frac{Q_2}{T_2}$$

Note that while writing the above equation, we have used absolute values of heat added or leaving the system. Let us now use the sign convention, thus taking Q_1 positive and Q_2 negative (heat rejected). Considering the sign, we can write

$$\frac{Q_1}{T_1} + \frac{Q_2}{T_2} = 0$$

where Q_1 is a positive number and Q_2 is a negative number. Applying similar arguments for cycle EFGH, we can write

$$\frac{Q_3}{T_3} + \frac{Q_4}{T_4} = 0$$

If all equations corresponding to all zigzag cycles are added, then we obtain

$$\frac{Q_1}{T_1} + \frac{Q_2}{T_2} + \frac{Q_3}{T_3} + \frac{Q_4}{T_4} = 0$$

Further, note that heat transferred during the adiabatic portion of the zigzag cycle is zero. We may write for all zigzag cycles (i in numbers)

$$\sum_i \frac{Q_i}{T_i} = 0 \qquad\qquad (4.5)$$

where Q_i is the net heat exchange at temperature T_i. Now, if we divide the cycles in such a way so that dQ_i be the infinitesimal small amount of heat exchanges at temperature T_i, then summation in above equation can be replaced with an integral and above equation reduces to

$$_R \oint \frac{dQ_i}{T_i} = 0 \tag{4.6}$$

the circle through the integral sign indicates that integration takes place over the complete cycle and the letter R emphasize the fact that it is only valid for a reversible cyclic process. This result is known as *Clausius theorem* and is the first part of Caluisius's mathematical statement of the second law of thermodynamics. The second part is valid for an irreversible process and will be discussed in Sect. 4.1.2.

4.1.2   Irreversible Case, Clausius Inequality

Clausius gave a very important theorem for cyclic processes which leads to the concept of entropy. This theorem is known as the Clausius inequality. To prove this theorem, we will consider two engines working between same reservoirs; see Fig. 4.3. We also assume that one (E) of the engines is reversible and the other (E$'$) one is irreversible. If η and η'_{irr} be the efficiencies of these two engines, then from Carnot's theorem

$$\eta > \eta'_{irr}$$

$$\frac{W_{rev}}{Q_1} > \frac{W_{irrev}}{Q_1}, \quad W_{rev} > W_{irrev}$$

Fig. 4.3 Schematic illustration of a reversible and irreversible engine connected across two same reservoirs

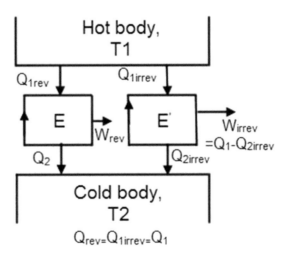

$Q_{rev} = Q_{1irrev} = Q_1$

Applying the first law and assuming $Q_{1rev} = Q_{1irrev} = Q_1$, gives

$$W_{rev} = Q_1 - Q_2, \quad W_{irrev} = Q_1 - Q_{2irrev}$$

Therefore

$$Q_1 - Q_2 > Q_1 - Q_{2irrev} \text{ implying } Q_{2irrev} > Q_2$$

Making use of the above result and the fact that for a reversible cyclic process, we have

$$_R \oint \frac{dQ_i}{T_i} = 0$$

Therefore, for an irreversible process

$$_{Irrev} \oint \frac{dQ}{T} = \frac{Q_1}{T_1} - \frac{Q_{2irrev}}{T_2} < 0$$

or

$$\oint \frac{dQ}{T} \leq 0 \tag{4.7}$$

This result is known as Clausius inequality. Note that equality holds for a reversible process, whereas inequality hold for an irreversible process.

4.2   Entropy and Mathematical Form of the Second Law

Let us consider a thermodynamic system with two equilibrium states i and f as shown in the generalized work diagram in Fig. 4.4. Since i and f are the equilibrium states, it is possible to take the system from initial state i to final state f and back to state i through a large number of reversible paths. Let R_1 and R_2 be two such reversible paths. We can take the system from i to f through path R_1 and then bring it back to initial equilibrium state through path R_2.

These two paths form a reversible cycle and application of Clausius theorem implies

$$_{R_1 R_2} \oint \frac{dQ}{T} = 0$$

The above integral can also be written as

$$_{R_1} \int_i^f \frac{dQ}{T} +_{R_2} \int_f^i \frac{dQ}{T} = 0$$

Fig. 4.4 Two different reversible paths joined by two equilibrium points

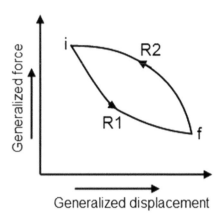

as R_2 is a reversible path, we can write

$$-_{R_2}\int_f^i \frac{dQ}{T} =_{R_2} \int_i^f \frac{dQ}{T}$$

Thus, finally we get

$$_{R_1}\int_i^f \frac{dQ}{T} =_{R_2} \int_i^f \frac{dQ}{T} \tag{4.8}$$

Equation 4.8 implies that the quantity $_R\int_i^f \frac{dQ_R}{T}$ is path independent. Therefore, $\exists$ a state function whose value at final state minus its value at initial state equals the integral $_R\int_i^f \frac{dQ_R}{T}$. *This state function is called as Entropy.* Therefore, the change in entropy between the initial and final equilibrium state is

$$\Delta S = S_f - S_i =_R \int_i^f \frac{dQ_R}{T} \tag{4.9}$$

This equation holds for any reversible path connecting initial and final states. Thus, entropy change between points i and f is independent of the choice of reversible paths connecting i and f. Therefore, similar to internal energy function U, entropy S is also a state function. Note that if the two states are infinitesimally close to each other, the above equation can also be written as

$$dS = \frac{dQ_R}{T} \quad or \quad dQ_R = TdS$$

This equation gives the mathematical representation of the second law of thermodynamics. This does not mean that a reversible heat transfer must take place for the entropy of a system to change. Indeed, there are many situations in which the entropy of a system changes when there is no transfer of heat (see for instance Sect. 4.2.6).

4.2.1   The Principle of Increasing Entropy, Irreversible Change

Most of the naturally occurring processes are irreversible in nature and accompanied by a net increase (or no change) in the entropy of the universe. This point can be easily understood with the help of Clausius inequality. In order to understand it in more detail, let us consider a cycle which contains an irreversible section (A to B) and a reversible section (B to A), as shown in Fig. 4.5. This makes entire cycle as irreversible one. The Clausius inequality for this cyclic process gives

$$\oint \frac{dQ}{T} \leq 0 \tag{4.10}$$

For the complete cycle, we have

$$\int_A^B \frac{dQ_{Irrev}}{T} + \int_B^A \frac{dQ_R}{T} \leq 0$$

$$\int_A^B \frac{dQ_{Irrev}}{T} - \int_A^B \frac{dQ_R}{T} \leq 0$$

Where dQ_{Irrev} and dQ_R represent heat exchange during the irreversible and reversible parts of the process. The second step we have written by using the fact that for reversible part B$\longrightarrow$A of the cycle, we can write

$$\int_B^A \frac{dQ_R}{T} = - \int_A^B \frac{dQ_R}{T}$$

It follows that

$$\int_A^B \frac{dQ_{Irrev}}{T} \leq \int_A^B \frac{dQ_R}{T} = dS \tag{4.11}$$

The equality in this expression holds if the process on the right-hand side is reversible. This equation implies that in an infinitesimal irreversible process between

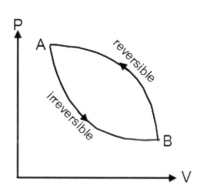

Fig. 4.5 Example of an irreversible cycle that consists of an irreversible (A-B) process followed by a reversible (B-A) process back to the initial state again

a pair of equilibrium states (A and B), there is a definite entropy change dS, but this is larger than the heat supplied in that irreversible process divided by the temperature of the external heat source. One must not confuse this heat with the heat supplied in any imaginary reversible process used to calculate dS. In the next subsection, we will take an example of an irreversible cycle and evaluate respective entropy change.

4.2.2 Entropy Change in a Reversible Cycle

Having defined the entropy through Eq. 4.9 and making use of Clausius theorem for a reversible cyclic process (Eq. 4.6)

$$\oint \frac{dQ}{T} = 0$$

Therefore, in any reversible cycle, the net change in entropy is zero. We can arrive at this result by taking an example of Carnot's cycle which is reversible in nature. For a Carnot cycle, Q_1 heat enters into the working substance and Q_2 exits through it. Therefore, we can write for one cycle, $\frac{Q_1}{T_1} = \frac{Q_2}{T_2}$ and Q_2 being negative, therefore

$$\oint \frac{dQ}{T} = \frac{Q_1}{T_1} + \frac{Q_2}{T_2} = 0$$

4.2.3 Entropy Change in an Irreversible Cycle

As we noted earlier, Carnot cycle is an example of a reversible cycle and such an engine possesses maximum efficiency given by:

$$\eta = 1 - \frac{Q_2}{Q_1} = 1 - \frac{T_2}{T_1}$$

Further, we also have $\frac{Q_2}{Q_1} = \frac{T_2}{T_1}$. However, if the cycle is irreversible, the efficiency η' will be lowered and given by

$$\eta' = 1 - \frac{Q_2}{Q_1} < 1 - \frac{T_2}{T_1} \quad or$$
$$\frac{Q_2}{Q_1} > \frac{T_2}{T_1} \quad or \quad \frac{Q_2}{T_2} > \frac{Q_1}{T_1}$$

During the complete irreversible cycle, the entropy of the source decreases by $\frac{Q_1}{T_1}$ and that of the sink increases by $\frac{Q_2}{T_2}$. The net entropy change of the working substance in this case also will be zero (similar to the reversible cycle in a Carnot engine). This is because, when the cycle is completed, the working substance recovers its initial state. Hence, the total increase in the entropy of the system (working substance) plus the surroundings (source and sink) is given by

$$\Delta S = \Delta S_{system} + \Delta S_{surrounding} = \frac{Q_2}{T_2} - \frac{Q_1}{T_1}$$

This is a positive quantity since $\frac{Q_2}{T_2} > \frac{Q_1}{T_1}$. *Therefore, in an irreversible cycle the entropy of the system plus its surroundings always increases.* In general, we can write

$$\Delta S_{Universe} \geq 0$$

where equality holds for a reversible cycle and inequality for an irreversible cycle. Here the word universe consists of the system and surroundings.

4.2.4 Entropy Changes and Second Law

It was pointed out in the beginning that the second law deals with the direction of heat flow. It has emerged out from experience, and one cannot prove it theoretically. The discussion from previous sections pertaining to entropy change in a reversible and irreversible process allows us to put the second law as follows:

All naturally occurring (or irreversible) processes proceeding from one equilibrium state to another equilibrium state of a system takes place in a direction which causes the entropy of the system plus surroundings to increase. Only in the limiting (or ideal) case of a reversible process, the entropy of the system plus surroundings remain constant (see Chap. 3, entropy change for a Carnot cycle which is reversible), that is, in general, we can write $\Delta S_{Universe} \geq 0$. And a process with $\Delta S_{Universe} < 0$ is impossible.

4.2.5 Entropy Change of an Isolated System

Now, consider that the system is thermally isolated. Then, $dQ = 0$ and

$$dS \geq 0 \qquad\qquad (4.12)$$

This is another statement of the second law of thermodynamics. It shows that for any thermally isolated (i.e., it can neither exchange heat nor work with its surroundings, then the entropy of the system alone is the entropy of the universe) system, the entropy either stays the same (for a reversible change) or increases (for an irreversible change).

For a reversible process in a thermally isolated system, $dS = dQ_R/T = 0$ because no heat can flow in or out. Therefore, we conclude that "*The entropy of a thermally isolated system increases in any irreversible process and is unaltered in a reversible process*". This is the principle of increasing entropy. Another way of putting this statement is that processes involving $\Delta S_{universe} < 0$ are impossible, that is, the entropy of an isolated system can never decrease.

4.2.6 Free Expansion and Corresponding Entropy Change

Consider a double-sectioned container (Fig. 4.6), with gas in the left part. The state of the gas is characterized by well defined values of the thermodynamic variables P, V and T. Let the container is surrounded by a adiathermal wall so that heat exchange between the double-sectioned container and the reservoir (surrounding) is absent. We also assume that the air from right-hand part is removed and there is a vacuum. Let the volumes of each part is equal to V and the intervening partition be removed so that the gas molecules rush into the right side before eventually settling down to a new equilibrium state. This process is known as free expansion. We wish to calculate the work done by the gas in this expansion. An immediate application of the equation $W = -\int_{V_i}^{V_f} P dV$ (which holds for a reversible change) would give a finite number, as the volume is changing during the expansion. But the relation $W = -\int_{V_i}^{V_f} P dV$ cannot be applied here as this process is irreversible. In this special case, the work done by the gas on the surrounding is zero, as there is no interaction between the system (confined chamber) and the surrounding.

The following two points should be clear while dealing with the thermodynamic processes:

1. It is important to be clear as to what is the system. Here it is the chamber as a whole and not just the left-hand part initially containing all the gas.
2. $W = -\int_{V_i}^{V_f} P dV$ is applicable only to reversible processes and to those special irreversible processes where there is no finite pressure drop across the piston.

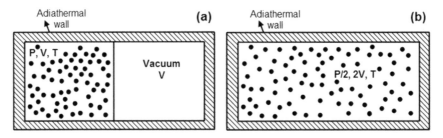

Fig. 4.6 Schematic diagram displaying free expansion of gas inside a chamber that is thermally isolated from surrounding **a** Gas enclosed in volume V only **b** Free expansion of gas, gas confined to volume 2V

If a system, say a gas, expands in such a way that heat exchanges are absent (adiabatic process) and also no work is done by or on the system, then the expansion is called the '*free expansion*'. For this case, applying the first law of thermodynamics, we get

$$U_f - U_i = Q + W = 0 + 0$$

This implies that $U_f = U_i$ and $T_f = T_i$, i.e., the initial and the final internal energies as well as temperatures are equal in free expansion. Therefore, the internal energy and the temperature of the gas do not change during the free expansion.

The Entropy Change in a Free Expansion

The process of free expansion just described above is an example of an irreversible process, involving no change in temperature, internal energy of the gas and absence of heat exchange. The entropy change for a reversible process are defined as:

$$\Delta S = S_f - S_i =_R \int_i^f \frac{dQ_R}{T} \tag{4.13}$$

In order to apply the above equation to the process of free expansion, we imagine a reversible isothermal doubling of the volume and calculate the entropy change for this. Such an expansion could be achieved by allowing the gas to expand slowly but being in thermal contact with a reservoir at T, as shown in Fig. 4.6. Applying the first law to this process, we have.

$$dQ = dU + PdV = 0 + PdV \tag{4.14}$$

Because dU $= 0$, as T $=$ constant. Therefore

$$dS = \frac{dQ}{T} = \frac{P}{T}dV = \frac{nR}{V}dV \tag{4.15}$$

Here, we have assumed that gas is ideal (PV $=$ nRT) and n moles are confined in the container. Therefore, the net change in entropy is

$$\Delta S = nR \int_V^{2V} \frac{dV}{V} = nR \ln 2 \tag{4.16}$$

Thus, we see that, in an irreversible process, without any heat exchange, the net change in entropy is non-zero. It is often mistakenly thought that heat has to flow into a system for there to be an entropy change and, conversely, that any adiabatic process takes place at constant entropy, or isentropically. This example shows this not to be so. As $dQ_R = TdS$ applies to a reversible process. only, a process has to be both adiabatic and reversible to be isentropic. Although our free expansion is adiabatic, it is not isentropic because it is irreversible.

There are several points to note from this result:

(a) Here $\Delta S_{system} = \int_V^{2V} \frac{dQ}{T} = nR \ln 2 > 0$, the equality between ΔS and $\frac{dQ}{T}$
holds for a reversible process

(b) $\Delta S_{total} = \Delta S_{system} + \Delta S_{surronding} > 0$, implies that process of free expansion is not reversible.

(c) A direct connection exists between the work needed to restore the system to the original state and the entropy change: The work required to bring the system back into its original position, i.e., $(V_2$ to $V_1)$ is

$$W = -\int_{V_2}^{V_1} P \, dV$$

$$= -nRT \int_{V_2}^{V_1} \frac{dV}{V} = nRT \ln \frac{V_2}{V_1} = T\Delta S \qquad (4.17)$$

Therefore, $W = T\Delta S = \dfrac{PV\Delta S}{nR}$.

4.2.7   Entropy Change for an Inelastic Collision

In an inelastic collision, the mechanical energy is converted into thermal energy. Such a process is clearly irreversible in nature. The entropy of the universe must, therefore, increase. Consider an object of mass m falling from a height h and making an inelastic collision with the ground [Fig. 4.10b]. Assuming that after the process, the object, ground, and atmosphere are all at a temperature T. If we consider the object, ground, and atmosphere as our isolated system, there is no heat conducted into or out of the system. The state of the isolated system has changed as its internal energy has been increased by an amount mgh. This change is identical as if we add heat Q = mgh to the system at constant temperature T. Since it is an irreversible process, we replace it with an imaginary reversible process in which heat Q_R = mgh is reversibly added to the system at temperature T, so that entropy change $\Delta S = \dfrac{dQ_R}{T}$.

4.3   Entropy–Temperature Diagrams

A Carnot cycle consists of two reversible isothermal processes and two reversible adiabatic processes as shown in Fig. 4.7a. Therefore, it forms a rectangle on a T-S diagram as shown in Fig. 4.7b, irrespective of the working substance. During the isothermal expansion AB at constant temperature T_1, the system utilizes its energy. In order to keep it at a constant temperature, heat has to be supplied externally. During isothermal expansion, the entropy of the working substance increases from S_1 to S_2. The adiabatic expansion BC occurs at the cost of energy of the working substance (no heat exchange between the system and the surrounding). Therefore, during this step, the temperature falls from T_1 to T_2, but the entropy remains constant. During

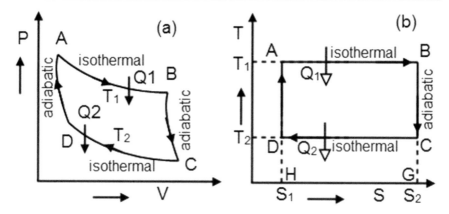

Fig. 4.7 a Carnot cycle and corresponding **b** temperature–entropy (T-S) diagram

the isothermal compression CD at constant temperature T_2, the entropy decreases from S_2 to S_1. Finally, during the adiabatic compression DA, the temperature rises to T1, and the entropy remains constant. Let the working substance absorb heat Q_1 (at T_1) during isothermal expansion AB and rejects heat Q_2 (at T_2) during isothermal compression CD. Using the relation dQ = TdS, we have

$$Q_1 = T_1(S_2 - S_1)$$
$$= AH \times AB = Area \ ABGH$$

Similarly, the heat Q_2 rejected at T_2 is

$$Q_2 = T_2(S_2 - S_1)$$
$$= DH \times DC = Area \ DCGH$$

The difference $Q_1 - Q_2$ is converted into useful work and is, therefore, the available energy per cycle. It is given by

$$Q_1 - Q_2 = area \ ABGH - \quad area \ DCGH = area ABCD$$

Therefore, the area on T-S diagram represents the energy available for performing useful work.

4.3.1   Utility of T-S Diagram

Instead of P-V diagrams, the T-S diagrams are very useful for the study of engines. The area of the cycle can be easily computed. Let us evaluate the efficiency of the

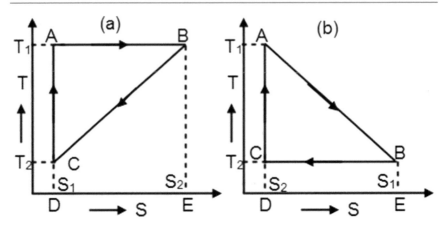

Fig. 4.8 a and **b** T-S diagrams for two different processes

Carnot Engine from T-S diagram. The efficiency of the engine is defined as

$$\eta = \frac{W}{Q_1} = \frac{Q_1 - Q_2}{Q_1} = \frac{area\ ABCD}{area\ ABGH} = \frac{AB \times BC}{AB \times AH}$$
$$= \frac{BC}{AH} = \frac{T_1 - T_2}{T_1} = 1 - \frac{T_2}{T_1}$$

Which is the same as obtained earlier. Let us take an example to understand the utility
of T-S diagrams.

Ex:4.1 **Two cyclic processes are shown in Fig. 4.8. Which one is more efficient?**
   Sol: Let Q_1 be the heat absorbed during the process in which entropy increases
      and Q_2 be the heat rejected during the process in which entropy decreases,
      then the efficiency is given by

$$\eta = \frac{Q_1 - Q_2}{Q_1} = \frac{area\ ABC}{Q_1}$$

   Here, $Q_1 - Q_2$ is the energy available for performing useful work and this is
   equal to area ABC (similar in both cases) of T-S diagrams. However, in each
   cycle, Q_1 is given by area ABED, which is larger for cycle (a). Therefore,
   cycle (b) is more efficient.

4.3.2   Slope of T-S Curve

If heat transfer occurs in a reversible process, we have the relation

$$dQ = TdS$$

If 1 mole of a gas is heated at constant volume (i.e., an isochoric process) at temperature T, then we can write

$$dQ = C_V dT$$

Above two equations yield

$$\frac{dT}{dS} = \frac{T}{C_V}$$

here dT/dS is the slope of the isochoric curve on T-S diagram. On the other hand, if the process is carried out at constant pressure (isobaric process), then

$$dQ = C_P dT$$

and we get

$$\frac{dT}{dS} = \frac{T}{C_P}$$

Therefore, the slopes of isochoric and isobaric curves on T-S diagram are T/C_V and T/C_P, respectively. Since $C_P > C_V$, the slope of the isochoric curve is larger. The ratio of the slopes is

$$\frac{\text{slope (isochoric)}}{\text{slope (isobaric)}} = \frac{\frac{T}{C_V}}{\frac{T}{C_P}} = \frac{C_P}{C_V} = \gamma$$

4.4   Central Equation of Thermodynamics

By combining the first and second law of thermodynamics, we can obtain a more elegant and useful statement of the first law of thermodynamics. In differential form, the law is

$$dU = dQ + dW$$

Here dQ and dW are not perfect differentials and are, therefore, not individually defined for a given change of state. In order to separate the contributions from heat and work to U, the constraints on the system have to be known so that the path of the change can be found. If the change takes place reversibly, the work done can be expressed as dW = −PdV. That is in terms of the system's parameters of state and only when the path is known, this can be integrated. The above equation is in fact the statement of the law of conservation of energy and holds good for reversible as well as irreversible processes. Therefore, taking an ideal gas as our model

$$dU = dQ + dW \qquad \text{always} \qquad (4.18)$$

Further, for an infinitesimal reversible process

$$dW = -PdV \qquad \text{reversible} \tag{4.19}$$

and we have defined entropy such that

$$dQ = TdS \qquad \text{reversible only} \tag{4.20}$$

Therefore, the above equations lead to

$$dU = TdS - PdV \qquad \text{reversible} \tag{4.21}$$

Note that this equation contains the variables that are functions of state so that all the differentials are perfect. As a result, the integration of this equation must be independent of the path of integration and the equation may be applied to any change of state, however, accomplished. To use the equation, we only require that initial and final states be defined and that there is some reversible path between them. To find the change in internal energy accompanying an irreversible change, we choose any convenient reversible path between initial and final states and integrate Eq. 4.21 along it. Thus, by expressing ΔU in terms of state functions only, we have

$$\boxed{dU = TdS - PdV} \qquad \text{always} \tag{4.22}$$

This equation is true for reversible as well as irreversible processes. All quantities in the equation are state functions, whose values are fixed by the end points and not on the path joining the end points.

For irreversible changes, the Eqs. 4.19 and 4.20 do not hold. For irreversible case, we have $dQ \leq TdS$ so that for Eq. 4.22 to remain valid, we should have $dW \geq -PdV$. One would expect this as in presence of irreversibility (for instance when there is friction), the total work done is greater than that which would be required to bring the same change in volume of the system without the irreversibility.

Equation 4.22 implies that internal energy changes whenever S or V changes. Therefore, U can be written in terms of the thermodynamic variables S and V, which are its so-called *natural variables*. Both of these variables are extensive (scale with the dimensions of the system). The variables P and T are intensive and behave like forces, since they show how the internal energy change w.r.t. some parameter. Thus, we can write U as $U \to U(S, V)$. Mathematically, for a small change in U, we can write

$$dU = \left(\frac{\partial U}{\partial S}\right)_V dS + \left(\frac{\partial U}{\partial V}\right)_S dV \tag{4.23}$$

Compairing Eqs. 4.22 and 4.23, we can identify P and T as

$$P = -\left(\frac{\partial U}{\partial V}\right)_S \quad \text{and} \quad T = \left(\frac{\partial U}{\partial S}\right)_V$$

The ratio of P and T can also be written in terms of the variables U, S and V, as follows:

$$\frac{P}{T} = -\left(\frac{\partial U}{\partial V}\right)_S \cdot \left(\frac{\partial S}{\partial U}\right)_V = \left(\frac{\partial S}{\partial V}\right)_U \tag{4.24}$$

where the last equation has been written using the reciprocity theorem. This means that if we know U(S,V) we can always find T and P by using these relations. In other words, T and P are, strictly speaking, redundant. However, it is highly inconvenient in practice to measure S or U than T and P, which can be easily measured.

Ex:4.2 **The internal energy U of a system is given by** $U = \dfrac{bS^3}{VN}$, **where b is a constant and other symbols have their usual meaning. The temperature and pressure of this system are equal to, respectively,**

[NET-2011]

(A) $T = \dfrac{bS^3}{kVN}, P = \dfrac{bS^3}{NV^2}$ (C) $T = \dfrac{bS^3}{V^2N}, P = \dfrac{2bS^2}{NV^2}$

(B) $T = \dfrac{3bS^2}{VN}, P = \dfrac{bS^3}{NV^2}$ (D) $T = \left(\dfrac{S}{N}\right), P = \dfrac{bS^3}{NV^2}$

Sol: We will make use of the equations we have just developed in the previous section. Here, the internal energy is $U = \dfrac{bS^3}{VN}$. The temperature and pressure are

$$T = \left(\frac{\partial U}{\partial S}\right)_V = \frac{3bS^2}{VN} \quad \text{and} \quad P = -\left(\frac{\partial U}{\partial V}\right)_S = \frac{bS^3}{NV^2}$$

Therefore, B is the right option.

Ex:4.3 **The entropy of an ideal gas is given by**
$$S = \frac{n}{2}\left[\sigma + 5R\ln\frac{U}{n} + 2R\ln\frac{V}{n}\right].$$
The internal energy 'U' of the ideal gas is

(A) $\dfrac{1}{2}nRT$ (C) $\dfrac{5}{2}nRT$

(B) $\dfrac{3}{2}nRT$ (D) $\dfrac{7}{2}nRT$

Sol: From definition of temperature

$$\frac{1}{T} = \left(\frac{\partial S}{\partial U}\right)_V = \frac{n}{2}5R\frac{1}{U} \quad Or \quad U = \frac{5}{2}nRT$$

Therefore, **C** is the right option.

Ex.4.4 **The entropy of an ideal gas is given by**
$$S = \frac{n}{2}\left[\sigma + 5R\ln\frac{U}{n} + 2R\ln\frac{V}{n}\right].$$
The specific heat C_V and C_P of the ideal gas are, respectively

(**A**) $\frac{3}{2}nR$ and $\frac{5}{2}nR$ (**C**) $\frac{3}{2}nR$ and $\frac{7}{2}nR$

(**B**) $\frac{5}{2}nR$ and $\frac{7}{2}nR$ (**D**) $\frac{5}{2}nR$ and $\frac{3}{2}nR$

Sol: Temperature can be evaluated as

$$\frac{1}{T} = \left(\frac{\partial S}{\partial U}\right)_V = \frac{n}{2}5R\frac{1}{U} \quad Or \quad U = \frac{5}{2}nRT$$

$$C_V = \left(\frac{\partial U}{\partial T}\right)_V = \frac{5}{2}nR \quad \text{and} \quad C_P = C_V + nR = \frac{7}{2}nR$$

Therefore, **B** is the right option.

Ex.4.5 **A large reservoir at temperature T_R is placed in thermal contact with a small system at temperature T_S s.t. $T_R > T_S$. They both end up at the temperature of the reservoir, T_R. Calculate (a) the change in entropy of the reservoir (b) the change in entropy of the system (c) total change in entropy of the universe OR the composite system and reservoir (d) repeat the calculations when $T_S > T_R$**

Sol: The heat transferred from the reservoir to the system is $\Delta Q = (T_R - T_S)$, where C is the heat capacity of the system. The heat ΔQ is transferred from reservoir to the system at fixed temperature T_R of the reservoir. Therefore, entropy change in reservoir is

$$\Delta S_R = -\frac{\Delta Q}{T_R} = C\frac{T_S - T_R}{T_R} \qquad (4.25)$$

Because $T_R > T_S$, therefore, ΔS_R is negative. The temperature of the system changes from T_S to T_R. Therefore, entropy change for the system is

$$\Delta S_S = \int \frac{dQ}{T} = \int_{T_S}^{T_R} C\frac{dT}{T} = C\ln\frac{T_R}{T_S} \qquad (4.26)$$

Because $T_R > T_S$, therefore, ΔS_S is a positive quantity. Hence, total change in entropy of the universe is

$$\Delta S_{universe} = \Delta S_S + \Delta S_R = C\left[\ln\frac{T_R}{T_S} + \frac{T_S}{T_R} - 1\right] \quad (4.27)$$

One can easily see that total entropy $\Delta S_{universe} \geq 0$. This example demonstrates that even though ΔS_R and ΔS_S can each be positive or negative, we always have that $\Delta S_{universe} \geq 0$.

4.5   A Few Other Examples of Entropy Change Calculations

Below, we will examine a few examples where heat exchanges are involved between system and reservoir. We begin with the entropy change of an ideal gas followed by various other practical examples.

4.5.1   Entropy of a Perfect Gas

In general, the internal energy of a gas will be the function of temperature and volume, i.e., $U \approx U(T, V)$. For small change in U, we can write

$$dU = \left(\frac{\partial U}{\partial T}\right)_V dT + \left(\frac{\partial U}{\partial V}\right)_T dV \quad (4.28)$$

Now the first law gives

$$dQ = dU - dW = dU + PdV$$

The above equations give

$$dQ = \left(\frac{\partial U}{\partial T}\right)_V dT + \left[P + \left(\frac{\partial U}{\partial V}\right)_T\right]dV \quad (4.29)$$

Using the definition of heat capacity at constant volume $C_V = \left(\frac{\partial Q}{\partial T}\right)_V = \left(\frac{\partial U}{\partial T}\right)_V$ and the fact that for an ideal gas $\left(\frac{\partial U}{\partial V}\right)_T = 0$, Eq. 4.29 gives us

$$dQ = C_V dT + PdV \quad (4.30)$$

For an ideal gas, all equilibrium states are connected by ideal gas equation, i.e., PV = nRT. For an infinitesimal quasi-static process, we can always write

$$PdV + VdP = nRdT \quad \text{Or} \quad PdV = nRdT - VdP$$

By using the last equation and Eq. 4.29

$$dQ = [C_V + nR]dT - VdP = C_PdT - VdP \tag{4.31}$$

Equations 4.30 and 4.31 can now be used to evaluate the entropy change of an ideal gas if C_P or C_V is given. Let us begin with the latter equation.

Let us consider a system of an ideal gas that undergoes a reversible process and absorbs an infinitesimal amount of heat dQ_R. The entropy change of gaseous system is

$$dS = \frac{dQ_R}{T}$$

Once we know dQ_R in terms of thermodynamic variables, the corresponding entropy change can be evaluated using above equation. From Eq. 4.31, we can write

$$\frac{dQ_R}{T} = C_P\frac{dT}{T} - \frac{V}{T}dP$$

or

$$dS = C_P\frac{dT}{T} - nR\frac{dP}{P} \tag{4.32}$$

The above equation can be used to find the entropy change ΔS of the ideal gas between any two equilibrium states i and f.

$$\Delta S = S_f - S_i = \int_i^f dS = C_P \int_{T_i}^{T_f} \frac{dT}{T} - nR \int_{P_i}^{P_f} \frac{dP}{P} \tag{4.33}$$

or

$$S_f = C_P \ln T_f - nR \ln P_f + [S_i - C_P \ln T_i + nR \ln P_i]$$
$$S_f = C_P \ln T_f - nR \ln P_f + S_0 \tag{4.34}$$

where S_0 represents term in brackets. Equation 4.34 indicates that by assigning different values to T_f and P_f (for fixed S_0), corresponding value of S_f can be calculated. Note that any value S_f or S_i taken alone carries no meaning. The actual change in entropy will be given by the difference in entropy values and S_0 disappears when difference is taken.

In the same way, we can evaluate entropy change of an ideal gas as a function of temperature and volume by using Eq. 4.30. For a reversible change,

$$\frac{dQ_R}{T} = C_V\frac{dT}{T} + \frac{P}{T}dV$$

and

$$dS = C_V\frac{dT}{T} + nR\frac{dV}{V}$$

where C_V is the heat capacity at constant volume. After integration, we obtain

$$S = C_V \ln T_f + nR \ln V_f + S_0 \tag{4.35}$$

where S_0 has similar meaning as above.

4.5.2   The Entropy Change on Heating a Substance

Let us suppose that a liquid of mass m and specific heat c is heated from T_i to T_f. We can imagine that the initial (T_i) and final states (T_f) are connected by a reversible process of heating, making use of an infinite number of heat-reservoirs with temperature ranging from T_i to T_f. Let dQ be the amount of heat taken in by the liquid (which is at temperature T) for an infinitesimally small change dT in temperature, then we can write:

$$dQ = mcdT \tag{4.36}$$

The corresponding entropy change is

$$dS = \frac{dQ}{T} = mc\frac{dT}{T} \tag{4.37}$$

The net entropy change between initial and final states is

$$\Delta S_{liquid} = \int_{T_i}^{T_f} mc\frac{dT}{T} = mc \ln \frac{T_f}{T_i} \tag{4.38}$$

Here, ΔS_{liquid} is a positive number, since $T_f > T_i$. Therefore, the entropy of the liquid increases. Since, entropy is a state function, its value depends upon the initial and final equilibrium states and not upon the way how these states have been achieved. Thus, in the actual irreversible heating, the entropy change would be the same.

In the actual irreversible heating, the liquid is placed in contact with a single reservoir. In this case, the decrease in entropy of the reservoir (surroundings) is less than the increase in entropy of the liquid (system). Thus, in the actual irreversible process, the entropy of the universe increases. The following example will elaborate this point more clearly.

Ex:4.6 **Consider a beaker of water is heated at atmospheric pressure between room temperature $27°$ and $100°C$ in three different ways as below.**

   (a.)   **Water beaker is placed directly on a reservoir at $100°C$.**
   (b.)   **The same water beaker is heated in two steps, first the temperature is raised from $27°$ to $50°$ and in the subsequent step from $50°$ to $100°$.**
   (c.)   **In last case, the water beaker is first heated from $27°C$ to $50°C$ and then to $80°C$ followed by a final heating upto $100°C$. Comment upon the results obtained**

Sol:

(a) Let us calculate the net entropy change when the water beaker is heated from 27 °C to 100 °C. We have shown in Eq. 4.38 that the entropy change of the water is

$$\Delta S_{liquid} = mc \ln \frac{T_f}{T_i} = C_P \ln \frac{T_f}{T_i} = C_P \ln \left\{ \frac{373}{300} \right\}$$

Next, let us calculate the entropy change of the reservoir. The reservoir loses an amount of heat $Q = C_P \left(T_f - T_i \right)$ irreversibly in this process. To calculate its entropy change, imagine the reservoir losing this heat reversibly. Then

$$\Delta S_{reservoir} = \int \frac{dQ}{T_f} = \frac{1}{T_f} \int dQ$$

$$= -C_P \frac{\left(T_f - T_i \right)}{T_f} = -\frac{73 C_P}{373} = -0.196 C_P$$

The reservoir temperature is constant and entropy change is negative as heat flows out from the reservoir. Therefore, the entropy change for the universe is

$$\Delta S_{universre} = \Delta S_{liquid} + \Delta S_{reservoir}$$

$$= C_P \left[\ln \left\{ \frac{373}{300} \right\} - 0.196 \right] = 0.022 C_P$$

which is positive, as it should be for an irreversible process.

(b) Let us evaluate the entropy change $\Delta S_{universe}$ if the water is heated in two stages by placing it first on a reservoir at 50 °C and, when it has reached that temperature, transferring it to a second reservoir at 100 °C for the final heating. The initial and final states for water are still same, irrespective of the way they have been achieved. The entropy change for water is the same as given by Eq. 4.39, i.e.

$$\Delta S_{liquid} = C_P \ln \left\{ \frac{373}{300} \right\} \tag{4.39}$$

The net entropy change for the reservoirs is found using the same method that we have employed, in previous case, except that we have taken the water to final temperature in two steps.

$$\Delta S_{reservoir} = -C_P \left[\frac{23}{323} + \frac{50}{373} \right]$$

Therefore, for the universe

$$\Delta S_{universre} = C_P \left[\ln\left\{ \frac{373}{300} \right\} - \frac{23}{323} - \frac{50}{373} \right]$$
$$= 0.013 C_P$$

This is again positive, but less than the entropy change occurring when a single reservoir was employed.

(c) For the third case, initial and final states are the same, therefore, entropy change for liquid is still same. For the reservoir, we have

$$\Delta S_{reservoir} = -C_P \left[\frac{23}{323} + \frac{30}{353} + \frac{20}{373} \right] = -0.210 C_P$$

Therefore, for the universe

$$\Delta S_{universre} = C_P \left[\ln\left\{ \frac{373}{300} \right\} - 0.210 \right] = 0.008 C_P$$

Again a positive number, but less than the entropy change occurring when a single or double reservoir was employed. This is reasonable as the use of two or three reservoirs is closer to a reversible heating, employing a large number of reservoirs, than the use of just one. Thus, we conclude that when a large number of reservoir with intermediate temperature are employed to achieve the same initial and final states, the entropy change for the universe is close to zero

4.5.3   The Entropy Change on Cooling a Substance

When a substance is cooled from temperature T_i to T_f, its entropy decreases, because the substance will loose heat during cooling. Let the temperature of the substance drop by dT, when it looses dQ amount of heat. We assume that mass of liquid is m and specific heat capacity is c. Therefore, $dQ = mcdT$. Small entropy change can be written as

$$dS = \frac{dQ}{T} = mc\frac{dT}{T}$$
$$\Delta S = mc \int_{T_i}^{T_f} \frac{dT}{T} = mc \ln\left\{ \frac{T_f}{T_i} \right\} \tag{4.40}$$

Since, $T_f < T_i$ (cooling), ΔS is negative. Note that, the total entropy of the universe, however, can never decrease, it always increases. Note that for heating or cooling, if temperature change from T_i to T_f and process occurs at either constant pressure or at constant volume then

$$\Delta S = \int \frac{dQ}{T} = mc_p \ln\left\{ \frac{T_f}{T_i} \right\} \tag{4.41}$$

and

$$\Delta S = \int \frac{dQ}{T} = mc_v \ln \left\{ \frac{T_f}{T_i} \right\} \tag{4.42}$$

Ex:4.7 **A mug of coffee cools from 100 °C to room temperature, 20 °C. The mass of the coffee is m = 0.25 kg and its specific heat capacity may be assumed to be equal to that of water, = 4190 $Jkg^{-1}K^{-1}$. Calculate the change in entropy**

   (i)   of the coffee
   (ii)   of the surroundings
   (iii) of the universe

Sol: The temperature of the coffee drops from 100 °C (373 K) to 20 °C (293 K). Therefore, the entropy change of coffee is

$$\Delta S = mc \ln \left\{ \frac{T_f}{T_i} \right\}$$

$$= 0.25 \times 4190 \times \ln \left(\frac{293}{373} \right) = -253 \, \text{JK}^{-1}$$

The entropy of coffee decrease, as the temperature has reduced. Heat flows from coffee to the surroundings, while the surrounding temperature remains at 293 K (large heat sink). We calculate $\Delta S_{surrounding}$ by knowing how much heat the coffee loses
The heat (Q) lost by the coffee = Heat gained by the surrounding. Therefore

$$Q = mc\Delta T$$
$$= -0.25 \times 4190. \, (373 - 293) = -8.38 \times 10^4 \, \text{J}$$

Q is negative for the coffee, as it loses this amount of heat, while for surrounding it is positive. The entropy change of the reservoir is

$$\Delta S_{reservoir} = \frac{Q}{T}$$
$$= mc\frac{\Delta T}{T} = \frac{8.38 \times 10^4}{293} = 286 \, \text{JK}^{-1}$$

which is positive, and larger than the entropy decrease of the coffee. The total entropy change of the universe is

$$\Delta S_{universe} = \Delta S_{coffee} + \Delta S_{reservoir}$$
$$= -253 + 286 = 33 \, \text{JK}^{-1}$$

Which is a positive number, as expected for an irreversible process.

4.5.4   Entropy Change of Melting Ice

In some physical processes, one can put heat into the system without changing its temperature at all. This occurs at phase transitions, such as melting of ice and boiling of water. Let us assume that ice of mass m and latent heat L, melts at a constant temperature T (in Kelvin). While melting, it takes an amount of heat $Q = mL$. We can imagine a reversible process that connects the initial and final states by bringing ice in contact with a heat reservoir at a temperature only slightly higher than T. As the temperature stays constant, the entropy change during the process of melting is

$$\Delta S = \frac{Q}{T} = \frac{mL}{T},$$ (4.43)

this is a positive quantity. This equation tells us that entropy change for the process (melting of ice) is positive. There is an equal decrease in entropy of the surroundings (heat-reservoir), so that in reversible melting the entropy change of the universe is zero. In the actual irreversible melting, however, the entropy of the universe increases.

4.5.5   Entropy Change When Ice is Converted into Steam

Let m be the mass of ice which melts into water at temperature T_1 (in Kelvin) and L_1 be the latent heat of fusion of ice. Let T_2 be the temperature at which water is boiled to steam. Therefore, the entropy change when ice is converted into water is

$$\Delta S_1 = \frac{Q}{T} = \frac{mL_1}{T_1},$$ (4.44)

In the next stage, water is heated from T_1 to T_2. The entropy change in heating water from T_1 to T_2 is

$$\Delta S_2 = mc \ln \left\{ \frac{T_2}{T_1} \right\} = C \ln \left\{ \frac{T_2}{T_1} \right\}$$

where c is the specific heat of the water. Now at T_2, the water is converted into steam. Let L_2 be the latent heat of vaporization of water. Then the entropy change when water is converted into steam is

$$\Delta S_3 = \frac{mL_2}{T_2}$$ (4.45)

Thus, the total entropy change (for ice alone or system alone) is given

$$\Delta S_{system} = \Delta S_1 + \Delta S_2 + \Delta S_3$$
$$= \frac{mL_1}{T_1} + mc \ln \left\{ \frac{T_2}{T_1} \right\} + \frac{mL_2}{T_2}$$ (4.46)

Which is entropy change for system (sublimation of ice) only.

4.5.6 Entropy Change on Mixing of Same Liquids at Different Temperatures

Consider a liquid of mass m at temperature T_1, which is mixed isobarically (i.e., at constant pressure) with an equal mass of same liquid at temperature T_2 $(T_1 > T_2)$ in a thermally insulated container. Let the specific heat of liquid is 'c'. After mixing, the heat flows from hotter to colder liquid. Let T be the common temperature after mixing. The principle of calorimetry implies that heat lost by one liquid is equivalent to heat gain by the other liquid. Thus

$$mc\,(T_1 - T) = mc\,(T - T_2)$$

Therefore,

$$T = \frac{T_1 + T_2}{2} \tag{4.47}$$

In order to calculate the entropy change during the actual irreversible process, we replace it with an imaginary process involving an infinite number of heat reservoirs with temperatures varying from T_1 to T. Since entropy is a state function, it depends upon the initial and final states of the process and not on the path followed. Therefore, for the liquid whose temperature changes from T_1 to T $(T_1 > T)$, the entropy change is

$$\Delta S_1 = \int_{T_1}^{T} \frac{dQ}{T} = mc \int_{T_1}^{T} \frac{dT}{T} = mc \ln \left\{ \frac{T}{T_1} \right\} \tag{4.48}$$

Similarly, for the liquid whose temperature changes from T_2 to T $(T_2 < T)$, the entropy change is given by

$$\Delta S_2 = \int_{T_2}^{T} \frac{dQ}{T} = mc \int_{T_2}^{T} \frac{dT}{T} = mc \ln \left\{ \frac{T}{T_2} \right\} \tag{4.49}$$

The net entropy change for the liquid is

$$\Delta S = \Delta S_1 + \Delta S_2 = mc \left[\ln \left\{ \frac{T}{T_1} \right\} + \ln \left\{ \frac{T}{T_2} \right\} \right]$$

$$= mc \ln \frac{T^2}{T_1 T_2} = mc \ln \left\{ \frac{T}{\sqrt{T_1 T_2}} \right\}^2$$

$$= 2mc \ln \left\{ \frac{T}{\sqrt{T_1 T_2}} \right\} = 2mc \ln \left\{ \frac{(T_1 + T_2)/2}{\sqrt{T_1 T_2}} \right\} \tag{4.50}$$

Here ΔS is positive as the arithmetic mean $(\frac{T_1 + T_2}{2})$ is always larger than the geometric mean $\sqrt{T_1 T_2}$. Thus, the entropy of the system increases in the irreversible process of mixing.

Because the system is thermally insulated, for the surroundings, the entropy change is zero. ΔS, that we calculated, represents the entropy change of the universe. We conclude that the entropy of the universe increases.

4.5.7   Entropy Change on Mixing of Two Different Liquids at Different Temperature

Let us consider the mixing of two different liquids of masses m_1, m_2, specific heat capacities c_1 and c_2, which are at different temperatures T_1 and T_2 s.t $T_1 > T_2$. In such a case, after mixing, the heat will flow from hotter to cooler liquid. Let T be the final equilibrium temperature after irreversible mixing. Then applying the principle of calorimetry, we get

$$m_1 c_1 (T_1 - T) = m_2 c_2 (T - T_2)$$

$$T = \frac{m_1 c_1 T_1 + m_2 c_2 T_2}{m_1 c_1 + m_2 c_2} \tag{4.51}$$

To evaluate the change in entropies of two liquids, we replace the actual irreversible process by reversible processes involving infinite succession of reservoirs, so that we have same final states as in case of irreversible mixing of two liquids. Therefore, for liquid one (which cools from T_1 to T), the entropy change is

$$\Delta S_1 = m_1 c_1 \int_{T_1}^{T} \frac{dT}{T}$$

$$= m_1 c_1 \ln\left\{\frac{T}{T_1}\right\} < 0, \because T < T_1 \tag{4.52}$$

For liquid which gains heat

$$\Delta S_2 = m_2 c_2 \int_{T_2}^{T} \frac{dT}{T}$$

$$= m_2 c_2 \ln\left\{\frac{T}{T_2}\right\} > 0, \because T > T_2 \tag{4.53}$$

Thus, total entropy change for system is

$$\Delta S = \Delta S_1 + \Delta S_2$$

$$= \ln\left(\frac{T}{T_1}\right)^{m_1 c_1} + \ln\left(\frac{T}{T_2}\right)^{m_2 c_2}$$

$$= \ln\left[\left(\frac{T}{T_1}\right)^{m_1 c_1} \left(\frac{T}{T_2}\right)^{m_2 c_2}\right] > 0 \tag{4.54}$$

4.5.8   Entropy Change in Expanding a Spring

This is a non-conventional example of entropy change calculation. Let us try to understand this problem. We imagine a mass-spring system shown in Fig. 4.9. When mass m is released, it produces an extension x in a thermally isolated spring in air at temperature T. After releasing the mass, the mass-spring system starts oscillating with decreasing amplitude (damped oscillations), and eventually comes to rest. Note that this process is irreversible. What is corresponding entropy change of the universe?. Let k be the spring constant and spring has been displaced form its equilibrium position by an amount x. The potential energy of the spring when it is extended by an amount x is

$$V = \frac{1}{2}kx^2$$

When the mass falls by a distance x against gravity, corresponding change in gravitational potential energy will be -mgx (reduced from previous value). Hence, total potential energy of the mass on the spring is given by

$$V(x) = \frac{1}{2}kx^2 - mgx$$

Note that the position of mechanical equilibrium is obtained when $\dfrac{\partial V(x)}{\partial x} = 0$. This gives $x_e = \dfrac{mg}{k}$ and the potential energy corresponding to equilibrium position is

$$V(x = x_e) = \frac{1}{2}kx_e^2 - mgx_e$$
$$= \frac{1}{2}kx_e^2 - kx_e^2 = -\frac{1}{2}kx_e^2 = -\frac{1}{2}mgx_e$$

Which is half the change in gravitational potential energy. Therefore, as the mass falls and eventually comes to rest at equilibrium position x_e, the energy $-mgx_e/2 =$

Fig. 4.9 The extension produced in the loaded spring. In **a**, spring is not extended x = 0, in **b** mass is released, causing the spring to extend by an amount x

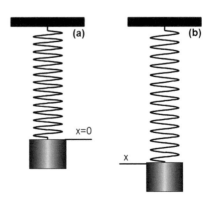

$kx_e^2/2$ is liberated as heat to the surroundings at temperature T. Therefore, the energy dissipated in the form of heat is

$$mgx_e - \frac{1}{2}mgx_e = \frac{1}{2}mgx_e$$

Provided the temperature of the surrounding does not change, the corresponding entropy change (increase) for the universe is

$$\Delta S_{universe} = \frac{kx_e^2}{2T}$$

Entropy Change in Expanding a Spring by Fixed Length in N Steps, Reversibility

Now we will consider a similar example of expanding a spring by fixed length H in N equal steps. Let when spring is loaded by a mass M but in N equal steps, such that m = M/N and extension produced in it each time be h = H/N. As discussed in previous article, when spring is loaded by a mass m, it not only undergoes extension, but also performs damped oscillations about the new equilibrium position. As a result, it gradually dissipates its energy into heat. If we assume that h and H are static extensions in the spring due to masses m and M, respectively, then it follows that loss of gravitational potential energy of the spring when loaded with mass m = M/N is mgh and 2 mgh when it is loaded with 2 m. Therefore, total loss of gravitational potential energy of mass is

$$mgh + 2mgh + \cdots + Nmgh = mgh\frac{N(N+1)}{2}$$
$$= \frac{1}{2}MgH\left[1 + \frac{1}{N}\right]$$

The potential energy in the spring corresponding to equilibrium position s

$$V(x = H) = \frac{1}{2}kH^2 - mgH$$
$$= -\frac{mgH}{2}$$

Therefore, the energy dissipated in the form of heat is given by

$$\frac{1}{2}MgH\left[1 + \frac{1}{N}\right] - \frac{1}{2}MgH = \frac{1}{2}\frac{MgH}{N}$$

If surrounding temperature does not change, the corresponding entropy change of the universe will be

$$\Delta S_{universe} = \frac{1}{2}\frac{MgH}{NT}$$

This relation shows that entropy change is inversely proportional to the number of steps. As N is very large, correspondingly $\Delta S_{universe} \longrightarrow 0$, as required for a reversible process. However, in actual irreversible process, there will be a net entropy change of the universe as proved in previous case.

4.5.9 Entropy Change in Charging a Capacitor

Before we proceed with this problem, we make an important assumption that, battery, capacitor and resistor collectively form an isolated system. Let a capacitor of capacitance C be charged to a potential difference V by connecting it (in series with a resistor) to a battery of voltage V. Note that when capacitor is charged to a voltage V, the energy stored in the capacitor is

$$C_E = \frac{1}{2}CV^2$$

While charging it, the current developed through the resistor is

$$I(t) = \frac{V}{R}e^{-t/RC}$$

This current produces a heat across the resistor, which is given by

$$H = \int_0^\infty I^2(t)dt$$
$$= \frac{1}{2}CV^2$$

Therefore, energy delivered by the voltage source is (energy across the capacitor + energy losses across the resistor) $2 \times \frac{1}{2}CV^2$. This is the case when the process is carried out in single step. Out of the energy delivered (CV^2) by the voltage source, $\frac{1}{2}CV^2$ is dissipated as heat across the resistor and remaining $\frac{1}{2}CV^2$ is stored in the capacitor. Therefore, state of the resistor has changed in this process. Assuming, this energy is received by the resistor at temperature T (surrounding also at same T), then corresponding entropy change of resistor will be

$$\Delta S = \frac{CV^2}{2T}$$

Since this is an isolated system, whatever entropy change is there for the system (resistor, capacitor and battery), the same entropy change will occur for the universe.

However, when the process (charging) is carried out in N steps, so that in each step, the potential across the capacitor is raised by V/N when it is connected to a cell. The heat produced across resistor will be

$$H_{resistor} = N \times \frac{1}{2}C\left(\frac{V}{N}\right)^2 = \frac{1}{2}\frac{CV^2}{N}$$

Therefore, energy delivered by the voltage source is

$$E_{source} = E_{capacitor} + H_{resistor}$$

$$= \frac{1}{2}CV^2 + \frac{1}{2}\frac{CV^2}{N} = \frac{1}{2}CV^2\left[1 + \frac{1}{N}\right]$$

Assuming no change in the temperature of the environment, the corresponding entropy change is given by

$$\Delta S = \frac{H_{resistor}}{T} = \frac{1}{N}\frac{CV^2}{2T}$$

The above relation indicates that, ΔS reduces with increasing N. In a limiting case when N approaches ∞, ΔS tends to zero as one expects for a quasi-static process.

4.6 Nernest Theorem, Third Law of Thermodynamics

The second law of thermodynamics defines the entropy change between two equilibrium states as (Eq. 4.9)

$$dS = \frac{dQ_R}{T} \tag{4.55}$$

If we integrate this equation from absolute zero to a temperature T, we obtain

$$S = \int_0^T \frac{dQ}{T} + S_0 \tag{4.56}$$

There is no way to determine the entropy S_0 by using second law. This is where the third law of thermodynamics is concerned and enables us to determine the entropy of a system as the temperature is reduced to absolute zero. After examining the data on chemical thermodynamics, W. H. Nernst gave the first statement of the third law of thermodynamics. On the basis of the data, Nernst postulated that $\Delta S \longrightarrow 0$ as T $\longrightarrow 0$. This is rewritten as "*Near absolute zero, all reactions in a system in internal equilibrium take place with no change in entropy*". Max Planck modified it as *The entropy of all systems in internal equilibrium is the same at absolute zero, and may be taken to be zero*. The essential point in the Planck statement is that the entropies of material (perfect crystals, liquids or gases) crystals are equal at T $= 0$ and it is a matter of convenience to put entropy $S_0 = 0$ at T $= 0$.

4.6.1 Consequences of Third Law

After acquainting ourselves with various statements of the third law, it is now important to understand some of its consequences.

4.6.2   Heat Capacity at Absolute Zero

The heat capacity is given by

$$C = T \left(\frac{\partial S}{\partial T} \right) = \frac{\partial S}{\partial \ln T} \longrightarrow 0$$

Because, as $T \Longrightarrow 0$, $\ln T \Longrightarrow -\infty$ and $S \Longrightarrow 0$. Therefore, $C \Longrightarrow 0$. Note that this result disagrees with the classical prediction of C = R/2 (or $E_{av} = \frac{1}{2}kT$) per mole per degree of freedom. This is a valid statement, in a sense that this observation emphasizes the fact that the equipartition theorem is valid at high temperature and does not provide accurate results at low temperature.

Thermal Expansion
As noted above, $S \Longrightarrow 0$ as $T \Longrightarrow 0$. This allows us to write

$$\left(\frac{\partial S}{\partial P} \right) \longrightarrow 0$$

As $T \Longrightarrow 0$. With the use of Maxwell's relation

$$\left(\frac{\partial S}{\partial P} \right)_T = - \left(\frac{\partial V}{\partial T} \right)_P$$

Therefore, as $T \Longrightarrow 0$

$$\frac{1}{V} \left(\frac{\partial V}{\partial T} \right) \Longrightarrow 0 \tag{4.57}$$

Hence, the isobaric expansivity $\beta_P \Longrightarrow 0$

Deviation from Ideal Gas Behaviour at Absolute Zero
For an ideal gas, we note that $C_P - C_V = R$ per mole. However, as $T \Longrightarrow 0$, both C_P and C_V tend to zero, and this condition cannot be satisfied. Moreover, for an ideal monoatomic gas, we expect that $C_V = \frac{3}{2}R$ per mole, and as we noticed above, this also does not hold down to absolute zero. Yet another problem is linked with the entropy expression for an ideal gas

$$S = C_V \ln T + R \ln V + constant$$

As the $T \Longrightarrow 0$, this equation predicts $S \Longrightarrow -\infty$, which is not the case. Therefore, we notice that the third law forces us to abandon the ideal gas formalism when thinking about gases at low temperature.

Unattainability of Absolute Zero
Another important statement of the third law is "*it is impossible to reach absolute zero using a finite number of processes*". This can be understood as follows: A

refrigerator is used for producing cold environment. Its performance is measured in terms of coefficient of performance which is given by

$$COP = \frac{Q_2}{Q_1 - Q_2} = \frac{T_2}{T_1 - T_2} \tag{4.58}$$

As the temperature T_2 of the reservoir from which heat is extracted approaches absolute 0 K, COP becomes vanishingly small. In other words, as the reservoir becomes colder, the refrigerator finds it more difficult to run further. Hence, it is impossible to take the reservoir down to 0 K by a finite number of cycles.

4.7 Unavailability of Energy; Heat Death of Universe

When an irreversible process takes place, energy is conserved, but some of the energy is 'wasted,' meaning it becomes unavailable to do work. Let us take two examples to understand this point. In first case, we consider the process of heat conduction from a hot to cold object. Separate hot and cold objects can serve as high and low temperature reservoirs to run a heat engine and extract useful work. After two objects are put in contact with each other, they reach an equilibrium uniform temperature and no useful work can be extracted from them. With regard to being able to do some useful work, order has changed to disorder during this process. This means during this entire process (try to correlate it with solved problem 10) entropy increases and order goes to disorder (Fig. 4.10).

In another example, we consider a block falling to the ground. When the block was at a height h, one can use its potential energy mgh to do some useful work. After hitting the ground, due to inelastic collision, this energy is no longer available (dissipated in

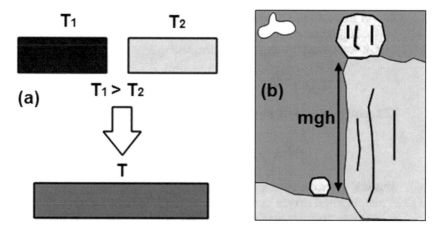

Fig. 4.10 a Two reservoirs at different temperatures can be used to extract useful work. When they acquire common temperature, it is no more possible to extract work. **b** Example of a rock on hill, the potential energy mgh of a rock can be used to extract the useful work. Once the rock hits the ground, the potential energy changes into thermal energy and transferred to the ground

the form of heat) because it has become the disordered internal energy of the block and its surroundings. The energy that has become unavailable (to do useful work) is equal to $mgh = T \Delta S_{Universe}$. Where, $\Delta S_{Universe}$ is entropy change of the system (falling rock) and surrounding (the ground where energy is dissipated). In simple words, we can call the unavailable energy as work lost, i.e., $W_{lost} = T \Delta S_{Universe}$. Both of these examples illustrate another important aspects of second law: *In an irreversible process, energy equal to $T \Delta S_{Universe}$ becomes unavailable for doing useful work. Here T is the temperature of the coldest available reservoir.* Note that in any natural process, no energy is ever lost (it is always conserved). Rather, the energy becomes less useful to extract useful work. As the time passes, energy is degraded in a sense that it goes from more orderly form (such as mechanical energy) to less orderly form (internal or thermal energy) and cannot be used to extract useful work. Entropy has a connection here as the amount of energy that becomes unavailable to do work is proportional to the entropy change during that process. A natural consequence of this degradation of energy is that as time goes on, the universe will approach a state of maximum disorder with uniform temperature. In such a case, no useful work can be done. All energy of the universe will degrade to thermal energy and all kind of changes will cease. This prediction is called as heat death of the universe.

4.8   Solved Problems

Q.1 **A 20 Ω resistor is held at a temperature of 300 K. A current of 10 mA is passed through the resistor for 1 min. Ignoring the changes in the source of the current, evaluate the entropy change in (a) the Resistor (b) the universe**

Sol: (a) Since, the resistor is held at fixed temperature T, there is no change in the state of the resistor. Therefore, for a resistor $\Delta S = 0$.
(b) As the current of 10 mA passes through the resistor (20 Ω) for 60 s. The heat produced and hence given to surrounding is

$$I^2 Rt = 0.01^2 \times 20 \times 60 = 0.12\,J$$

The entropy change for surrounding is

$$\Delta S_{surrounding} = \frac{0.12}{300} = 0.4\,mJK^{-1}$$

This is also the entropy change for the universe.

Q.2 **A paddle wheel driven by an electric motor is fixed in a large water reservoir whose temperature is maintained at 500 K. Estimate the entropy change of the reservoir if the paddle wheel is operated for two hours by a 250 W motor.**

Sol: The energy consumed s^{-1} by the paddle wheel $= 250\,\mathrm{Js}^{-1} \times 2 \times 3600 = 18 \times 10^5\,\mathrm{J}$. This much energy is transferred into the reservoir at a temperature of $500\,\mathrm{K}$. Therefore, the entropy change is

$$\Delta S = \frac{dQ}{T} = \frac{18 \times 10^5}{500} = 3.6\,\mathrm{JK}^{-1}$$

Q.3 **Experimental measurements of heat capacity per mole of Aluminium at low temperatures show that the data can be fitted to the formula, $C_V = aT + bT^3$, where a = 0.00135 JK^{-2}mol^{-1}, b = 2.48$\times$ 10^{-5} JK^{-4}mol^{-1} and T is the temperature in Kelvin. The entropy of a mole of Aluminium at such temperatures is given by the formula**

[JAM-2007]

(A) $aT + \dfrac{b}{3}T^3 + c, c > 0$ a constant (C) $aT + \dfrac{b}{3}T^3$

(B) $\dfrac{aT}{2} + \dfrac{b}{4}T^3 + c,\ c > 0,$ a constant (D) $\dfrac{aT}{2} + \dfrac{b}{4}T^3$

Sol: The change in entropy

$$\Delta S = \int \frac{dQ}{T} = \int \frac{C_V dT}{T} = \int \frac{aT + bT^3}{T} dT = aT + \frac{bT^3}{3} + c$$

Q.4 **Consider free expansion of one mole of an ideal gas in an adiabatic container from volume V_1 to V_2. The entropy change of the gas, calculated by considering a reversible process between the original state (V_1, T) to the final state (V_2, T) where T is the temperature of the system is denoted by ΔS_1. The corresponding change in the entropy of the surrounding is ΔS_2. Which of the following combinations is correct?**

[JAM-2011]

(A) $\Delta S_1 = R \ln\left(\dfrac{V_1}{V_2}\right),\ \Delta S_2 = -R \ln\left(\dfrac{V_1}{V_2}\right)$

(B) $\Delta S_1 = -R \ln\left(\dfrac{V_1}{V_2}\right),\ \Delta S_2 = R \ln\left(\dfrac{V_1}{V_2}\right)$

(C) $\Delta S_1 = R \ln\left(\dfrac{V_2}{V_1}\right),\ \Delta S_2 = 0$

(D) $\Delta S_1 = -R \ln\left(\dfrac{V_2}{V_1}\right),\ \Delta S_2 = 0$

Sol: This is an example of free expansion. The container is adiabatic and the gas
     expands from volume V_1 to V_2 without exchanging heat with the surrounding.
     Therefore, the internal energy U and temperature T of the system do not
     change. This is an irreversible process. The net entropy change for the system
     (gas) is

$$\Delta S_1 = R \ln \frac{V_2}{V_1}$$

Because no heat exchange is involved for surrounding, the entropy change is
$\Delta S_2 = 0$. Therefore, 'C' is the right option.

Q.5 **A rigid and thermally isolated tank is divided into two compartments of
     equal volume V, separated by a thin membrane. One compartment con-
     tains one mole of an ideal gas A and the other compartment contains one
     mole of a different ideal gas B. The two gases are in thermal equilibrium
     at a temperature T. If the membrane ruptures, the two gases mix. Assume
     that the gases are chemically inert. The change in the total entropy of the
     gases on mixing is**

[JAM-2015]

(A) 0   (C) $\frac{3}{2} R \ln 2$
(B) $R \ln 2$ (D) $2R \ln 2$

Sol: Since the container is thermally insulated, there is no heat exchange between
     the system and the surrounding. Each gas expands from volume V to 2V.
     Therefore, for gas A, the change in entropy is

$$\Delta S_A = R \ln \left(\frac{2V}{V} \right) = R \ln 2$$

Similarly for gas B

$$\Delta S_B = R \ln \left(\frac{2V}{V} \right) = R \ln 2$$

The net change in entropy of the gases is

$$\Delta S_{gases} = \Delta S_A + \Delta S_B = 2R \ln 2$$

Therefore, D is the right option. Note that, entropy change for the surrounding
is **ZERO**. Thus, the entropy change for the universe is

$$\Delta S_{universe} = \Delta S_{gases} + \Delta S_{surrounding} = 2R \ln 2 + 0 = 2R \ln 2$$

Q.6 **An ideal gas is originally confined to a volume V_1 in an insulated container of volume $V_1 + V_2$. The remainder of the container is evacuated. The partition is then removed and the gas expands to fill the entire container. If the initial temperature of the gas was T, what is the final temperature?.**

Sol: This is an example of free expansion. Ideal gas expands from volume V_1 to $V_1 + V_2$ but it is isolated from its surrounding through a adiathermal wall. Hence, $\Delta Q = 0$. Therefore, the internal energy of the gas doesn't change during this process. Hence, the temperature of the gas stays fixed at its initial temperature.

Q.7 **An insulated chamber is divided into two halves of volumes. The left half contains an ideal gas at temperature T_0 and the right half is evacuated. A small hole is opened between the two halves, allowing the gas to flow through, and the system comes to equilibrium. No heat is exchanged between the walls. Find the final temperature of the system.**

Sol: This is free expansion of gas. The final temperature is T_0

Q.8 **Consider the Figure shown below. The liquid is in contact with the reservoir at temperature T. When the object of mass 'm' falls through a height 'h', the paddle inside the liquid moves and this causes an increase in the temperature of the liquid. The adiathermal contact between the system and the reservoir causes the heat flow from the former to the latter, as a result, the state of system is not affected. Then evaluate the following**
(i) the entropy change of the water
(ii) the entropy change of paddle
(iii) the entropy change of mass 'm'
(iv) the entropy change of reservoir

Sol: The Figure shows a Joule paddle wheel arrangement. Here the system is water at room temperature. It is surrounded by adiabatic cylindrical walls and top. The bottom is conducting wall in contact with heat reservoir at temperature T. When the mass falls through height 'h', it causes the paddle to move. Thus, water temperature rises. The system (water) shares this added energy with the reservoir at temperature T. At the end, the water is in the same initial state. Hence, no change in water temperature, and hence in the entropy. (i) Therefore, the entropy change in water (Fig. 4.11)

$$\Delta S_{water} = 0$$

(ii) The paddle is also a part of the system and its state also does not change. Therefore

$$\Delta S_{paddle} = 0$$

(iii) When mass falls through a height h, its potential energy reduces by an amount mgh. This is accompanied by increase in its kinetic energy so that its total internal energy stays fixed. In other words, the state of mass m is not

Fig. 4.11 Joule paddle
wheel arrangement, also
known as entropy generator

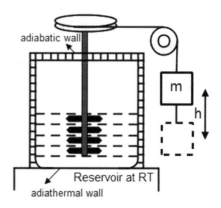

affected by the its vertical motion. As entropy is a state variable, corresponding
entropy change is

$$\Delta S_m = 0$$

(iv) The reservoir receives a net amount of heat from the system. The heat
transferred is equal to the reduction in the potential energy of mass m. Therefore, entropy change

$$\Delta S_{reservoir} = \frac{mgh}{T}$$

Q.9 **The temperature–entropy diagram of a reversible engine cycle is given
below in Figure. Its efficiency is (Fig. 4.12)**

(A) $\dfrac{2}{3}$ (C) $\dfrac{1}{4}$

(B) $\dfrac{1}{3}$ (D) $\dfrac{1}{2}$

Sol: Let Q_1 be the heat entering during process AB and Q_2 be the heat rejected
during BC. Step CA occurs at constant entropy (adiabatic), hence $Q = 0$ for
this case. Now

$$Q_1 = T_0 S_0 + \frac{1}{2} T_0 S_0 = \frac{3}{2} T_0 S_0$$

and $Q_2 = T_0 S_0$ and the efficiency is

$$\eta = \frac{W}{Q_1} = \frac{Q_1 - Q_2}{Q_1} = 1 - \frac{2}{3} = \frac{1}{3}$$

Fig. 4.12 Temperature–
entropy
diagram

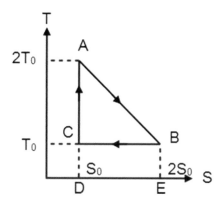

Q.10 **One gram of ice at $0\,^\circ$C is melted and heated to water at $39\,^\circ$C. Assume that the specific heat remains constant over the entire process. The latent heat of fusion of ice is 80 Calories/gm. The entropy change in the process (in Calories per degree) is ....**

[JAM-2015]

Sol: Let ΔS_1 be the entropy change when converting ice into water and ΔS_2 be the entropy change corresponding to heating water at $39\,^\circ$C. Therefore

$$\Delta S_1 = \frac{mL}{T} = \frac{1 \times 80}{273} = 0.29 \, \text{Cal/K}$$

and

$$\Delta S_2 = mc \ln \frac{T_f}{T_i} = 1 \times 1 \ln \frac{302}{273} = 0.1 \, \text{Cal/K}$$

$$\Delta S = \Delta S_1 + \Delta S_2 = 0.29 + 0.1 = 0.39 \, \text{Cal/K}$$

Q.11 **Two blocks of copper, each of mass 850 g, are put into thermal contact in an insulated box. Their initial temperatures are $52\,^\circ$C and $12\,^\circ$C and the specific heat of copper is 0.1 cal (g^{-1}C). What is the change in the entropy of the system and of the universe?**

Sol: Let the final equilibrium temperature of the blocks be T_f°C. The heat lost by the hotter block must be absorbed by the cooler one, that is, we can write

$$mc(52 - T_f) = mc(T_f - 12)$$
$$T_f = 32\,^\circ\text{C}$$

The entropy change corresponding to hotter block (cooling from $52\,°C$, $325\,K$ to $32\,°C$ $305\,K$) is

$$\Delta S_{hot} = mc \ln \frac{T_f}{T_i} = 850 \times 0.1 \ln \frac{305}{325} = -5.4\,Cal/K$$

For cooler block, the entropy change corresponding to the temperature change from $12\,°C$, $285\,K$ to $32\,°C$ $305\,K$ is

$$\Delta S_{cold} = mc \ln \frac{T_f}{T_i} = 850 \times 0.1 \ln \frac{305}{285} = 5.77\,Cal/K$$

Therefore, the net change in the entropy of universe is

$$\Delta S = \Delta S_{hot} + \Delta S_{cold} = -5.4 + 5.77 = 0.37\,Cal\ K^{-1}$$

Q.12 If n mole of an ideal gas expands isothermally and reversibly to twice its original volume. Calculate the change in entropy of the gas, of the universe. What happens if the gas undergoes free expansion?

Sol: As $dT = dU = 0$, Hence, $dQ = -dW$, where

$$dW = -PdV$$
$$W = -nRT \int_{V_i}^{V_f} \frac{dV}{V} = -nRT \ln \frac{V_f}{V_i}$$

Therefore, $Q = -W = nRT \ln \frac{V_f}{V_i}$, so that The entropy change corresponding to gas expansion is

$$\Delta S_{gas} = \frac{Q}{T} = nR \ln \frac{2V}{V} = nR \ln 2$$

This is an entropy increase. There is also an equal decrease in the entropy of the heat source which supplies heat to the gas for isothermal expansion. Therefore,

$$\Delta S_{reservoir} = -nR \ln 2$$

so that the entropy change for the universe is

$$\Delta S_{universe} = \Delta S_{gas} + \Delta S_{reservoir} = nR \ln 2 - nR \ln 2 = 0$$

Thus, the entropy of the universe(gas-plus reservoir) remains unchanged.
The case of FREE EXPANSION
This is an example of an irreversible adiabatic process ($dQ = 0$). If the gas undergoes free expansion, then the entropy of the reservoir (surrounding) does not change because the expanding gas is thermally isolated from its reservoir (surrounding). The entropy change for gas undergoing free expansion is $nR \ln 2$. The same will be the entropy change for the universe, i.e., $nR\ln 2$.

Q.13 **One mole of gas at $0\,^\circ$C is expanded reversibly and adiabatically to twice its initial volume. What is the entropy change for the universe?**

Sol: For reversible and adiabatic expansion (or compression), dQ = 0 (no heat exchange between gas and surrounding). Therefore

$$\Delta S_{gas} = \Delta S_{surrounding} = 0$$

Therefore

$$\Delta S_{universe} = \Delta S_{gas} + \Delta S_{surrounding} = 0 + 0 = 0$$

Q.14 **The temperature at the surface of the Sun is approximately 5 700 K, and the temperature at the surface of the Earth is approximately 300 K. What entropy change occurs when 1000 J of energy is transferred by radiation from the Sun to the Earth**

Sol: The corresponding entropy change is

$$\Delta S_{Earth} = \frac{Q}{T_{Earth}} = \frac{1000}{300} = 3.3\,\text{JK}^{-1}$$

Q.15 **An ideal gas is taken from an initial temperature T_i to a higher final temperature T_f along two different reversible paths. Path A is at constant pressure and path B is at constant volume. Let ΔS_A and ΔS_B be the respective entropy changes along paths A and B. Then correct relationship between ΔS_A and ΔS_B is**

(A) $\Delta S_A < \Delta S_B$ **(C)** $\Delta S_A = \Delta S_B$
(B) $\Delta S_A > \Delta S_B$ **(D)** $\Delta S_A \geq \Delta S_B$

Sol: For first case (path A) i.e., expansion during constant pressure. The reversible heat is

$$dQ_r = dU - dW$$

Here, dW $= -$PdV, is negative. For constant volume process, dW $= 0$. Therefore, dQ_r is larger for constant pressure process, leading to a larger value for the change in entropy. Therefore, $\Delta S_A > \Delta S_B$.

Q.16 **One mol of H_2 gas is contained in the left side of the container shown in Fig. 4.13 , which has equal volumes left and right. The right side is evacuated. When the valve is opened, the gas streams into the right side. Assuming that container is perfectly insulated,**
(i) What is the final entropy change of the gas?
(ii) Does the temperature of the gas change? Assume the container is so large that the hydrogen behaves as an ideal gas

Fig. 4.13 Illustration for
Q.16

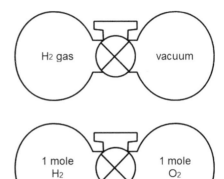

Fig. 4.14 Illustration for
Q.17

Sol: (i) This is an example of free expansion. Therefore, the corresponding entropy
change for one mole of H_2 is

$$\Delta S_{H_2} = nR \ln \frac{V_f}{V_i} = 1 \times 8.314 \ln 2 = 5.76 \, \text{JK}^{-1}$$

(ii) There will be no temperature change during free expansion of ideal gas,
there will be no change in internal energy. Because, $dQ = 0$, $U = $ constant,
and hence T stays fixed.

Q.17 **A 4 L perfectly insulated container has a centre partition that divides it
into two equal parts as shown in Fig. 4.14. The left side contains 1 mole
of H_2 gas, and the right side contains 1 mole of O_2 gas. Both gases are
at room temperature and at atmospheric pressure. The partition is then
removed and the gases are allowed to mix. What is the net entropy change
of the system?**

Sol: The entropy change for H_2 gas

$$\Delta S_{H_2} = nR \ln \frac{V_f}{V_i} = 1 \times 8.314 \ln \frac{4}{2} = 5.76 \, \text{JK}^{-1}$$

Similarly, corresponding entropy change for O_2 is

$$\Delta S_{O_2} = nR \ln \frac{V_f}{V_i} = 1 \times 8.314 \ln \frac{4}{2} = 5.76 \, \text{JK}^{-1}$$

Net entropy change is

$$\Delta S_{system} = \Delta S_{H_2} + \Delta S_{O_2} = 5.76 + 5.76 = 11.52 \, \text{JK}^{-1}$$

Q.18 **A solid metallic cube of heat capacity S is at temperature 300 K. It is brought in contact with a reservoir at 600 K. If the heat transfer takes place only between the reservoir and the cube, the entropy change of the universe after reaching the thermal equilibrium is**

[JAM June 2014]

(A)   0.69 S   (C)   0.27 S
(B)   0.54 S   (D)   0.19 S

Sol: Answer is **D**. The heat absorbed by cube $= C\Delta T = S\,(600 - 300) = 300\,S$ J. The entropy change of reservoir is

$$\Delta S_{reservoir} = -\frac{dQ}{T} = -\frac{300S}{600}\,\text{J/K} = -0.5S\,\text{J/K}$$

Similarly, entropy change of cube is

$$\Delta S_{cube} = C\ln\frac{T_f}{T_i} = S\ln\frac{600}{300} = 0.69S$$

Therefore, entropy change for universe

$$\Delta S_{universe} = \Delta S_{reservoir} + \Delta S_{cube} = -0.5S + 0.69S = 0.19S$$

Q.19 **A capacitor of capacitance $1\,\mu F$ is connected to a battery of e.m.f. 100 V at $0\,^\circ$C. What is the entropy change of the universe?**

[GNDU-2019]

Sol: We can assume, battery, capacitor and series resistor to be an isolated system. During charging of a capacitor, the energy stored in it is

$$E_{capacitor} = \frac{1}{2}CV^2 = \frac{1}{2}1\,\mu\text{F}100^2 = 5000\,\mu\text{J}$$

The remaining $\frac{1}{2}CV^2$ of energy is dissipated as heat in the resistor. This will raise the entropy of the resistor (a part of the system). Note that only the energy dissipated as heat will be taken into account while doing the entropy calculation. The energy stored in capacitor can still be used for performing useful work. The entropy change of charging resistor is

$$\Delta S_{resistor} = \frac{5000\,\mu\text{J}}{273\,\text{K}} = 18.3\,\mu\text{JK}^{-1}$$

Since system is isolated, the entropy change for resistor will be the corresponding entropy change for the universe.

$$\Delta S_{Universe} = 18.3\,\mu\text{JK}^{-1} + 0 = 18.3\,\mu\text{JK}^{-1}$$

Q.20 **A capacitor of capacitance $1\,\mu F$ is connected to a battery of e.m.f. 100 V at $0\,^\circ$C. The capacitor after being charged to 100 V, is discharged through a resistor at $0\,^\circ$C. What is the entropy change of the universe?**

[GNDU-2019]

Sol: When the capacitor is discharged through the same resistor, the energy stored in the capacitor becomes heat in the resistor. Therefore

$$\Delta S_{Universe} = \Delta S_{resistor} = 18.3\,\mu JK^{-1} + 0 = 18.3\,\mu JK^{-1}$$

Q.21 **What is the entropy change for universe when one mole of gas at $0\,^\circ$C is expanded reversibly and isothermally to twice its initial volume?**

Sol: Here $dU = 0, dW = -dQ = -RT \ln 2$, Since gas receives heat durng isothermal expansion, therefore, $\Delta S_{gas} = R \ln 2$ and $\Delta S_{surrounding} = -R \ln 2$. Hence

$$\Delta S_{universe} = \Delta S_{gas} + \Delta S_{surrounding} = 0$$

Q.22 **What is the entropy change for universe when One mole of gas at $0\,^\circ$C is expanded reversibly and adiabatically to twice its initial volume?**

Sol: As the process is adiabatic, $dQ = 0$, so

$$\Delta S_{gas} = \Delta S_{surrounding} = 0$$

Therefore $\Delta S_{universe} = 0$.

4.9   Multiple Choice Questions

Q.1 **For any process, the second law of thermodynamics requires that the change in entropy of the universe is**

[GATE-2004]

(A)	positive only	**(C)**	zero only
(B)	positive or zero	**(D)**	negative or zero

Q.2 **The internal energy U of a system is given by $U = \dfrac{bS^3}{VN}$, where b is a constant and other symbols have their usual meaning. The temperature and Pressure of this system is equal to, respectively,**

[NET-2011]

(A) $T = \dfrac{bS^3}{kVN}, P = \dfrac{bS^3}{NV^2}$ **(C)** $T = \dfrac{bS^3}{V^2N}, P = \dfrac{2bS^2}{NV^2}$

(B) $T = \dfrac{3bS^2}{VN}, P = \dfrac{bS^3}{NV^2}$ **(D)** $T = \left(\dfrac{S}{N}\right), P = \dfrac{bS^3}{NV^2}$

Q.3 **A sample of ideal gas with initial pressure P and volume V is taken through an isothermal expansion process during which the change in entropy is found to be ΔS. The universal gas constant is R. Then the work done by the gas is**

[GATE-2003]

(A) $\dfrac{PV\Delta S}{nR}$

(B) $nR\Delta S$

(C) pV

(D) $\dfrac{P\Delta S}{nRV}$

Q.4 **Which one of the following is false for a reversible process**

(A) $dQ = TdS$

(B) $dW = -PdV$

(C) $dU = TdS - PdV$

(D) $\oint \dfrac{dQ}{T} < 0$

Q.5 **A thermally insulated ideal gas of volume V_1 and temperature T expands to another enclosure of volume V_2 through a porous plug. What is the change in the temperature of the gas?**

[JEST-2013]

(A) 0

(B) $T \ln\left(\dfrac{V_1}{V_2}\right)$

(C) $T \ln\left(\dfrac{V_2}{V_1}\right)$

(D) $T \ln\left(\dfrac{V_2 - V_1}{V_2}\right)$

Q.6 **In Q.5 above, what is the change in internal energy of the gas?**

(A) 0

(B) $T \ln\left(\dfrac{V_1}{V_2}\right)$

(C) $T \ln\left(\dfrac{V_2}{V_1}\right)$

(D) $T \ln\left(\dfrac{V_2 - V_1}{V_2}\right)$

Q.7 **In Q.5 above, what is the entropy change of the gas, (assume that n mole of an ideal gas are there)?**

(A) 0

(B) $2 \ln 2$

(C) $n \ln \dfrac{V_2}{V_1}$

(D) $nR \ln \dfrac{V_2}{V_1}$

Q.8 **Consider an ideal gas of mass m at temperature T_1 which is mixed iso-barically (i.e. at constant pressure) with an equal mass of same gas at temperature T_2 in a thermally insulated container. What is the change of entropy of the universe?**

[JEST-2012]

(A) $2mc_p \ln \left(\dfrac{T_1 + T_2}{2\sqrt{T_1 T_2}} \right)$ (C) $2mc_p \ln \left(\dfrac{T_1 + T_2}{2T_1 T_2} \right)$

(B) $2mc_p \ln \left(\dfrac{T_1 - T_2}{2\sqrt{T_1 T_2}} \right)$ (D) 0

Q.9 **The value of entropy at absolute zero of temperature would be**
[JAM-2005]

(A)   zero for all the materials
(B)   finite for all the materials
(C)   zero for some materials and non-zero for others
(D)   unpredictable for any material

Q.10 **Consider the Fig. 4.15 shown below. The chamber is divided into two equal parts of volume V each. In all three cases, the container has an adiabatic wall. In (a), gas is confined to volume V while other part of the chamber has vacuum. In (b), two different gases are contained in two compartments. In (c), same gas is contained in both parts of the chamber. Now suppose that inner partition is removed so that gases are allowed to expand. Then entropy change for (a), (b) and (c), respectively is**

(A) $NK_B \ln 2, NK_B \ln 2, 0$ (C) $2NK_B \ln 2, 2NK_B \ln 2, 0$
(B) $NK_B \ln 2, 2NK_B \ln 2, 0$ (D) $NK_B \ln 2, NK_B \ln 2, NK_B \ln 2$

Q.11 **Which one of the following is not true?**

(A)   the heat capacity tends to zero as $T \longrightarrow 0$
(B)   all gases behaves like real gas as $T \longrightarrow 0$
(C)   thermal expansion stops as $T \longrightarrow 0$
(D)   Curie's law breaks down as $T \longrightarrow 0$

Q.12 **Which one of the following is not true regarding system (s) in internal equilibrium?**

(A)   Near absolute zero, all reactions in a system take place with no change in entropy

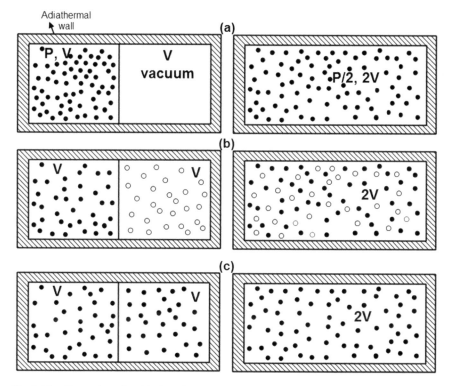

Fig. 4.15 a Expansion of an ideal gas **b** Mixing of two different gases **c** Mixing of two identical gases

(B)   The entropy of all systems is same at absolute zero, and may be taken to be zero

(C)   The contribution to the entropy of a system(s) by each aspect of the system tends to zero as T⟶0

(D)   Near absolute zero, all reactions in a system take place with non-zero entropy

Q.13 **Which one of these is an intensive variable?**

(A)   area   (C)   internal energy
(B)   volume   (D)   temperature

Q.14 **The entropy change for a mole of gas, when it undergoes free expansion is**

(A)   positive   (C)   No change
(B)   negative   (D)   cannot predicted

Q.15 In a reversible adiabatic process, the entropy

(A)   increases   (C)   remain same
(B)   decreases   (D)   none

Q.16 In a reversible process, the entropy of the system

(A)   increases   (C)   remain constant
(B)   decreases   (D)   none

Q.17 In an irreversible process, the entropy of the system

(A)   increases   (C)   remain constant
(B)   decreases   (D)   none

Q.18 Which of the following represents a reversible process?

(A) $dS < 0$ (C) $dS = 0$
(B) $dS > 0$ (D)   none

Q.19 In a natural process, the entropy

(A)   increases   (C)   remains same
(B)   decreases   (D)   cannot say

Q.20 Net entropy change of the system in a Carnot cycle is

(A)   zero   (C)   negative
(B)   positive   (D) >1 J/K

Q.21 In which state the entropy is maximum?

(A)   solid   (C)   gas
(B)   liquid   (D)   cannot say

Q.22 Which relation represents the Clausius theorem (R stands for reversible process)

(A) $\oint \dfrac{dQ}{T} = 0$ (C) $\oint \dfrac{dQ}{T} > 0$

(B) $\oint \dfrac{dQ}{T} < 0$ (D) $\oint \dfrac{dQ}{T} \neq 0$

Q.23 Which relation represents the Clausius theorem for any process

(A) $\oint \dfrac{dQ}{T} = 0$ (C) $\oint \dfrac{dQ}{T} > 0$

(B) $\oint \dfrac{dQ}{T} \leq 0$ (D) $\oint \dfrac{dQ}{T} \neq 0$

Q.24 When water vapour condenses into water, its entropy

(A)   increase   (C)   remain same
(B)   decrease   (D)   cannot say

Q.25 Consider free expansion of an ideal gas. Which one of the following is false?

(A)   the internal energy of the gas doesn't change
(B)   the temperature of the gas doesn't change
(C)   the entropy of the gas increases as the gas expands
(D)   the entropy of the gas does not change

Keys and Solution to MCQ Type Questions

Q.1 B	Q.6 A	Q.11 B	Q.16 C	Q.21 C
Q.2 B	Q.7 D	Q.12 D	Q.17 A	Q.22 A
Q.3 A	Q.8 A	Q.13 D	Q.18 C	Q.23 B
Q.4 D	Q.9 C	Q.14 A	Q.19 A	Q.24 B
Q.5 A	Q.10 B	Q.15 C	Q.20 A	Q.25 D

Hint.1 Answer is **B**. Now, consider that the system is thermally isolated. Then, dQ = 0 and

$$dS \geq 0 \qquad\qquad (4.59)$$

This is another statement of the second law of thermodynamics. It shows that for any thermally isolated system, the entropy either stays the same (for a reversible change)or increases (for an irreversible change). For a reversible process in a thermally isolated system, $TdS = dQ_R = 0$ because no heat can flow in or out. Therefore, we conclude that "The entropy of a thermally isolated system increases in any irreversible process and is unaltered in a reversible process. This is the principle of increasing entropy"

Sol.7 Answer is **D**. As the gas expands, adiabatically dU = 0, therefore, dQ = −dW = PdV. The entropy change for the gas is

$$\Delta S = \int_{V_1}^{V_2} \frac{dQ}{T}$$

$$= \int_{V_1}^{V_2} \frac{PdV}{T} = nR \int_{V_1}^{V_2} \frac{dV}{V} = nR \ln\left(\frac{V_2}{V_1}\right)$$

Hint.9 Answer is **C**. If the system attains absolute zero then it becomes a perfectly ordered system with unique configuration. The entropy change for such a system will be zero. However, there are a few exceptions, for instance CO may exist either as CO or as OC. Entropy change for CO per molecule at absolute zero is $S = k_B \ln 2$.

4.10   Exercises

1. Show that for a reversible cyclic process, $\oint \dfrac{dQ}{T} = 0$.

2. Show that for an irreversible cyclic process, $\oint \dfrac{dQ_{irrev}}{T} < 0$.

3. State and prove Clausius theorem.

4. What is entropy? Explain the principle of increase of entropy.

5. What do you mean by entropy? Show that entropy remains constant in reversible process but increases in irreversible process.

6. Develop an expression for entropy of an ideal gas in terms of its pressure, volume and specific heat.

7. What is a T-S diagram? What is the importance of T-S diagrams? Obtain an expression for efficiency of a reversible Carnot engine with the help of T-S diagram.

8. Show that T-S diagram for a Carnot cycle is a rectangle.

9. Show that entropy change depends only on initial and final states and independent of path followed.

10. Explain why unavailable energy of the universe is increasing.

11. Show that entropy of the universe increases in any natural process.

12. What is the meaning of unavailable energy? How it is related to entropy of a system?

13. A capacitor of capacitance C is charged from a battery to voltage V. If charging is performed at 300 K using a resistor, what is corresponding entropy change of the universe?

14. What is free expansion? Obtain an expression for entropy change in free expansion.

15. Entropy change in a reversible process is defined as $dS = \dfrac{dQ_r}{T}$, does it mean that energy exchange is essential for an entropy change? Explain your answer.

Thermodynamic Potentials and Maxwell Relations

5

In the previous chapter, we discussed that the internal energy U of a system is a function of the state. This means that the change ΔU when the system moves between two equilibrium states is independent of the route followed to arrive at these final equilibrium states. This property makes it a very useful quantity, but not easy to determine and not well suited for the analysis of a few thermodynamic processes and it is convenient to introduce additional state functions by using various other combinations of state functions such as T, P, V and S in such a way that the resulting quantity has dimensions of energy. These new state functions are known as *thermodynamic potentials*. Various state functions like U+TS, U-PV, U+2PV-TS, etc. may be constructed from these state variables. However, not all of them are really useful. But three of them are very useful as they provide a direct link with the experiment and have got a very special status in thermodynamics. These additional thermodynamic potentials include enthalpy, H = U + PV, Helmholtz free energy, F = U − TS and Gibbs free energy, G = H − TS. In this chapter, we will discuss these four functions U, H, F and G and explore why these are so important. First, let us again review some properties pertaining to the internal energy function U.

5.1 Thermodynamic Potentials

In this section, we will learn about four thermodynamic potentials. We continue our discussion with internal energy.

5.1.1 The Internal Energy, U

Following the discussion from the previous chapter, the change in internal energy for a system between two equilibrium states can be written as

$$dU = TdS - PdV \tag{5.1}$$

© The Author(s) 2022
S. Sharma, *Thermal and Statistical Physics*,
https://doi.org/10.1007/978-3-031-07685-5_5

This equation indicates that a change in internal energy U is due to a change in S or V (both extensive). We can always write T and P in terms of change dS and dV in S and V. This makes T and P redundant variables. This shows that S and V are the natural variables to describe any change in U, i.e., U = U(S, V). And for a small change in U, we can write

$$dU = \left(\frac{\partial U}{\partial S}\right)_V dS + \left(\frac{\partial U}{\partial V}\right)_S dV \tag{5.2}$$

Equations 5.1 and 5.2 imply that T and P can also be written as

$$T = \left(\frac{\partial U}{\partial S}\right)_V \quad \text{and} \quad P = -\left(\frac{\partial U}{\partial V}\right)_S \tag{5.3}$$

For an isochoric (constant volume)

$$dU = TdS \tag{5.4}$$

and for an isochoric and reversible process

$$dU = TdS = dQ_R = C_V dT \tag{5.5}$$

Therefore,

$$\Delta U = \int_{T_1}^{T_2} C_V dT \tag{5.6}$$

Equation 5.5 also shows that

$$C_V = \left(\frac{\partial U}{\partial T}\right)_V = \left(\frac{\partial Q_R}{\partial T}\right)_V = T\left(\frac{\partial S}{\partial T}\right)_V \tag{5.7}$$

This is valid only for a system held at constant volume. Now we would like to know if a system is held at constant pressure (a bit easier constraint to apply) and how the energy function behaves. For this case, the energy function (or so-called thermodynamic potential) has been given the name Enthalpy and is described in the next section.

5.1.2   The Enthalpy, H

Enthalpy is defined as

$$H = U + PV \tag{5.8}$$

As the quantities on the right-hand side take unique values for each equilibrium state, H is also a state function. Differentiating,

$$
\begin{aligned}
dH &= dU + PdV + VdP \\
&= TdS - PdV + PdV + VdP = TdS + VdP
\end{aligned}
\tag{5.9}
$$

The above equation holds for a reversible process ($\because dQ_R = TdS$). For a process which is reversible and isobaric,

$$dH = dQ_R = TdS \tag{5.10}$$

Therefore,

$$C_P = \left(\frac{\partial Q_R}{\partial T}\right)_P = \left(\frac{\partial H}{\partial T}\right)_P \tag{5.11}$$

$$dH = C_P dT \quad \text{so that} \quad \Delta H = \int_{T_1}^{T_2} C_P dT \tag{5.12}$$

This equation shows that for reversible isobaric processes, the enthalpy represents the heat absorbed by the system. An experiment in open air inside a laboratory occurs at constant pressure as pressure is provided by the atmosphere. Thus, isobaric conditions are relatively easy to obtain. From Eq. 5.9, we notice that S and P are the natural variables for H. We can obtain T and V in terms of change in H, S and P as below.

$$T = \left(\frac{\partial H}{\partial S}\right)_P \quad \text{and} \quad V = \left(\frac{\partial H}{\partial P}\right)_S \tag{5.13}$$

Fundamental difficulty with both U and H is that one of their natural variables is entropy S, which is not straightway determined. It would be more convenient to replace it with another variable that controls the entropy. A better choice for this will be temperature T, which is easier to control and measure. This is accomplished for the other two thermodynamic potentials, namely the Helmholtz and Gibbs functions.

5.1.3   The Helmholtz Function, F

The Helmholtz function has been designed to address the problems in which temperature and volume are the important variables. It is defined as

$$F = U - TS \tag{5.14}$$

As quantities on the right side acquire unique values for each equilibrium state, F is also a state function. For an infinitesimal small change in F,

$$dF = dU - TdS - SdT = TdS - PdV - TdS - SdT$$
$$= -PdV - SdT \tag{5.15}$$

For an isothermal process (dT = 0),

$$dF = -PdV \text{ so that } \Delta F = -\int_{V_1}^{V_2} PdV \tag{5.16}$$

Thus, a positive change in F means reversible work is done by the surrounding on the system ($dV = V_f - V_i$ negative). On the other hand, a negative change in F represents a reversible work done by the system on the surrounding (positive dV). Equation 5.15 reveals that V and T are natural variables for F and we can write F = F(V, T). For an infinitesimal small change in F

$$dF = \left(\frac{\partial F}{\partial V}\right)_T dV + \left(\frac{\partial F}{\partial T}\right)_V dT \tag{5.17}$$

Equations 5.17 and 5.15 on comparison imply that P and S can be written as

$$P = -\left(\frac{\partial F}{\partial V}\right)_T, \quad \text{and} \quad S = -\left(\frac{\partial F}{\partial T}\right)_V \tag{5.18}$$

In the end, we also notice that if T and V are constant, then dF = 0, or F is a constant for the isothermal and isochoric processes.

5.1.4   The Gibbs Function, G

Gibbs free energy is defined as

$$G = H - TS \tag{5.19}$$

For a small change in G,

$$dG = dH - TdS - SdT = TdS + VdP - TdS - SdT$$
$$= VdP - SdT \tag{5.20}$$

The above equation indicates that P and T are the natural variables for G. Hence we can write G as G = G(P, T). In a process where T and P are constant, dG = 0. Therefore, G is conserved in any isothermal isobaric process.

For an infinitesimal small change in G, we can write

$$dG = \left(\frac{\partial G}{\partial P}\right)_T dP + \left(\frac{\partial G}{\partial T}\right)_P dT \qquad (5.21)$$

The above equations enable us to write V and S as

$$V = -\left(\frac{\partial G}{\partial P}\right)_T \quad \text{and} \quad S = -\left(\frac{\partial G}{\partial T}\right)_P \qquad (5.22)$$

5.1.5   Physical Meaning of Free Energy

So far, we have defined four functions of the state known as thermodynamic potentials. Out of these, it appears that internal energy U is the only one that is more useful. However, we will see that depending upon the constraints imposed on the system, other functions are also important. Before we arrive at the discussion, let us first understand the physical meaning of the *free energy*.

Consider a big object lying on the top of a cliff [Fig. 5.1a]. There is potential energy associated with this object sitting at a certain height w.r.t. ground. If one can connect the object to a pulley system, then by lowering it down the cliff, useful mechanical work can be extracted. In other words, this system is capable of providing useful work. Once the object is down the cliff, no more useful work can be extracted from this system. It reduces its potential to provide useful work (lack of enough free energy). Thus, when the object is at the edge of a cliff it is capable of providing useful mechanical work whereas when it is at bottom of the cliff, no more useful work can be extracted. The quantity that describes the amount of available useful work a system can provide is called the *free energy*. Depending upon the constraints imposed on the system, free energy is of two types, Helmholtz free energy [Fig. 5.1c] and the Gibbs free energy [Fig. 5.1d]. The example below will clear the difference between these two. Consider the case of petrol, which stores free energy that is released when it is burned. There are two options with us. First, consider the burning of petrol in open air at constant atmospheric pressure. Here, combustion occurs at constant pressure and the released energy is known as the *Gibbs free energy*. In the second case, petrol is burned in a container of a fixed volume and has a fixed amount of petrol and air. Here, combustion takes place at constant volume and the associated free energy is called *Helmholtz free energy*.

Another example will be that of mustard oil. When burned as a lamp in open air at constant pressure, the free energy is called the Gibbs free energy. On the other hand, when consumed as edible oil, it provides energy to the human body and is defined as another form of free energy.

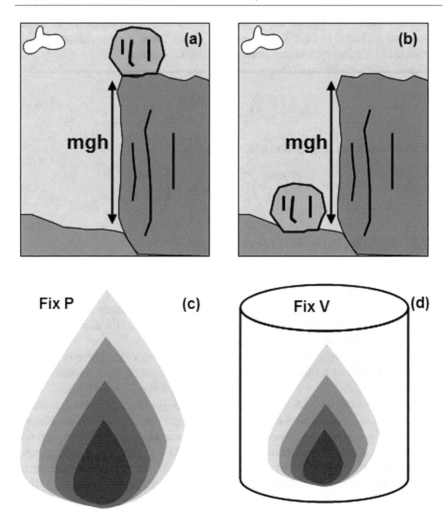

Fig. 5.1 a An object at height can provide useful mechanical work. **b** Once the object is at ground, its ability to provide useful work reduces (free energy reduced). **c** combustion of oil in open at constant pressure P (Gibbs free energy). **d** combustion of same oil at constant volume (Helmholtz free energy)

5.2   Maxwell's Relations

In this section, we will derive four important equations used in solving various problems in thermodynamics. These equations are known as Maxwell's relations. These relations correlate the partial differential of quantities that can be hard to measure (directly) with the partial differential of the quantities that are relatively easy to measure. The basic approach to developing these relations is described below. Let us consider a state function (f) of variables x and y. Therefore, df will be a perfect differential, where df is the small change in f defined as follows:

$$df = \left(\frac{\partial f}{\partial x}\right)_y dx + \left(\frac{\partial f}{\partial y}\right)_x dy \tag{5.23}$$

Because df is a perfect differential, we can write

$$\left(\frac{\partial^2 f}{\partial x \partial y}\right) = \left(\frac{\partial^2 f}{\partial y \partial x}\right) \tag{5.24}$$

If we define

$$F_x = \left(\frac{\partial f}{\partial x}\right)_y \quad and \quad F_y = \left(\frac{\partial f}{\partial y}\right)_x$$

then we have

$$\left(\frac{\partial F_y}{\partial x}\right) = \left(\frac{\partial F_x}{\partial y}\right) \tag{5.25}$$

5.2.1   First Maxwell Relation Derived from Internal Energy, U

Since all four thermodynamic potentials are perfect differentials and extensive quantities, we can apply the procedure developed above to find the four relations. Let us begin with the first thermodynamic potential, the internal energy U. A small change in U is

$$dU = TdS - PdV \tag{5.26}$$

Since S and V are natural variables for U, we can write U = U(S, V) and

$$dU = \left(\frac{\partial U}{\partial S}\right)_V dS + \left(\frac{\partial U}{\partial V}\right)_S dV \tag{5.27}$$

$$T = \left(\frac{\partial U}{\partial S}\right)_V \quad and \quad P = -\left(\frac{\partial U}{\partial V}\right)_S \tag{5.28}$$

∵ dU is a perfect differential, we can write

$$\left(\frac{\partial^2 U}{\partial S \partial V}\right) = \left(\frac{\partial^2 U}{\partial V \partial S}\right) \tag{5.29}$$

or equivalently

$$\frac{\partial}{\partial S}\left(\frac{\partial U}{\partial V}\right)_S = \frac{\partial}{\partial V}\left(\frac{\partial U}{\partial S}\right)_V \tag{5.30}$$

or

$$-\left(\frac{\partial P}{\partial S}\right)_V = \left(\frac{\partial T}{\partial V}\right)_S \tag{5.31}$$

This is our first Maxwell's thermodynamic relation.

5.2.2   Second Maxwell Relation Derived from Enthalpy, H

$$dH = TdS + VdP \tag{5.32}$$

Since S and P are natural variables for H, we can write H = U(S, P) and

$$dH = \left(\frac{\partial H}{\partial S}\right)_P dS + \left(\frac{\partial H}{\partial P}\right)_S dP \tag{5.33}$$

$$T = \left(\frac{\partial H}{\partial S}\right)_P \quad \text{and} \quad V = \left(\frac{\partial H}{\partial P}\right)_S \tag{5.34}$$

∵ dH is a perfect differential, we can write

$$\left(\frac{\partial^2 H}{\partial S \partial P}\right) = \left(\frac{\partial^2 H}{\partial P \partial S}\right) \tag{5.35}$$

or equivalently

$$\frac{\partial}{\partial S}\left(\frac{\partial H}{\partial P}\right)_S = \frac{\partial}{\partial P}\left(\frac{\partial H}{\partial S}\right)_P \tag{5.36}$$

or

$$\left(\frac{\partial V}{\partial S}\right)_P = \left(\frac{\partial T}{\partial P}\right)_S \tag{5.37}$$

This is the second Maxwell's thermodynamic relation.

5.2.3   Third Maxwell Relation Derived from Helmholtz Free Energy, F

$$dF = -SdT - PdV \tag{5.38}$$

Since T and V are natural variables for F, we can write F = F(T, V) and

$$dF = \left(\frac{\partial F}{\partial T}\right)_V dT + \left(\frac{\partial F}{\partial V}\right)_T dV \tag{5.39}$$

$$S = -\left(\frac{\partial F}{\partial T}\right)_V \quad \text{and} \quad P = -\left(\frac{\partial F}{\partial V}\right)_T \tag{5.40}$$

∵ dF is a perfect differential, we can write

$$\left(\frac{\partial^2 F}{\partial T \partial V}\right) = \left(\frac{\partial^2 F}{\partial V \partial T}\right) \tag{5.41}$$

or equivalently

$$\frac{\partial}{\partial T}\left(\frac{\partial F}{\partial V}\right)_T = \frac{\partial}{\partial V}\left(\frac{\partial F}{\partial T}\right)_V \tag{5.42}$$

or

$$\left(\frac{\partial P}{\partial T}\right)_V = \left(\frac{\partial S}{\partial V}\right)_T$$

or

$$\left(\frac{\partial S}{\partial V}\right)_T = \left(\frac{\partial P}{\partial T}\right)_V \tag{5.43}$$

This is the third Maxwell's thermodynamic relation. This equation tells us that for a thermodynamic system, the entropy change corresponding to the change in volume at constant temperature is equal to change of pressure per unit change in temperature at constant volume. Indeed, this equation can be applied to understand the equilibrium between two states of the same substance. For instance, the phase change corresponds to the vaporization of liquid or melting of a solid. This equation will be further used to develop the Clausius–Clapeyron equation.

5.2.4   Fourth Maxwell Relation Derived from Gibbs Free Energy, G

$$dG = -SdT + VdP \tag{5.44}$$

Since T and P are natural variables for G, we can write G = G(T, P) and

$$dG = \left(\frac{\partial G}{\partial T}\right)_P dT + \left(\frac{\partial G}{\partial P}\right)_T dP \tag{5.45}$$

$$S = -\left(\frac{\partial G}{\partial T}\right)_P \quad \text{and} \quad V = \left(\frac{\partial G}{\partial P}\right)_T \tag{5.46}$$

∵ dG is a perfect differential, we can write

$$\left(\frac{\partial^2 G}{\partial T \partial P}\right) = \left(\frac{\partial^2 G}{\partial P \partial T}\right) \tag{5.47}$$

or equivalently

$$\frac{\partial}{\partial T}\left(\frac{\partial G}{\partial P}\right)_T = \frac{\partial}{\partial P}\left(\frac{\partial G}{\partial T}\right)_P \tag{5.48}$$

or

$$\left(\frac{\partial V}{\partial T}\right)_P = -\left(\frac{\partial S}{\partial P}\right)_T \tag{5.49}$$

This is the fourth Maxwell's thermodynamic relation.

5.2.5   A Thermodynamic Mnemonic Diagram

One should not memorize these equations; rather it is better to remember how to develop them. All these equations can be derived by following the same approach as done above. There is a convenient way to remember them with the help of a simple mnemonic diagram shown below in Fig. 5.2.

The following points can be remembered to recall the Maxwell relations:

(i)   Cross multiplication of the variables always gives the form TS = PV (dimensions of energy).
(ii) Take conjugate pairs of variables as constant.
(iii) The sign is positive if T appears with P (in one of the partial derivatives).

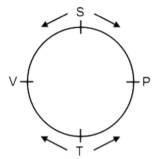

Clockwise from S (+ve sign)

S → P
↑□↓ $\left(\dfrac{\partial S}{\partial V}\right)_T = \left(\dfrac{\partial P}{\partial T}\right)_V$
V ← T

Anticlockwise from S (-ve sign)

S → V
↑□↓ $\left(\dfrac{\partial S}{\partial P}\right)_T = -\left(\dfrac{\partial V}{\partial T}\right)_P$
P ← T

Clockwise from T (+ve sign)

T → V
↑□↓ $\left(\dfrac{\partial T}{\partial P}\right)_S = \left(\dfrac{\partial V}{\partial S}\right)_P$
P ← S

Anticlockwise from T (-ve sign)

T → P
↑□↓ $\left(\dfrac{\partial T}{\partial V}\right)_T = -\left(\dfrac{\partial P}{\partial S}\right)_P$
V ← S

Fig. 5.2 A mnemonic diagram to recall Thermodynamic Maxwell's equations. While moving clockwise (anticlockwise) take positive (negative) sign

How to write Maxwell relations with the help of a mnemonic diagram?

These four relations can be recalled by remembering the phrase **S**pecial **P**rogramme on **TV** on the mnemonic diagram as shown in Fig. 5.2. The first two relations can be obtained by starting from S. While starting from S and going clockwise, write the four variables as shown in Fig. 5.2. This way we obtain one of the four equations. Similarly, while moving anticlockwise (don't forget to include a negative sign) write the variables as shown to obtain the second equation. Note that irrespective of clockwise or anticlockwise movement from S or T, variables are always written following a clockwise pattern.

5.3 Clausius–Clapeyron Equation

As we noted earlier, the Maxwell third equation can be used to study the equilibrium between two phases of the same substance. This can be understood from the discussion below. The Maxwell third equation is

$$\left(\frac{\partial S}{\partial V}\right)_T = \left(\frac{\partial P}{\partial T}\right)_V$$

This gives

$$T\left(\frac{\partial S}{\partial V}\right)_T = T\left(\frac{\partial P}{\partial T}\right)_V$$

or

$$\left(\frac{\partial Q}{\partial V}\right)_T = \left(\frac{\partial P}{\partial T}\right)_V$$

This equation implies that under isothermal expansion (e.g., vaporization of water), the heat absorbed per unit volume equals the product of absolute temperature and rate of change in pressure with the temperature at constant volume (isochoric process). Here, ∂Q represents the quantity of heat absorbed per unit increase of volume ∂V at constant temperature T. This quantity of heat is called latent heat L $(= \partial Q/dm)$ or more precisely specific latent heat.

Let us consider a cylinder containing a liquid in equilibrium with its vapours. (Note that pressure exerted by vapours is called *saturated vapour pressure*. This pressure is independent of the quantity of liquid and the vapours present, as long as they are in equilibrium with each other. The only parameter that changes saturated vapour pressure is the temperature.) Now, suppose we allow the liquid to expand isothermally. Under these conditions, the vapour pressure will not change and liquid will evaporate to fill the empty space with the vapour. Since evaporation occurs at a constant temperature, we can write $\delta Q = L\mathrm{d}m$. Here, L is the specific latent heat of evaporation. The corresponding volume change will be $(v_{vap} - v_{liq})\mathrm{d}m$, where v_{vap} and v_{liq} correspond to specific volumes for vapour and liquid, respectively. Therefore, we can write

$$\left(\frac{\partial Q}{\partial V}\right)_T = \frac{L}{v_{vap} - v_{liq}}$$
$$= T\left(\frac{\partial P}{\partial T}\right)_V$$

If the system volume is fixed and temperature is increased, the liquid evaporates till it reaches a new equilibrium state and therefore a new value for saturated vapour pressure. We can write

$$\left(\frac{\partial P}{\partial T}\right)_V = \frac{L}{T\left[v_{vap} - v_{liq}\right]} \tag{5.50}$$

This is known as Clausius–Clapeyron's latent heat equation and holds for both the changes of state, i.e., from liquid to vapour and solid to liquid. In the latter case, L will represent the latent heat of fusion; v_{vap} and v_{liq} are the specific volume of a substance in vapour and liquid states, respectively. It may be noted that L is expressed in Joules/kg.

5.3.1   Effect of Pressure on Boiling Points of Liquids

Whenever a liquid boils, i.e., changes from a liquid state to a gaseous state, its volume increases from say V_1 to V_2 so that $V_2 - V_1$ is positive. Therefore, dP/dT becomes a positive quantity. This implies that the boiling point of a liquid rises with an increase in pressure or vice versa. Therefore, a liquid will boil at a lower temperature when pressure is lowered. It indicates that water will boil at a temperature less than $100\,^{\circ}$C, when the pressure is lower than one atmospheric pressure.

5.3.2   Effect of Pressure on Melting Points of Solids

When a solid melts, two situations arise depending upon the material we are investigating. Melting may be accompanied by an increase in volume as in the case of certain substances like wax and sulphur, otherwise there may be a decrease in volume as in the case of certain substances like melting of ice and gallium
(i) When $V_2 > V_1$ (for wax and sulphur), the slope dP/dT is a +ve quantity. This means that the melting point of such substances rises with an increase in pressure.
(ii) When $V_2 < V_1$ (ice or gallium), $(V_2 - V_1)$ is a negative quantity. Hence, dP/dT is also negative which means that the melting point of such substances decreases with an increase in pressure. Thus, ice will melt at a temperature lower than $0\,^{\circ}$C when the pressure is higher than the normal atmospheric pressure.

5.4   Phase Transformation in Pure Substances, Triple Point of Water

A phase transformation reflects a discontinuous change in the properties of a substance. It is discontinuous in the sense that its environment changes infinitesimally. A known example is the melting of ice, and boiling of water, either of which can be accomplished by a slight variation in temperature. In this case, different forms of substance, for instance here ice, water and vapour, are called phases. A graph showing the equilibrium phases as a function of temperature and pressure is known as a phase diagram. Figure 5.3a gives a qualitative phase diagram for water. The Clausius–Clapeyron equation indicates that the slope dP/dT of the vapour pressure curve and sublimation curve is positive. The melting process however may be accompanied by $\Delta V = V_2 - V_1$ as positive or negative (Fig. 5.3a and b, respectively) depending upon whether a solid expands or contracts upon melting. Ice contracts upon melting, and the ice–liquid boundary has a negative dP/dT. In panel a, as we see, the phase diagram is divided into three regions indicating the conditions under which ice, water and steam can exist as a most stable phase. The three lines on a phase diagram represent conditions under which two different phases can co-exist in equilibrium. Note that the pressure at which a gas can co-exist with its liquid phase is called *vapour pressure*. Experimental data indicates that at T $= 0.01\,^{\circ}$ C and P $= 0.006$ bar, all three

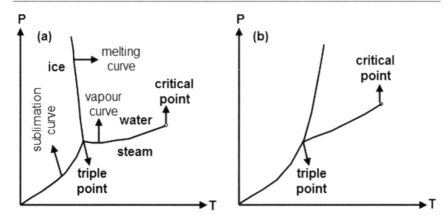

Fig. 5.3 a Phase diagram of water. **b** Phase diagram of CO_2

phases can co-exist; and this point is known as ***triple point***. At low pressure (left side of the triple point), water cannot exist in equilibrium, and ice directly sublimates into vapours.

As mentioned, the liquid–vapour boundary will always have a positive slope. For a liquid in equilibrium with vapours (gas), if we raise the temperature, we must apply more pressure to keep the liquid from vaporizing and maintain the equilibrium between these two phases. As the pressure increase, the gaseous phase becomes more and more dense, thereby decreasing the difference between liquid and vapours. Eventually, a point is reached in the liquid–vapour line where there is no longer any change from liquid to the gaseous phase. This point is called ***critical point***. For water, the critical point occurs at 374 °C and 221 bars. A solid–liquid boundary does not have a critical point. For a substance which expands on melting, the slope dP/dT will be positive and such a qualitative example of carbon dioxide is shown in Fig. 5.3b.

5.5   Thermodynamic Potential with More Than Two Natural Variables, Development of Maxwell Relations

When a thermodynamic potential consists of more than two natural variables, the procedure used earlier can be used to set up new thermodynamic Maxwell's equations. Earlier we discussed examples where thermodynamic potentials consist of two natural variables and yielded one Maxwell relation each. In some circumstances, thermodynamic potentials may consist of more than two natural variables and corresponding Maxwell equations will then be determined by the available number of natural variables. For instance, if thermodynamic potential consists of (t+1) natural variables, there will be t(t+1)/2 separate pairs of mixed second derivatives or a set of Maxwell's relations.

Internal Energy

Let us try to understand it with the help of thermodynamic potential, internal energy U. A small change in internal energy between a pair of equilibrium states is given by

$$dU = TdS - PdV + \mu dN$$

Here, the last term originates from the fact that addition or removal of particles from the system strongly influences the total energy of the system. Here, μ is the chemical potential and will be discussed in more detail in one of the coming chapters. As U is a perfect differential and has three natural variables S, V and N. We can write $U = U(S, V, N)$. For a small change in U,

$$dU = \left(\frac{\partial U}{\partial S}\right)_{V,N} dS + \left(\frac{\partial U}{\partial V}\right)_{S,N} dV + \left(\frac{\partial U}{\partial N}\right)_{V,S} dN$$

Comparing the above equations, we get

$$T = \left(\frac{\partial U}{\partial S}\right)_{N,V}, \quad P = -\left(\frac{\partial U}{\partial V}\right)_{S,N}, \quad \mu = \left(\frac{\partial U}{\partial N}\right)_{V,S} \qquad (5.51)$$

Because dU is a perfect differential, the first pair of natural variables, i.e., **(S, V)** gives

$$\left(\frac{\partial^2 U}{\partial S \partial V}\right)_N = \left(\frac{\partial^2 U}{\partial V \partial S}\right)_N, \quad or \quad \frac{\partial}{\partial S}\left(\frac{\partial U}{\partial V}\right)_{S,N} = \frac{\partial}{\partial V}\left(\frac{\partial U}{\partial S}\right)_{N,V}$$

thus, finally giving

$$-\left(\frac{\partial P}{\partial S}\right)_{N,V} = \left(\frac{\partial T}{\partial V}\right)_{S,N} \qquad (5.52)$$

This is the first Maxwell's relation.

Second pair of natural variables, i.e., **(S, N)** gives

$$\left(\frac{\partial^2 U}{\partial S \partial N}\right)_V = \left(\frac{\partial^2 U}{\partial N \partial S}\right)_V \quad or \quad \frac{\partial}{\partial S}\left(\frac{\partial U}{\partial N}\right)_{S,V} = \frac{\partial}{\partial N}\left(\frac{\partial U}{\partial S}\right)_{V,N}$$

thus, finally giving

$$\left(\frac{\partial \mu}{\partial S}\right)_{V,N} = \left(\frac{\partial T}{\partial N}\right)_{S,V} \tag{5.53}$$

This is the second Maxwell's relation.

Third pair of natural variables, i.e., **(V, N)** gives

$$\left(\frac{\partial^2 U}{\partial V \partial N}\right)_{S} = \left(\frac{\partial^2 U}{\partial N \partial V}\right)_{S} \quad or \quad \frac{\partial}{\partial V}\left(\frac{\partial U}{\partial N}\right)_{S,V} = \frac{\partial}{\partial N}\left(\frac{\partial U}{\partial V}\right)_{S,N}$$

thus, finally giving

$$\left(\frac{\partial \mu}{\partial V}\right)_{S,N} = -\left(\frac{\partial P}{\partial N}\right)_{S,V} \tag{5.54}$$

This is the third Maxwell's relation.

Enthalpy

Similarly, for other thermodynamic potentials, one can develop a similar set of equations. The final results are written below for the convenience of the reader. A small change in enthalpy can be written as

$$dH = TdS + VdP + \mu dN$$

Following the procedure illustrated above, the first pair of natural variables, i.e., **(S, P)** gives the first equation as below

$$\left(\frac{\partial T}{\partial P}\right)_{S,N} = \left(\frac{\partial V}{\partial S}\right)_{P,N} \tag{5.55}$$

The second pair of natural variables, i.e., **(S, N)** gives the second equation as below

$$\left(\frac{\partial T}{\partial N}\right)_{S,P} = \left(\frac{\partial \mu}{\partial S}\right)_{P,N} \tag{5.56}$$

The third pair of natural variables, i.e., **(P, N)** gives the third equation as below

$$\left(\frac{\partial V}{\partial N}\right)_{S,P} = \left(\frac{\partial \mu}{\partial P}\right)_{S,N} \tag{5.57}$$

Helmholtz Free Energy

A small change in *Helmholtz free energy* can be written as

$$dF = -SdT - PdV + \mu dN$$

The first pair of natural variables, i.e., **(T, V)** gives the first equation as below

$$\left(\frac{\partial S}{\partial V}\right)_{T,N} = \left(\frac{\partial P}{\partial T}\right)_{V,N} \tag{5.58}$$

The second pair of natural variables, i.e., **(T, N)** gives the second equation as below

$$-\left(\frac{\partial S}{\partial N}\right)_{T,V} = \left(\frac{\partial \mu}{\partial T}\right)_{V,N} \tag{5.59}$$

The third pair of natural variables, i.e., **(V, N)** gives the third equation as below

$$-\left(\frac{\partial P}{\partial N}\right)_{T,V} = \left(\frac{\partial \mu}{\partial V}\right)_{T,N} \tag{5.60}$$

Gibbs Free Energy

A small change in the *Gibbs free energy* can be written as

$$dG = -SdT + VdP + \mu dN$$

The first pair of natural variables, i.e., **(T, P)** gives the first equation as below

$$\left(\frac{\partial V}{\partial T}\right)_{P,N} = -\left(\frac{\partial S}{\partial P}\right)_{T,N} \tag{5.61}$$

The second pair of natural variables, i.e., **(T, N)** gives the second equation as below

$$\left(\frac{\partial \mu}{\partial T}\right)_{P,N} = -\left(\frac{\partial S}{\partial N}\right)_{P,T} \tag{5.62}$$

The third pair of natural variables, i.e., **(P, N)** gives the third equation as below

$$\left(\frac{\partial \mu}{\partial P}\right)_{N,T} = \left(\frac{\partial V}{\partial N}\right)_{P,T} \tag{5.63}$$

5.6   Use of Maxwell's Relations in Solving Various Thermodynamic Problems

Now we will go through various examples of how Maxwell's relations can be used to solve various problems in thermodynamics. Let us begin by evaluating quantities $\left(\frac{\partial C_V}{\partial V}\right)_T$ and $\left(\frac{\partial C_P}{\partial P}\right)_T$. Before we evaluate these quantities, let us examine what appears to be a contradiction in these terms. As C_V gives the heat capacity at constant volume, what do we really want to know by asking for its variation with volume?. The answer lies in the fact that C_V measurements are taken at constant volume by measuring the amount of heat put in divided by the corresponding temperature rise. Now suppose we want to repeat the (C_V) measurement at a different volume V. We will get a different value for C_V in this case. It is in this sense that C_V could depend on V and so it is reasonable to investigate the behaviour of $\left(\frac{\partial C_V}{\partial V}\right)_T$ and $\left(\frac{\partial C_P}{\partial P}\right)_T$.

By definitions of C_V and C_P, we have

$$C_V = \frac{dQ_V}{dT} = T\left(\frac{\partial S}{\partial T}\right)_V \tag{5.64}$$

$$C_P = \frac{dQ_P}{dT} = T\left(\frac{\partial S}{\partial T}\right)_P \tag{5.65}$$

With the help of Maxwell's relations, certain differentials of the heat capacities can be expressed in simpler forms which consist of directly observable functions of the state. For instance, let us take an example of C_P.

$$\left(\frac{\partial C_P}{\partial P}\right)_T = \left(\frac{\partial}{\partial P}\right)_T \left\{ T\left(\frac{\partial S}{\partial T}\right)_P \right\}$$
$$= T\left(\frac{\partial}{\partial P}\right)_T \left(\frac{\partial S}{\partial T}\right)_P \tag{5.66}$$

We can do this as T remains constant under the partial differential with respect to P. Reversing the order of differentiation, we obtain

$$\left(\frac{\partial C_P}{\partial P}\right)_T = T\left(\frac{\partial}{\partial T}\right)_P \left(\frac{\partial S}{\partial P}\right)_T \tag{5.67}$$

$$= -T\left(\frac{\partial}{\partial T}\right)_P \left(\frac{\partial V}{\partial T}\right)_P \tag{5.68}$$

where in the last step we have used Maxwell's relation

$$\left(\frac{\partial S}{\partial P}\right)_T = -\left(\frac{\partial V}{\partial T}\right)_P \tag{5.69}$$

Thus, we can write

$$\left(\frac{\partial C_P}{\partial P}\right)_T = -T\left(\frac{\partial^2 V}{\partial T^2}\right)_P \tag{5.70}$$

Following the same procedure, we can evaluate $\left(\frac{\partial C_V}{\partial V}\right)_T$ as below

$$\left(\frac{\partial C_V}{\partial V}\right)_T = \left(\frac{\partial}{\partial V}\right)_T \left\{T\left(\frac{\partial S}{\partial T}\right)_V\right\}$$

$$= T\left(\frac{\partial}{\partial V}\right)_T \left(\frac{\partial S}{\partial T}\right)_V \tag{5.71}$$

We can do this as T remains constant under the partial differential with respect to V. Reversing the order of differentiation, we obtain

$$\left(\frac{\partial C_V}{\partial V}\right)_T = T\left(\frac{\partial}{\partial T}\right)_V \left(\frac{\partial S}{\partial V}\right)_T \tag{5.72}$$

$$= T\left(\frac{\partial}{\partial T}\right)_V \left(\frac{\partial P}{\partial T}\right)_V \tag{5.73}$$

where in the last step we have used Maxwell's relation

$$\left(\frac{\partial S}{\partial V}\right)_T = \left(\frac{\partial P}{\partial T}\right)_V \tag{5.74}$$

Thus, we can write

$$\left(\frac{\partial C_V}{\partial V}\right)_T = T\left(\frac{\partial^2 P}{\partial T^2}\right)_V \tag{5.75}$$

Case of Ideal Gas

For ideal gas obeying the equation of state $PV = nRT$, the differentials $\left(\frac{\partial^2 P}{\partial T^2}\right)_V$ and $\left(\frac{\partial^2 V}{\partial T^2}\right)_P$ are zero. Therefore, both these coefficients vanish for an ideal gas.

To Establish Relationship Between Heat Capacities

With the help of Maxwell's relations, it is also possible to establish a relationship between C_P and C_V in terms of expansivity and compressibility. This is obtained directly by expanding entropy S as a function of T and V, i.e., S = S(T, V). This allows us to write immediately

$$dS = \left(\frac{\partial S}{\partial T}\right)_V dT + \left(\frac{\partial S}{\partial V}\right)_T dV \qquad (5.76)$$

so that

$$\left(\frac{\partial S}{\partial T}\right)_P = \left(\frac{\partial S}{\partial T}\right)_V + \left(\frac{\partial S}{\partial V}\right)_T \left(\frac{\partial V}{\partial T}\right)_P \qquad (5.77)$$

or in terms of heat capacities

$$C_P - C_V = T \left(\frac{\partial S}{\partial V}\right)_T \left(\frac{\partial V}{\partial T}\right)_P \qquad (5.78)$$

The Maxwell relation

$$\left(\frac{\partial S}{\partial V}\right)_T = \left(\frac{\partial P}{\partial T}\right)_V = -\left(\frac{\partial P}{\partial V}\right)_T \left(\frac{\partial V}{\partial T}\right)_P \qquad (5.79)$$

And hence, we have

$$C_P - C_V = \frac{\alpha_P^2}{\kappa_T} V T \qquad (5.80)$$

where we have used isobaric expansivity (α_P) and isothermal compressibility κ_T as

$$\alpha_P = \frac{1}{V} \left(\frac{\partial V}{\partial T}\right)_P, \quad \text{and} \quad \kappa_T = -\frac{1}{V} \left(\frac{\partial V}{\partial P}\right)_T \qquad (5.81)$$

For one mole of an ideal gas, Eq. 5.78 gives

$$C_P - C_V = R \qquad (5.82)$$

Ratio Between Isothermal and Adiabatic Compressibilities

$$\frac{\kappa_T}{\kappa_S} = \frac{-\dfrac{1}{V}\left(\dfrac{\partial V}{\partial P}\right)_T}{-\dfrac{1}{V}\left(\dfrac{\partial V}{\partial P}\right)_S} = \frac{\left(\dfrac{\partial V}{\partial T}\right)_P \left(\dfrac{\partial T}{\partial P}\right)_V}{\left(\dfrac{\partial V}{\partial S}\right)_P \left(\dfrac{\partial S}{\partial P}\right)_V} \quad \text{reciprocity theorem}$$

$$= \frac{\left(\dfrac{\partial S}{\partial V}\right)_P \left(\dfrac{\partial V}{\partial T}\right)_P}{\left(\dfrac{\partial S}{\partial P}\right)_V \left(\dfrac{\partial P}{\partial T}\right)_V} \quad \text{rearranging}$$

$$= \frac{\left(\dfrac{\partial S}{\partial T}\right)_P}{\left(\dfrac{\partial S}{\partial T}\right)_V} = \frac{C_P/T}{C_V/T} = \gamma \tag{5.83}$$

Difference Between Adiabatic and Isothermal Compressibilities

Starting with the relation

$$C_P - C_V = \frac{V T \alpha_P^2}{\kappa_T} \tag{5.84}$$

show that difference between adiabatic and isothermal compressibilities is given by

$$\kappa_S - \kappa_T = -\frac{V T \alpha_P^2}{C_P}$$

Since the ratio of isothermal to adiabatic compressibilities is

$$\frac{\kappa_T}{\kappa_S} = \frac{C_P}{C_V} = \gamma \tag{5.85}$$

dividing both sides by C_V in Eq. 5.84

$$\frac{C_P}{C_V} - 1 = \frac{V T \alpha_P^2}{C_V \kappa_T} \tag{5.86}$$

or

$$\gamma - 1 = \frac{VT\alpha_P^2}{C_V \kappa_T}$$

$$\kappa_T - \kappa_S = \frac{\kappa_S}{\kappa_T} \frac{\alpha_P^2 VT}{\gamma C_V} \tag{5.87}$$

after rearranging these terms, we get

$$\kappa_S - \kappa_T = -\frac{\alpha_P^2 VT}{C_P} \tag{5.88}$$

5.7   The TdS Equations

The TdS equations allow entropy changes to be evaluated during various reversible processes in terms of either dV and dT, or dP and dT, or dV and dP, and even in terms of directly measurable quantities such as the coefficient of expansion and the bulk modulus. These TdS equations are developed by considering a pure, simple compressible system which undergoes an internally reversible process. One can express entropy in terms of any two of T, V or P variables.

5.7.1   First TdS Equation

To continue with the first TdS equation, we assume entropy S to be a function of T and V, i.e., S = S(TV). This allows us to write immediately

$$dS = \left(\frac{\partial S}{\partial T}\right)_V dT + \left(\frac{\partial S}{\partial V}\right)_T dV$$

$$\begin{aligned} TdS &= T\left(\frac{\partial S}{\partial T}\right)_V dT + T\left(\frac{\partial S}{\partial V}\right)_T dV \\ &= C_V dT + T\left(\frac{\partial S}{\partial V}\right)_T dV \\ &= C_V dT + T\left(\frac{\partial P}{\partial T}\right)_V dV \end{aligned} \tag{5.89}$$

While writing the above equation, we have used one of the thermodynamic Maxwell equations.

$$\left(\frac{\partial P}{\partial T}\right)_V = \left(\frac{\partial S}{\partial V}\right)_T$$

Equation 5.89 is known as the *first TdS equation*. This equation can be written in a more useful form in terms of isobaric volume expansivity and isothermal elasticity as below.

$$TdS = C_V dT + T \left(\frac{\partial P}{\partial T} \right)_V dV$$

$$= C_V dT - T \left(\frac{\partial P}{\partial V} \right)_T \left(\frac{\partial V}{\partial T} \right) P dV$$

$$= C_V dT + T \alpha_P E_T dV$$

where α_P and $E_T = 1/\kappa_T$ are, respectively, isobaric expansivity and isothermal elasticity.

5.7.2   Second TdS Equation

Take T and P as independent variables for entropy, i.e., S = S(T, P). Similar analysis gives

$$dS = \left(\frac{\partial S}{\partial T} \right)_P dT + \left(\frac{\partial S}{\partial P} \right)_T dP$$

Therefore,

$$TdS = T \left(\frac{\partial S}{\partial T} \right)_P dT + T \left(\frac{\partial S}{\partial P} \right)_T dP$$

Using Maxwell relation

$$\left(\frac{\partial S}{\partial P} \right)_T = - \left(\frac{\partial V}{\partial T} \right)_P$$

and for a process at constant pressure, we can write TdS = dH, so that

$$T \left(\frac{\partial S}{\partial T} \right)_P = \left(\frac{\partial H}{\partial T} \right)_P$$

$$= C_P$$

Therefore, we can write

$$TdS = C_P dT - T \left(\frac{\partial V}{\partial T} \right)_P dP \tag{5.90}$$

Equation 5.90 is known as the *second TdS equation*. Further, one can rewrite the above equation in terms of isobaric volume expansivity as below

$$TdS = C_P dT - TV \alpha_P dP$$

5.7.3   Third TdS Equation

A similar analysis taking P and V as independent variables for entropy, i.e., S = S(P, V) gives

$$dS = \left(\frac{\partial S}{\partial P}\right)_V dP + \left(\frac{\partial S}{\partial V}\right)_P dV$$

$$TdS = T\left(\frac{\partial S}{\partial P}\right)_V dP + T\left(\frac{\partial S}{\partial V}\right)_P dV$$

For a constant volume process, $TdS = C_V dT$, so that

$$T\left(\frac{\partial S}{\partial P}\right)_V = C_V \left(\frac{\partial T}{\partial P}\right)_V$$

and in a process that takes place at constant pressure, $TdS = C_P dT$, so that

$$T\left(\frac{\partial S}{\partial V}\right)_P = C_P \left(\frac{\partial T}{\partial V}\right)_P$$

$$TdS = C_V \left(\frac{\partial T}{\partial P}\right)_V dP + C_P \left(\frac{\partial T}{\partial V}\right)_P dV \tag{5.91}$$

This is the *third TdS equation*.

5.8   The Energy Equations

It is known that at a constant temperature, the internal energy of a real gas varies with volume as well as with pressure. For a pure substance, we can write in general, U = U(T, V) or U = U(T, P).

5.8.1   First Energy Equation

Take the first case, i.e., U = U(T, V), such that for an infinitesimal change in U, we can write

$$dU = \left(\frac{\partial U}{\partial T}\right)_V dT + \left(\frac{\partial U}{\partial V}\right)_T dV \tag{5.92}$$

Also, between a pair of equilibrium states, the internal energy change can be written as

$$dU = TdS - PdV$$

so that

$$\left(\frac{\partial U}{\partial V}\right)_T = T\left(\frac{\partial S}{\partial V}\right)_T - P = T\left(\frac{\partial P}{\partial T}\right)_V - P \qquad (5.93)$$

This equation is known as *first energy equation*. After combining Eqs. 5.92 and 5.93, we obtain

$$dU = C_V dT + \left[T\left(\frac{\partial P}{\partial T}\right)_V - P\right]dV$$

If C_V and equation of state are known, one can integrate this equation to evaluate the internal energy change associated with the change of state for a substance.

$$\Delta U = \int C_V dT + \int \left[T\left(\frac{\partial P}{\partial T}\right)_V - P\right]dV$$

The Case of Ideal Gas

For one mole of an ideal gas, $PV = RT$, so that

$$\left(\frac{\partial P}{\partial T}\right)_V = \frac{R}{V}$$

From the first energy equation (under isothermal conditions),

$$\left(\frac{\partial U}{\partial V}\right)_T = T\left(\frac{\partial P}{\partial T}\right)_V - P = \frac{RT}{V} - P = 0$$

Therefore, when the temperature doesn't change, the internal energy of an ideal gas is independent of volume. This allows us to write

$$\Delta U = \int_{T_1}^{T_2} C_V dT = C_V(T_2 - T_1)$$

While writing the above equation, it is assumed that heat capacity does not change over the temperature range of interest.

The Case of Real Gas

For a real gas that obeys the Van der Waals equation of state

$$P = \frac{RT}{V - b} - \frac{a}{V^2}$$

so that

$$\left(\frac{\partial P}{\partial T}\right)_V = \frac{R}{V - b} \tag{5.94}$$

Using this results in Eq. 5.93, we obtain

$$\left(\frac{\partial U}{\partial V}\right)_T = T\left(\frac{\partial P}{\partial T}\right)_V - P$$
$$= \frac{RT}{V - b} - P = \frac{a}{V^2}$$

Similar to the previous case, if C_V and the equation of state are known, one can evaluate the internal energy change associated with the change of state for a substance as below

$$\Delta U = \int C_V dT + \int_{V_1}^{V_2} \frac{a}{V^2} dV = \int C_V dT - a\left[\frac{1}{V_2} - \frac{1}{V_1}\right]$$

Therefore, for a real gas internal energy depends upon volume. As we see, with an increase in volume, the internal energy increases at a constant temperature.

5.8.2 Second Energy Equation

For a pure substance, we can write in general, $U = U(T, P)$, so that for an infinitesimal change in U, we can write

$$dU = \left(\frac{\partial U}{\partial T}\right)_P dT + \left(\frac{\partial U}{\partial P}\right)_T dP \tag{5.95}$$

Also, between a pair of equilibrium states, the internal energy change can be written as

$$dU = TdS - PdV$$

so that

$$\left(\frac{\partial U}{\partial P}\right)_T = T\left(\frac{\partial S}{\partial P}\right)_T - P\left(\frac{\partial V}{\partial P}\right)_T$$
$$= -T\left(\frac{\partial V}{\partial T}\right)_P - P\left(\frac{\partial V}{\partial P}\right)_T \tag{5.96}$$

While writing the last result, we have used the Maxwell equation. This equation is known as the *second energy equation*. Combining Eqs. 5.96 and 5.95, we obtain

$$dU = \left(\frac{\partial U}{\partial T}\right)_P dT - \left[T\left(\frac{\partial V}{\partial T}\right)_P + P\left(\frac{\partial V}{\partial P}\right)_T\right]dP$$

5.9   Applications to Various System

Till now, most of the discussion has been centred around the ideal gas. We have reformulated the first law of thermodynamics as

$$dU = TdS - PdV \tag{5.97}$$

However, in this chapter we will show that the methodology developed can be applied to other systems on equal footing. A few examples are given below in a table. However, we will only have three of these here. First of all, we will discuss an example of an elastic rod and then surface tension in a liquid. In the end, we will discuss a paramagnetic system. Let us define the generalized work as

$$dW = Xdx \tag{5.98}$$

where X is the generalized force and x is the generalized displacement (Table 5.1).

5.9.1   Stretching of an Elastic Rod by a Constant Force

Let us take an example of an elastic rod (metal) of length L, cross-sectional area A being held at temperature T. Under external infinitesimal small force $d\mathbb{F}$, the rod extends by an amount dL. Now we can apply the laws of thermodynamics to this system. For a gas, the first law reads

$$dU = TdS - PdV \tag{5.99}$$

Table 5.1 Examples of Generalized forces and Generalized displacements for various different systems

System	X	x	dW
Fluid	$-P$	V	$-PdV$
Elastic rod	$\mathbb{F}$	L	$\mathbb{F}dL$
Liquid film	γ	A	$\gamma\,dA$
Magnetic	B	m	$-m.B$
Dielectric	E	p_E	$-E.dp_E$

Fig. 5.4 Extension of rod by
applying a constant force $\mathbb{F}$

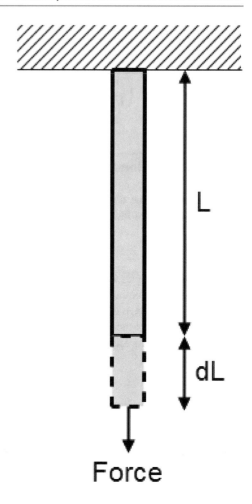

One can develop an identical expression for the system under discussion following identical steps used in the case of gases. Let us continue with internal energy first (Fig. 5.4).

Internal Energy

For an elastic rod, the modified form of the first law reads

$$dU = TdS + \mathbb{F}dL \qquad (5.100)$$

Note the positive sign for $dW = \mathbb{F}dL$ as the work is done on the elastic wire by external force and wire length is extended. Therefore it is taken positive. From here, we note that for the system under investigation, S and L are natural variables for internal energy U (which is an exact differential). Therefore,

$$\frac{\partial}{\partial S}\left(\frac{\partial U}{\partial L}\right) = \frac{\partial}{\partial L}\left(\frac{\partial U}{\partial S}\right)$$

$$\left(\frac{\partial \mathbb{F}}{\partial S}\right)_L = \left(\frac{\partial T}{\partial L}\right)_S$$

This is a required Maxwell equation describing the thermodynamics of the present system.

Enthalpy

For a gas, the enthalpy change can be written as

$$dH = TdS + VdP$$

For present system, it becomes

$$dH = TdS - Ld\mathbb{F} \tag{5.101}$$

so that we obtain

$$-\left(\frac{\partial L}{\partial S}\right)_{\mathbb{F}} = \left(\frac{\partial T}{\partial \mathbb{F}}\right)_S \tag{5.102}$$

Helmholtz Free Energy

$$dF = \mathbb{F}dL - SdT \tag{5.103}$$

so that we obtain

$$-\left(\frac{\partial S}{\partial L}\right)_T = \left(\frac{\partial \mathbb{F}}{\partial T}\right)_L \tag{5.104}$$

Gibbs Free Energy

Similarly, a change in the Gibbs free energy is

$$dG = -Ld\mathbb{F} - SdT \tag{5.105}$$

so that we obtain

$$\left(\frac{\partial S}{\partial \mathbb{F}}\right)_T = \left(\frac{\partial L}{\partial T}\right)_{\mathbb{F}} \tag{5.106}$$

NOTE

These equations can also be obtained by simply making a comparison between the modified energy Eqs. 5.1 and 5.100. After comparing these two, we see that if we replace P with $-\mathbb{F}$ and V with L in four Maxwell thermodynamic equations (for gaseous system), we obtain equations that are valid for the present case.

5.9.2 Isothermal Stretching of an Elastic Rod

Let us take an example of an elastic rod (metal) of length L, cross-sectional area A being held at temperature T. Under external infinitesimal small force d$\mathbb{F}$, the rod extends by an amount dL. Let E_T be the isothermal Young's modulus. Then,

$$E_T = \frac{Stress}{Strain} = \frac{d\mathbb{F}/A}{dL/L} = \frac{L}{A}\left(\frac{\partial \mathbb{F}}{\partial L}\right)_T \tag{5.107}$$

We also define *isothermal expansivity* ($\alpha_{\mathbb{F}}$) at constant applied force $\mathbb{F}$ as fractional change in length with temperature.

$$\alpha_{\mathbb{F}} = \frac{1}{L}\left(\frac{\partial L}{\partial T}\right)_{\mathbb{F}} \tag{5.108}$$

Now we can apply the laws of thermodynamics to this system. For a gas, the modified form of the first law reads

$$dU = TdS - PdV$$

For the system under investigation, it becomes

$$dU = TdS + \mathbb{F}dL \tag{5.109}$$

Note the positive sign for $dW = \mathbb{F}dL$ as the work is done on the elastic wire by external force and wire length is extended. Therefore it is taken positive. Similarly, we can write another thermodynamic potential, for instance, the Helmholtz free energy as

$$dF = dU - TdS - SdT$$
$$= \mathbb{F}dL - SdT \tag{5.110}$$

The above equation implies that

$$S = -\left(\frac{\partial F}{\partial T}\right)_L \tag{5.111}$$

and the applied force is

$$\mathbb{F} = \left(\frac{\partial F}{\partial L}\right)_T \tag{5.112}$$

As dF is a perfect differential and here L and T are natural variables for F, therefore

$$\frac{\partial}{\partial T}\left(\frac{\partial F}{\partial L}\right) = \frac{\partial}{\partial L}\left(\frac{\partial F}{\partial T}\right)$$

or

$$\left(\frac{\partial \mathbb{F}}{\partial T}\right)_L = -\left(\frac{\partial S}{\partial L}\right)_T \tag{5.113}$$

Now

$$\left(\frac{\partial \mathbb{F}}{\partial T}\right)_L = -\left(\frac{\partial \mathbb{F}}{\partial L}\right)_T \left(\frac{\partial L}{\partial T}\right)_{\mathbb{F}}$$
$$= -AE_T\alpha_{\mathbb{F}} \tag{5.114}$$

Equations 5.113 and 5.114 allow us to write

$$\left(\frac{\partial S}{\partial L}\right)_T = AE_T\alpha_{\mathbb{F}} \tag{5.115}$$

For an elastic rod (not made from rubber), E_T and $\alpha_{\mathbb{F}}$ are positive. Therefore, Eq. 5.115 indicates that the entropy of the rod increases when it is extended isothermally by an amount ΔL. In other words, extending the rod isothermally by an amount of ΔL would lead to the absorption of heat ΔQ, which is given by

$$\Delta Q = T\Delta S = AE_T\alpha_{\mathbb{F}}T\Delta L \tag{5.116}$$

Why Does Stretching a Wire (Metallic) Increase the Entropy?

A metallic wire contains a large number of small crystallites. By stretching the wire, it may distort those small crystallites and result in an increase in entropy together with the absorption of heat. For instance, crystallites may distort from cubic to tetragonal symmetry, thus increasing the entropy. In addition, after stretching the volume per atom may increase and this also results in an increase in entropy of the wire.

Case of Rubber Piece

In case of rubber $\alpha_{\mathbb{F}} < 0$. This means when a piece of rubber is extended isothermally, heat ΔQ is negative, that is heat is emitted. Stretching rubber at constant temperature results in the alignment of rubber molecules, thus reducing the entropy of the rubber piece with the release of heat ΔQ.

Change in Internal Energy of the Rod Under Isothermal Extension

The internal energy change under the isothermal extension of the rod is

$$dU = TdS + \mathbb{F}dL$$

or we can write

$$
\begin{aligned}
\left(\frac{\partial U}{\partial L}\right)_T &= T\left(\frac{\partial S}{\partial L}\right)_T + \mathbb{F} \\
&= TAE_T\alpha_{\mathbb{F}} + \mathbb{F}
\end{aligned}
\tag{5.117}
$$

The above relation contains two terms, the first term expresses the energy going into the rod due to its isothermal extension while the second expresses the energy going into the rod by work. Thus, the internal energy of the rod increases when it is extended isothermally.

5.9.3   Adiabatic Stretching of a Wire

When the wire is extended by an external constant force $\mathbb{F}$, the corresponding thermodynamic variables change from PVT to $\mathbb{F}LT$. Similar to TdS equations developed for gaseous system, one can follow identical strategy and rewrite equations for the present case. Let us assume entropy S = S(T, $\mathbb{F}$), so that

$$dS = \left(\frac{\partial S}{\partial T}\right)_{\mathbb{F}} dT + \left(\frac{\partial S}{\partial \mathbb{F}}\right)_T d\mathbb{F}$$

Therefore,

$$TdS = T\left(\frac{\partial S}{\partial T}\right)_{\mathbb{F}} dT + T\left(\frac{\partial S}{\partial \mathbb{F}}\right)_{T} d\mathbb{F}$$

$$= C_{\mathbb{F}} dT + T\left(\frac{\partial L}{\partial T}\right)_{\mathbb{F}} d\mathbb{F} \tag{5.118}$$

where we have used $C_{\mathbb{F}} = T\left(\frac{\partial S}{\partial T}\right)_{\mathbb{F}}$ is the heat capacity of the wire under constant applied force and used Maxwell's equation

$$\left(\frac{\partial S}{\partial \mathbb{F}}\right)_{T} = \left(\frac{\partial L}{\partial T}\right)_{\mathbb{F}}$$

When the wire is stretched under adiabatic (and reversible) conditions ($dS = 0$), therefore, the corresponding temperature change is given by

$$dT = -\frac{T}{C_{\mathbb{F}}}\left(\frac{\partial L}{\partial T}\right)_{\mathbb{F}} d\mathbb{F}$$

$$= -\frac{T}{C_{\mathbb{F}}}\alpha_{\mathbb{F}}L\frac{AdLE_{S}}{L}$$

$$= -\frac{AE_{S}\alpha_{\mathbb{F}}T}{C_{\mathbb{F}}}dL \tag{5.119}$$

Here, $\alpha_{\mathbb{F}}$ is the coefficient of linear expansion

$$\alpha_{\mathbb{F}} = \frac{1}{L}\left(\frac{\partial L}{\partial T}\right)_{\mathbb{F}}$$

and E_S is adiabatic elasticity defined below

$$E_S = \frac{dF/A}{dL/L} = \frac{LdF}{AdL}$$

Since E_S and α_L are positive for most of the substances (except rubber), dT will be negative when the wire is stretched adiabatically.

Case of Rubber Piece

For a piece of rubber α_L is negative, implying a positive change in temperature when a constant force is applied under adiabatic conditions. This means that rubber will heat up under adiabatic stretching by a constant force.

Isothermal Stretching of an Elastic Rod

One can arrive at the same conclusion for an elastic rod stretched by a constant force isothermally (the case we discussed in the previous article). For isothermal stretching, Eq. 5.119 implies

$$TdS = \Delta Q = T \left(\frac{\partial L}{\partial T} \right)_{\mathbb{F}} d\mathbb{F} = A T E_T \alpha_L dL \qquad (5.120)$$

where E_T is isothermal elasticity defined earlier. Therefore, under isothermal stretching via a constant force, heat will flow into the wire.

5.9.4   Temperature Change During Adiabatic Process

Here, we will discuss the temperature changes that take place during adiabatic expansion and compression of a gas. We note from Maxwell's first thermodynamic relation

$$\left(\frac{\partial T}{\partial V} \right)_S = - \left(\frac{\partial P}{\partial S} \right)_V$$

Multiplying and dividing by T on RHS of the above equation, we obtain

$$\left(\frac{\partial T}{\partial V} \right)_S = -T \left(\frac{\partial P}{\partial Q} \right)_V$$

At constant volume, an increasing amount of heat will result in an increase in pressure, hence the term $\frac{\partial P}{\partial Q}$ will be positive. Therefore, the term $\frac{\partial T}{\partial V}$ should be negative. Physically speaking, temperature should fall with an increase in volume at fixed entropy (adiabatic expansion). That is, an adiabatic expansion must be accompanied by fall in temperature.

Now consider the case of adiabatic compression. For this, we consider the second Maxwell's equation.

$$\left(\frac{\partial T}{\partial P} \right)_S = \left(\frac{\partial V}{\partial S} \right)_P$$

it reduces to

$$\left(\frac{\partial T}{\partial P} \right)_S = T \left(\frac{\partial V}{\partial Q} \right)_P$$

At constant pressure, an increasing quantity of heat will result in an increase in volume so that the term $\left(\frac{\partial V}{\partial Q} \right)_P$ is positive. Therefore, the term $\left(\frac{\partial T}{\partial P} \right)_S$ must be positive. That is, temperature must rise for an increase in pressure at constant entropy (adiabatic compression). In other words, an adiabatic compression must be accompanied by rise in temperature.

5.10   Magneto Caloric Effect

Consider a system of magnetic moments which cannot interact with each other. Under the influence of an external field, these moments line up and the system is said to exhibit paramagnetism. If the material is magnetized isothermally, usually a heat exchange with the surrounding occurs. Further, if magnetization changes adiabatically, the temperature changes. This interdependence of thermal and magnetic properties is known as the *magneto caloric effect*.

Assuming the material to be isotropic, the modified form of the first law becomes

$$dU = TdS - mdB \tag{5.121}$$

where m is the magnetic moment and B is the magnetic field. The last term on the right side of the equation is magnetic interaction energy ($-$m.dB) of the moments. The Helmholtz function for the system under consideration is

$$F = U - TS$$

so that

$$dF = dU - TdS - SdT = -SdT - mdB \tag{5.122}$$

The above equation implies

$$S = -\left(\frac{\partial F}{\partial T}\right)_B, \quad \text{and} \quad m = -\left(\frac{\partial F}{\partial B}\right)_T \tag{5.123}$$

as dF is a perfect differential,

$$\frac{\partial}{\partial T}\left(\frac{\partial F}{\partial B}\right) = \frac{\partial}{\partial B}\left(\frac{\partial F}{\partial T}\right)$$

giving

$$\left(\frac{\partial m}{\partial T}\right)_B = \left(\frac{\partial S}{\partial B}\right)_T \tag{5.124}$$

For the system under consideration, $m = MV$, M being intensity of magnetization. And the magnetic susceptibility is defined as

$$\chi = \frac{M}{H} = \frac{\mu_0 M}{B} = \frac{\mu_0 m}{VB} \tag{5.125}$$

Therefore,

$$\left(\frac{\partial m}{\partial T}\right)_B = \frac{VB}{\mu_0}\left(\frac{\partial \chi}{\partial T}\right)_B \tag{5.126}$$

For paramagnetic material, the quantity $\left(\dfrac{\partial \chi}{\partial T}\right)_B < 0$, thus Eqs. 5.124 and 5.126

$$\left(\frac{\partial S}{\partial B}\right)_T = \frac{VB}{\mu_0}\left(\frac{\partial \chi}{\partial T}\right)_B \tag{5.127}$$

This equation correlates the isothermal change of entropy with magnetic field B to a differential of χ. The heat exchanged under isothermal change of B is

$$\Delta Q = T\left(\frac{\partial S}{\partial B}\right)_T \Delta B = \frac{TVB}{\mu_0}\left(\frac{\partial \chi}{\partial T}\right)_B \Delta B \tag{5.128}$$

$\Delta Q < 0$, because $\left(\dfrac{\partial \chi}{\partial T}\right) < 0$ for paramagnetic materials. Since ΔQ is negative, heat is actually emitted under isothermal magnetization of the paramagnet.

Adiabatic Demagnetization

Now let us evaluate the change in T when magnetic field is reduced under adiabatic conditions.

$$\left(\frac{\partial T}{\partial B}\right)_S = -\left(\frac{\partial T}{\partial S}\right)_B\left(\frac{\partial S}{\partial B}\right)_T = -\frac{1}{T}\left(\frac{\partial T}{\partial S}\right)_B T\left(\frac{\partial S}{\partial B}\right)_T$$

$$= -\frac{T}{C_B}\left(\frac{\partial S}{\partial B}\right)_T = -\frac{TVB}{\mu_0 C_B}\left(\frac{\partial \chi}{\partial T}\right)_B \tag{5.129}$$

This equation indicates that $\left(\dfrac{\partial T}{\partial B}\right)_S > 0$, thus

$$\left(\frac{\partial T}{\partial B}\right)_S = \left(\frac{T_f - T_i}{B_f - B_i}\right)_S > 0$$

This implies that $T_f < T_i$. Therefore, a material can be cooled by using adiabatic demagnetization, that is, by keeping the sample at constant entropy and reducing the magnetic field. This process can yield temperatures as low as a few millikelvin for electronic systems and a few microkelvin for nuclear systems.

Fig. 5.5 Entropy change for
a paramagnetic salt as a
function of magnetic field
and temperature

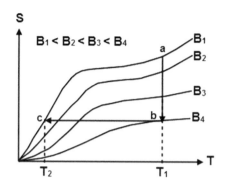

Adiabatic Demagnetization, Qualitative Discussion

For a typical paramagnetic material, the figure below displays the variation of entropy
with temperature and magnetic field. In the absence of a magnetic field, the entropy
falls close to the Curie temperature. It corresponds to the onset of spontaneous
ordering in paramagnetic salt. From the figure, it is also clear that entropy at higher
temperature can also be reduced by increasing the magnetic order by applying an
external magnetic field.

The process of cooling the paramagnetic salt is illustrated in the figure. It consists
of two major steps.

(i) Isothermal magnetization of the paramagnetic salt.
(ii) Adiabatic demagnetization of the salt at constant entropy (Fig. 5.5).

During the first step, salt is taken to some initial temperature T_1. This is usually
achieved by evaporating liquid ^4He or liquid ^3He. After attaining some initial temperature T_1, salt is magnetized isothermally at an external magnetic field B_1. During
this process, the heat is evolved which is taken away to the helium bath through
exchange gas, and the entropy falls. The salt goes from state a to state b during this
process. The next step is then performed by isolating the sample from the surrounding (pumping out the exchange gas), s.t. $\Delta Q = 0$ and the field is removed. Because
$\Delta Q = 0$, the sample follows the constant entropy path and this reduces the temperature T of the salt. This step is known as adiabatic demagnetization. By reducing
the field to zero, the final state of the salt corresponds to point c, with temperature
T_2. Therefore, the lowest temperature to which the paramagnetic salt can be cooled
by adiabatic demagnetization is effectively the Curie temperature. ∵ when $\Delta Q = 0$
and B change from B_1 to 0, the entropy increases (spin randomization in absence
of field). This increase in entropy is balanced by a decrease in entropy of phonons
(lattice vibrations). As a result, the sample cools to the final temperature T_2.

5.11   Solved Problems

Q.1 **The internal energy E(T) of a system at a fixed volume is found to depend on the temperature T as**

$$E(T) = aT^2 + bT^4$$

where a and b are constant. Then the entropy S(T), as a function of temperature, is

[NET-JRF-June 2016]

(A) $\dfrac{1}{2}aT^2 + \dfrac{1}{2}bT^4$ (C) $2aT + \dfrac{4}{3}bT^3$

(B) $2aT^2 + 4bT^4$ (D) $2aT + 4bT^3$

Sol: Consider the relation

$$dE = TdS - PdV$$

As dV = 0 (given), the above equation allows us to write

$$T = \frac{dE}{dS}$$

so that

$$dS = \frac{1}{T}dE$$

$$= \frac{1}{T}\left[2aTdT + 4bT^3dT\right] = \left[2a + 4bT^2\right]dT$$

Therefore,

$$\Delta S = \int_{Si}^{Sf} dS = \int_{Ti}^{Tf} \left(2a + 4bT^2\right) dT$$

or we can write

$$S_f - S_i = 2a\left(T_f - T_i\right) + \frac{4b}{3}\left(T_f^3 - T_i^3\right)$$

$$S_f = 2aT_f + \frac{4b}{3}T_f^3 + 2aT_i + \frac{4b}{3}T_i^3 + S_i$$

$$= 2aT_f + \frac{4b}{3}T_f^3 + S_0$$

where S_0 consists of all terms in square brackets. In general, we can write

$$S = 2aT + \frac{4b}{3}T^3 + S_0$$

S can be evaluated at different temperatures and while evaluating the change between two equilibrium states, S_0 will disappear. Therefore, (C) is the right option.

Q.2 **A thermodynamic function**

$$G(T, P, N) = U - TS + PV$$

is given in terms of the internal energy U, temperature T, entropy S, pressure P, volume V and the number of particles N. Which of the following relations is true? (In the following, μ is the chemical potential.)
[NET-JRF-June 2017]

(A) $S = -\left(\dfrac{\partial G}{\partial T}\right)_{N,P}$　　　(C) $V = -\left(\dfrac{\partial G}{\partial P}\right)_{N,T}$

(B) $S = \left(\dfrac{\partial G}{\partial T}\right)_{N,P}$　　　(D) $\mu = -\left(\dfrac{\partial G}{\partial N}\right)_{P,T}$

Sol. Since $G = U - TS + PV$, we can write for a small change dG

$$dG = -SdT + PdV$$

$$S = -\left(\frac{\partial G}{\partial T}\right)_{N,V} \quad \text{and} \quad P = \left(\frac{\partial G}{\partial V}\right)_{N,T}$$

Q.3 **The relation between the internal energy U, entropy S, temperature T, pressure P, volume V, chemical potential μ and number of particles N of a thermodynamic system is**

$$dU = TdS - PdV + \mu dN$$

That U is an exact differential implies that
[NET-2017]

(A) $-\left(\dfrac{\partial P}{\partial S}\right)_{V,N} = \left(\dfrac{\partial T}{\partial V}\right)_{S,N}$　　　(C) $P\left(\dfrac{\partial U}{\partial T}\right)_{S,N} = -\dfrac{1}{T}\left(\dfrac{\partial U}{\partial V}\right)_{S,\mu}$

(B) $P\left(\dfrac{\partial U}{\partial T}\right)_{S,N} = S\left(\dfrac{\partial U}{\partial V}\right)_{S,\mu}$　　　(D) $\left(\dfrac{\partial P}{\partial S}\right)_{V,N} = \left(\dfrac{\partial T}{\partial V}\right)_{S,N}$

Sol: Given that $dU = TdS - PdV + \mu dN$. As dU is a perfect differential, we can write U = U(S, V, N). For a small change in U,

$$dU = \left(\frac{\partial U}{\partial S}\right)_{V,N} dS + \left(\frac{\partial U}{\partial V}\right)_{S,N} dV + \left(\frac{\partial U}{\partial N}\right)_{V,S} dN$$

Comparing the above equations, we get

$$T = \left(\frac{\partial U}{\partial S}\right)_{N,V}, \quad P = -\left(\frac{\partial U}{\partial V}\right)_{S,N}, \quad \mu = \left(\frac{\partial U}{\partial N}\right)_{V,S}$$

Because dU is a perfect differential, the first pair gives

$$\left(\frac{\partial^2 U}{\partial S \partial V}\right) = \left(\frac{\partial^2 U}{\partial V \partial S}\right) \quad \text{or} \quad \frac{\partial}{\partial S}\left(\frac{\partial U}{\partial V}\right)_{S,N} = \frac{\partial}{\partial V}\left(\frac{\partial U}{\partial S}\right)_{N,V}$$

thus, finally giving

$$-\left(\frac{\partial P}{\partial S}\right)_{N,V} = \left(\frac{\partial T}{\partial V}\right)_{S,N}$$

Therefore, **A** is the correct option.

Q.4 **The relation between the internal energy E, entropy S, temperature T, force F and length L of a thermodynamic system is**

$$dE(S, L) = TdS + \mathbb{F}dL$$

That E is an exact differential implies that

(A) $\quad -\left(\frac{\partial T}{\partial L}\right)_S = \left(\frac{\partial \mathbb{F}}{\partial S}\right)_L$ (C) $\quad \frac{1}{L}\left(\frac{\partial T}{\partial L}\right)_S = \left(\frac{\partial \mathbb{F}}{\partial S}\right)_L$

(B) $\quad \left(\frac{\partial T}{\partial L}\right)_S = \left(\frac{\partial \mathbb{F}}{\partial S}\right)_L$ (D)   None of these

Sol: Given that $dE(S, L) = TdS + \mathbb{F}dL$ as dE is a perfect differential

$$dE = \left(\frac{\partial E}{\partial S}\right)_L dS + \left(\frac{\partial E}{\partial L}\right)_S dL$$

Thus, we can write after comparison

$$T = \left(\frac{\partial E}{\partial S}\right)_L \quad \text{and} \quad \mathbb{F} = \left(\frac{\partial E}{\partial L}\right)_S$$

As dE is a perfect differential

$$\left(\frac{\partial^2 E}{\partial S \partial L}\right) = \left(\frac{\partial^2 E}{\partial L \partial S}\right) \quad \text{or} \quad \left(\frac{\partial \mathbb{F}}{\partial S}\right)_L = \left(\frac{\partial T}{\partial L}\right)_S$$

Hence **B** is the right option.

Q.5 **The Gibbs free energy for a gas is given as**

$$G = RT \ln \left(\frac{P}{P_0} \right) - BP$$

where B is a function of T only. The pressure exerted by this gas is given by

(A) $\dfrac{RT}{V}$ (C) $\dfrac{RT}{V - B}$

(B) $\dfrac{RT}{V + B}$ (D) $\dfrac{RT}{V} + B$

Sol: We can write V as

$$V = \left(\frac{\partial G}{\partial P} \right)_T = RT \left[\frac{1}{P} - B \right] \quad \text{or} \quad P = \frac{RT}{V + B}$$

Q.6 **A Van der Waals gas has the equation of state**

$$\left(P + \frac{a}{V^2} \right) (V - b) = KT$$

The gas undergoes an isothermal expansion from volume V_1 to V_2. The change in Helmholtz free energy ΔF corresponding to this process is?

Sol: For an infinitesimal small change

$$dF = -PdV - SdT$$

For an isothermal process $dF = -PdV$, hence,

$$\Delta F = -\int_{V_1}^{V_2} \left[\frac{KT}{V - b} - \frac{a}{V^2} \right] dV = -KT \ln \left(\frac{V_2 - b}{V_1 - b} \right) + a \left[\frac{1}{V_1} - \frac{1}{V_2} \right]$$

Q.7 **Which among the following sets of Maxwell relations is correct? (U—internal energy, H—enthalpy, A—Helmholtz free energy and G—Gibbs free energy)?**

[GATE-2010]

(A) $T = \left(\dfrac{\partial U}{\partial V} \right)_S$ and $P = \left(\dfrac{\partial U}{\partial S} \right)_V$

(B) $V = \left(\dfrac{\partial H}{\partial P} \right)_S$ and $T = \left(\dfrac{\partial H}{\partial S} \right)_P$

(C) $P = -\left(\dfrac{\partial G}{\partial V} \right)_T$ and $V = \left(\dfrac{\partial G}{\partial P} \right)_S$

(D) $P = -\left(\dfrac{\partial A}{\partial S} \right)_T$ and $S = \left(\dfrac{\partial A}{\partial P} \right)_V$

Sol: Since dH = TdS + VdP, this allows us to write

$$T = \left(\frac{\partial H}{\partial S}\right)_P \quad \text{and} \quad V = \left(\frac{\partial H}{\partial P}\right)_S$$

Therefore, Answer is **B**.

Q.8 **A solid melts into a liquid via first order phase transition. The relationship between the pressure P and the temperature T of the phase transition is $P = -2T + P_0$ where P_0 is a constant. The entropy change associated with the phase transition is $1.0\,\text{JK}^{-1}\text{mol}^{-1}$. The Clausius–Clapeyron equation for the latent heat is**

$$L = T\left(\frac{\partial P}{\partial T}\right)\delta v$$

Here, $\delta v = v_{liquid} - v_{solid}$ is the change in molar volume at the phase transition. The correct statement relating the values of volumes is
[JAM-2006]

(**A**) $v_{liquid} = v_{solid}$

(**B**) $v_{liquid} = v_{solid} - 1$

(**C**) $v_{liquid} = v_{solid} - \dfrac{1}{2}$

(**D**) $v_{liquid} = v_{solid} + \dfrac{1}{2}$

Sol: Given $P = -2T + P_0$, thus $\dfrac{dP}{dT} = -2$. Also

$$L = T\left(\frac{\partial P}{\partial T}\right)\delta v = -2T\,\delta v$$

Therefore, $\dfrac{dL}{dT} = -2\delta v$. As

$$dS = \frac{dQ}{T} = m\frac{dL}{dT} = 1\,\text{JK}^{-1}\text{mol}^{-1}$$

thus $\delta v = -\dfrac{1}{2}$.

Q.9 **At atmospheric pressure (10^5 Pa), aluminium melts at 550 K. As it melts, its density decreases from $3 \times 10^3\,\text{kgm}^{-3}$ to $2.9 \times 10^3\,\text{kgm}^{-3}$. Latent heat of fusion of aluminium is 24×10^3 J/kg. The melting point of aluminium at a pressure of 10^7 Pa is closest to**
[JAM-2014]

(A)   551.3 K   (C)   558.7 K
(B)   552.6 K   (D)   547.4 K

Sol: Applying the Clausius–Clapeyron equation, we have

$$\frac{dP}{dT} = \frac{L}{T}\frac{1}{V_f - V_i} = \frac{24000}{T}\frac{1}{\left(\dfrac{1}{2900} - \dfrac{1}{3000}\right)}$$

This allows us to write dP as

$$dP = 2088 \times 10^6 \frac{dT}{T}$$

$$P = 2088 \times 10^6 \ln T + C$$

Applying the given condition, i.e., at $P = 10^5$ Pa, $T = 550$ K. This gives

$$10^5 = 2088 \times 10^6 \ln 550 + C$$

giving $C = -13.18 \times 10^9$. Therefore, at 10^7 Pa we can write

$$10^7 = 2088 \times 10^6 \ln T - 13.18 \times 10^9$$
$$T = antilog(6.31) = 551.3 \text{ K}$$

Q.10 **If there is a 10% decrease in the atmospheric pressure at a hill compared to the pressure at sea level, then the change in the boiling point of water is .....°C (Take latent heat of vaporization of water as 2270 kJ/cal and the change in the specific volume at the boiling point to be 1.2 m^3/kg)**
                                                                     [JAM-2016]
Sol: From the Clausius relation,

$$dT = dP \times T\frac{V_2 - V_1}{L} = \frac{0.1 \times 1.01 \times 10^5 \times 373 \times 1.2}{2270 \times 10^3} = 2\,^\circ\text{C}$$

Q.11 **Water freezes at $0\,^\circ$C at atmospheric pressure $(1.01 \times 10^5$ Pa). The densities of water and ice at this temperature and pressure are 1000 kg m^{-3} and 934 kg m^{-3}, respectively. The latent heat of fusion is 3.34×10^5 JK^{-1}. The pressure required for depressing the melting temperature of ice by $10\,^\circ$C is.... GPa (up to two decimal places)**
                                                                     [GATE-2017]
Sol: From the Clausius–Clapeyron equation

$$\frac{dP}{dT} = \frac{L}{T(V_2 - V_1)}$$
$$\int_{P_1}^{P_2} dP = \frac{L}{T(V_2 - V_1)}\int_{T_1}^{T_2}\frac{dT}{T}$$

$$P_2 - P_1 = \frac{L}{T(V_2 - V_1)} \ln \frac{T_2}{T_1}$$

After substituting the corresponding values, $P_2 = 0.01 \times 10^{-2}$ GPa.

Q.12 **The change of pressure by one atmosphere changes the melting point of ice by 0.0074 °C and when one gm of ice melts its volume changes by 0.0907c.c. The latent heat of ice is .... cal/gm (up to one decimal place)**

Sol: The Clausius–Clapeyron equation reads

$$\frac{dP}{dT} = \frac{L}{T(V_2 - V_1)} \quad \text{or} \quad L = \frac{T(V_2 - V_1)\,dP}{dT}$$

Given dP = 1 atmosphere = 1.01×10^6 dynes/cm², T = 0 °C = 273 K, dT = 0.0074 °C = 0.0074 K $(V_2 - V_1) = 0.0907$ cm³. Therefore

$$L = \frac{273 \times 0.0907 \times 1.013 \times 10^6}{0.0074}$$

$$= 3.39 \times 10^9 \text{ ergs/gm} = \frac{3.39 \times 10^9}{4.18 \times 10^7} \text{ cal/gm} = 81.1 \text{ cal/gm}$$

Q.13 **At atmospheric pressure, lead melts at 327 °C. As it melts, the density decreases from 1.101 $\times 10^4$ to 1.065 $\times 10^4$ kgm⁻³. Latent heat of fusion of lead is 24.5 kJ kg⁻¹. The melting point of lead at 100 atm pressure is close to ...**

Sol: Given L = 24500 Jkg⁻¹, T = 327 °C = 600 K, $\rho_i = 1.101 \times 10^4$, $\rho_f = 1.065 \times 10^4$ kgm⁻³.

$$\frac{dP}{dT} = \frac{L}{T(V_2 - V_1)}$$

$$= \frac{24500}{600\left(\frac{1}{1.065} - \frac{1}{1.101}\right) \times 10^{-4}} = 1.33 \times 10^7 \text{ PaK}^{-1}$$

When pressure changes from 1 to 100 atm (i.e., $dP = 99$ atm $= 99 \times 10^5$ Pa), the corresponding change in temperature will be

$$dT = \frac{dP}{1.33 \times 10^7} \text{ K} = \frac{99 \times 10^5}{1.33 \times 10^7} \text{ K} = 0.75 \text{ K}$$

Since dT is the difference, dT = 0.75 K = 0.75°. Therefore, the melting point of lead at 100 atm will be $T_f = 327.75$ °C.

Q.14 **Find the change in the boiling point of water when the pressure is increased from 1.0 to 10 atmospheres. Given specific volume of steam = 1.677 m³ kg⁻¹ and that of water is 10^{-3} m³ kg⁻¹, latent heat of steam 2268 kJ/kg and boiling point of water at one atmosphere pressure 100 °C and pressure of one atmosphere 10^5 Pa m².**

Sol: The Clausius–Clapeyron equation implies

$$\frac{dP}{dT} = \frac{L}{T\,(V_2 - V_1)}$$

Therefore,

$$dT = \frac{T\,(V_2 - V_1)\,dP}{L}$$

$$= \frac{373\,(1.677 - 0.001)\,9 \times 10^5}{2268 \times 10^3} = 248.07 \text{ K}$$

Because dT measures the change in T, therefore change in the boiling point of water is $dT = 248.07\,°C$.

Q.15 **Find the depression in the melting point of ice when pressure is increased by one atmospheric. Given that the latent heat of fusion of ice $L = 3.4 \times 10^5$ Jkg^{-1}. Specific volume of 1 kg of ice and water at 0°C are 1.091×10^{-3} m^3 and 1.0×10^{-3} m^3, respectively. 1 atmospheric pressure $= 10^5$ Nm^2.**

Sol: Given that $dP = 1$ atm $= 10^5$ Pa $= 10^5$ Nm^{-2}, T $= 0°C = 273$ K, L $= 3.4 \times 10^5$ Jkg^{-1}, Specific volume of 1 kg of ice $V_1 = 1.091 \times 10^{-3}$ m^3, Specific volume of 1 kg of water $V_2 = 1.0 \times 10^{-3}$ m^3. Therefore, $V_2 - V_1 = -0.091 \times 10^{-3}$ m^3. From the Clausius–Clapeyron equation

$$dT = \frac{T\,(V_2 - V_1)\,dP}{L}$$

$$= \frac{273\,(-0.091 \times 10^{-3})\,10^5}{3.4 \times 10^5} = -0.0073 \text{ K}$$

Since dT measures the change in temperature, $dT = -0.0073\,°C$. This means that the melting point of ice will be depressed by $0.0073\,°C$ per atmosphere increase of pressure.

Q.16 **If the Helmholtz free energy for radiation is given by**

$$F = -\frac{8\pi^5 K_B^4 T^4}{45c^3 h^3} V$$

(a) **What is radiation pressure?**
(b) **What is entropy (S) of the system?**

Sol. (a) $dF = -SdT - PdV$

$$P = -\left(\frac{\partial F}{\partial V}\right)_T = \frac{8\pi^5 K_B^4 T^4}{45c^3 h^3}$$

(b) Entropy S is

$$S = -\left(\frac{\partial F}{\partial T}\right)_V = \frac{32\pi^5 K_B^4}{45c^3 h^3}$$

Q.17 **In general, the specific volume of a liquid is much less compared to that of its vapours. Assuming that the vapour obeys the ideal gas equation, then prove that**

$$P = constant \times e^{-L/RT}$$

Sol: The Clausius–Clapeyron equation reads

$$\frac{dP}{dT} = \frac{L}{T(V_2 - V_1)}$$

Since $V_1 <<< V_2$, one can approximate $V_2 - V_1 \approx V_2 = V$ say. Therefore, the Clapeyron equation reduces to

$$\frac{dP}{dT} = \frac{L}{TV}$$

As the vapour obeys the ideal gas equation, PV = RT, giving V = RT/P. Therefore,

$$\frac{dP}{dT} = \frac{L}{T\left(\frac{RT}{P}\right)} = \frac{LP}{RT^2}$$

$$\frac{dP}{P} = \frac{L}{RT^2}dT$$

After integration,

$$\ln P = -\frac{L}{RT} + constant$$

thus giving

$$P = constant \times e^{-L/RT}$$

Q.18 Prove the following:

$$U = F - T\left(\frac{\partial F}{\partial T}\right)_V$$

$$= -T^2\left[\frac{\partial}{\partial T}\left(\frac{F}{T}\right)_V\right]$$

Sol: By definition, the *Helmholtz free energy* is defined as

$$F = U - TS$$

$$dF = -SdT - PdV$$

This gives

$$S = -\left(\frac{\partial F}{\partial T}\right)_V$$

This together with Eq. 5.14 gives

$$F = U + T\left(\frac{\partial F}{\partial T}\right)_V$$

or

$$U = F - T\left(\frac{\partial F}{\partial T}\right)_V$$

Further, we can write

$$\left[\frac{\partial}{\partial T}\left(\frac{F}{T}\right)\right]_V = \frac{1}{T}\left(\frac{\partial F}{\partial T}\right)_V - \frac{F}{T^2}$$

$$= -\frac{1}{T^2}\left[F - T\left(\frac{\partial F}{\partial T}\right)_V\right]$$

or

$$F - T\left(\frac{\partial F}{\partial T}\right)_V = -T^2\left[\frac{\partial}{\partial T}\left(\frac{F}{T}\right)\right]_V$$

Therefore,

$$U = F - T\left(\frac{\partial F}{\partial T}\right)_V = -T^2\left[\frac{\partial}{\partial T}\left(\frac{F}{T}\right)_V\right]$$

Q.19 Prove the following

$$H = G - T\left(\frac{\partial G}{\partial T}\right)_P$$

$$= -T^2\left[\frac{\partial}{\partial T}\left(\frac{G}{T}\right)_P\right]$$

Sol: By definition, the *Gibbs free energy* is defined as

$$G = H - TS$$
$$dG = VdP - SdT$$

This gives

$$S = -\left(\frac{\partial G}{\partial T}\right)_P$$

This together with Eq. 5.130 gives

$$G = H + T\left(\frac{\partial G}{\partial T}\right)_P$$

or

$$H = G - T\left(\frac{\partial G}{\partial T}\right)_P$$

Further, we can write

$$\left[\frac{\partial}{\partial T}\left(\frac{G}{T}\right)\right]_P = \frac{1}{T}\left(\frac{\partial G}{\partial T}\right)_P - \frac{G}{T^2}$$
$$= -\frac{1}{T^2}\left[G - T\left(\frac{\partial G}{\partial T}\right)_P\right]$$

$$G - T\left(\frac{\partial G}{\partial T}\right)_P = -T^2\left[\frac{\partial}{\partial T}\left(\frac{G}{T}\right)\right]_P$$

Therefore,

$$H = G - T\left(\frac{\partial G}{\partial T}\right)_P = -T^2\left[\frac{\partial}{\partial T}\left(\frac{G}{T}\right)\right]_P$$

Q.20 **Prove the following thermodynamic relation:**

$$C_V = -T\left(\frac{\partial^2 F}{\partial T^2}\right)_V$$

Sol: For a small change in *Helmholtz free energy* between two equilibrium states, we can write

$$dF = -PdV - SdT, \quad \text{so that} \quad S = -\left(\frac{\partial F}{\partial T}\right)_V$$

Again, performing partial differentiation w.r.t. T keeping volume fixed, we get

$$\left(\frac{\partial S}{\partial T}\right)_V = -\left(\frac{\partial^2 F}{\partial T^2}\right)_V$$

or

$$T\left(\frac{\partial S}{\partial T}\right)_V = -T\left(\frac{\partial^2 F}{\partial T^2}\right)_V$$

Therefore,

$$C_V = T\left(\frac{\partial S}{\partial T}\right)_V = -T\left(\frac{\partial^2 F}{\partial T^2}\right)_V$$

Q.21 Prove the following thermodynamic relation:

$$C_P = -T\left(\frac{\partial^2 G}{\partial T^2}\right)_P$$

Sol: For a small change in the *Gibbs free energy* between two equilibrium states, we can write

$$dG = VdP - SdT \quad \text{so that} \quad S = -\left(\frac{\partial G}{\partial T}\right)_P$$

Again, performing partial differentiation w.r.t. T keeping pressure fixed, we get

$$\left(\frac{\partial S}{\partial T}\right)_P = -\left(\frac{\partial^2 G}{\partial T^2}\right)_P$$

or

$$T\left(\frac{\partial S}{\partial T}\right)_P = -T\left(\frac{\partial^2 G}{\partial T^2}\right)_P$$

Therefore,

$$C_P = T\left(\frac{\partial S}{\partial T}\right)_P = -T\left(\frac{\partial^2 G}{\partial T^2}\right)_P$$

5.12   Multiple Choice Questions

Q.1 **The four thermodynamic potentials are (P—Pressure, V—Volume, T—Temperature, U—Internal energy function, H—Enthalpy, F—Helmholtz function and G—Gibbs function)**

   (A)   P, V, T and U   (C)   U, H, F and G
   (B)   P, V, T and F   (D)   None of these

Q.2 **From Maxwell's relations, the ratio '$\dfrac{\kappa_T}{\kappa_S}$' is**

   (A)   1   (C) $\dfrac{1}{\gamma}$
   (B) γ (D) $\gamma - 1$

Q.3 **Gibbs potential 'G' is**

   (A) $U-PV-TS$ (C) $U+PV-TS$
   (B) $U+PV+TS$ (D) $U-PV-TS$

Q.4 **For a thermodynamic system, the natural variables for the Gibbs function are**

   (A)   S, V   (C)   T, P
   (B)   V, T   (D)   S, P

Q.5 **For a thermodynamic system, the natural variables for the Helmholtz function are**

   (A)   S, V   (C)   T, P
   (B)   V, T   (D)   S, P

Q.6 **Maxwell's thermodynamic relation is**

   (A) $\left(\dfrac{\partial S}{\partial T}\right)_T = \left(\dfrac{\partial P}{\partial V}\right)_T$ (C) $\left(\dfrac{\partial T}{\partial P}\right)_P = \left(\dfrac{\partial V}{\partial P}\right)_T$

   (B) $\left(\dfrac{\partial T}{\partial V}\right)_S = -\left(\dfrac{\partial P}{\partial S}\right)_V$ (D)   none of these

Q.7 **A thermodynamic system is maintained at constant temperature and pressure. In thermodynamic equilibrium, its**

                                                                    [JAM-2009]

   (A)   Gibbs free energy is minimum
   (B)   enthalpy is maximum
   (C)   Helmholtz free energy is minimum
   (D)   internal energy is zero

Q.8 **When a system is held at constant temperature and pressure, in a state of equilibrium the system attains a minimum value of**

(A) internal energy   (C) Helmholtz free energy

(B) enthalpy   (D) Gibbs free energy

Q.9 Isothermal compressibility κ_T of a substance is defined as $\kappa_T = -\dfrac{1}{V}\left(\dfrac{\partial V}{\partial P}\right)_T$. Its value for n moles of an ideal gas will be

(A) $\dfrac{1}{P}$ (C) $-\dfrac{1}{P}$

(B) $\dfrac{n}{P}$ (D) $-\dfrac{n}{V}$

Q.10 **For an ideal gas, the adiabatic compressibility κ_S is**

(A) $\dfrac{1}{P}$ (C) $-\dfrac{1}{P}$

(B) $\dfrac{1}{\gamma P}$ (D) $-\dfrac{\gamma}{V}$

Q.11 **For an ideal gas, the coefficient of isobaric expansion is**

(A) 1/P   (C) 1/T

(B) 1/V   (D) none

Q.12 **The isothermal compressibility κ_T of a gas following the equation of state**

$$P\,(V_m - b) = RT$$

is given by

(A) $\dfrac{1}{P}$ (C) $\dfrac{1}{P} - \dfrac{b}{V_m}$

(B) $-\dfrac{1}{P}$ (D) $\dfrac{1}{P} - \dfrac{b}{V_m P}$

Q.13 **The isothermal compressibility κ_T of an ideal gas at temperature T and volume V is given by**

                                             [GATE-2012]

(A) $-\dfrac{1}{V}\left(\dfrac{\partial V}{\partial P}\right)_T$ (C) $-V\left(\dfrac{\partial P}{\partial V}\right)_T$

(B) $\dfrac{1}{V}\left(\dfrac{\partial V}{\partial P}\right)_T$ (D) $V\left(\dfrac{\partial P}{\partial V}\right)_T$

Q.14 **The adiabatic compressibility κ_S of an ideal gas is given by**

(A) $-\dfrac{1}{V}\left(\dfrac{\partial V}{\partial P}\right)_S$ (C) $-V\left(\dfrac{\partial P}{\partial V}\right)_S$

(B) $\dfrac{1}{V}\left(\dfrac{\partial V}{\partial P}\right)_S$ (D) $V\left(\dfrac{\partial P}{\partial V}\right)_S$

Q.15 **The adiabatic expansivity β_S is given by**

(A) $\dfrac{1}{V}\left(\dfrac{\partial V}{\partial T}\right)_S$ (C) $-V\left(\dfrac{\partial P}{\partial V}\right)_S$

(B) $\dfrac{1}{V}\left(\dfrac{\partial V}{\partial P}\right)_S$ (D) $V\left(\dfrac{\partial P}{\partial V}\right)_S$

Q.16 **The isobaric expansivity β_P is given by**

(A) $\dfrac{1}{V}\left(\dfrac{\partial V}{\partial T}\right)_P$ (C) $-V\left(\dfrac{\partial T}{\partial V}\right)_P$

(B) $-\dfrac{1}{V}\left(\dfrac{\partial V}{\partial P}\right)_P$ (D) $V\left(\dfrac{\partial T}{\partial V}\right)_P$

Q.17 **The maximum work that can be obtained from a system during a given process in which initial and final temperatures of the system are equal to the surrounding temperature is equal to the decrease in**

(A)   internal energy   (C)   Helmholtz free energy
(B)   enthalpy   (D)   Gibbs free energy

Q.18 **For a system held at constant T and P in an equilibrium state, which one of the following quantity is minimum**

[JAM-2008]

(A)   internal energy   (C)   Helmholtz free energy
(B)   enthalpy   (D)   Gibbs free energy

Q.19 **For a system at constant temperature and volume, which of the following statements is correct at equilibrium?**

[GATE-2016]

(A)   The Helmholtz free energy attains a local minimum
(B)   The Helmholtz free energy attains a local maximum
(C)   The Gibbs free energy attains a local minimum
(D)   The Gibbs free energy attains a local maximum

Q.20 If U, F, H and G represent internal energy, Helmholtz free energy, enthalpy and Gibbs free energy, respectively, then which one of the following is a correct thermodynamic relation?

[JAM-2016]

(A)   dU = PdV − TdS   (C)   dF = −PdV + SdT
(B)   dH = VdP + TdS   (D)   dG = VdP + SdT

Q.21 Consider an ensemble of thermodynamic systems each of which is characterized by the same number of particles, pressure and temperature. The thermodynamic function describing the ensemble is

[JAM-2018]

(A)   Enthalpy   (C)   Gibbs free energy
(B)   Helmholtz free energy   (D)   entropy

Q.22 Which of the following relations is (are) true for thermodynamic variables?

[JAM-2018]

(A) $TdS = C_V dT + T \left(\dfrac{\partial P}{\partial T} \right)_V dV$

(B) $TdS = C_P dT - T \left(\dfrac{\partial V}{\partial T} \right)_P dP$

(C) $dF = -SdT + PdV$
(D) $dG = -SdT + VdP$

Q.23 The differential of the thermodynamic potential H can be expressed in terms of the differentials of its natural variables as

(A) $dH = C_P dT + \left(\dfrac{\partial H}{\partial P} \right)_T dT$ (C) $dH = dU + PdV + VdP$
                                                                              (D) $dH = -TdS + PdV$
(B) $dH = TdS + VdP$

Q.24 The differential of the thermodynamic potential G can be expressed in terms of the differentials of its natural variables as

(A) $dG = dU + PdV + VdP - SdT - TdS$
(B) $dG = TdS + VdP$
(C) $dG = -SdT + PdV$
(D) $dG = -SdT + VdP$

Q.25 **The change in entropy associated with a change in volume at constant pressure is equal to**

(A) $\left(\dfrac{\partial P}{\partial T}\right)_S$

(B) $-\left(\dfrac{\partial S}{\partial P}\right)_T$

(C) $\left(\dfrac{\partial T}{\partial V}\right)_S$

(D) $\left(\dfrac{\partial P}{\partial T}\right)_V$

Q.26 **The total energy, E of an ideal non-relativistic Fermi gas in three dimensions is given by**

$$E \propto \dfrac{N^{5/3}}{V^{2/3}}$$

where N is the number of particles and V is the volume of the gas. Identify the CORRECT equation of state (P being the pressure)
                                                                [GATE-2012]

(A) $PV = \dfrac{1}{3}E$

(B) $PV = \dfrac{2}{3}E$

(C) $PV = E$

(D) $PV = \dfrac{5}{3}E$

Q.27 **For a gas under isothermal condition, its pressure p varies with volume V as**

$$P \propto V^{-2/3}$$

The bulk module B is proportional to
                                                                [GATE-2014]

(A) $V^{-1/2}$

(B) $V^{-2/3}$

(C) $V^{-3/5}$

(D) $V^{-5/3}$

Q.28 **A real gas has specific volume v at temperature T. Its coefficient of volume expansion and isothermal compressibility are 'β_P' and 'κ_T', respectively. Its molar specific heat at constant pressure C_P and molar specific heat at constant volume C_V are related as**
                                                                [JAM-2014]

(A) $C_P = C_V + R$

(B) $C_P = C_V + \dfrac{TV\beta_P}{\kappa_T}$

(C) $C_P = C_V + \dfrac{TV\beta_P^2}{\kappa_T}$

(D) $C_P = C_V$

Q.29 **If the ratio of isothermal and adiabatic elasticities is E_T/E_S, then which of the following is true?**

(A) $\dfrac{E_S}{E_T} = \dfrac{C_P}{C_V}$ (C) $\dfrac{E_T}{E_S} = \dfrac{C_P}{C_V}$

(B) $\dfrac{E_S}{E_T} = \dfrac{C_V}{C_P}$ (D) $\dfrac{E_S}{E_T} = C_P - C_V$

Q.30 **For a real gas, $C_P - C_V$ is**

(A) $C_P - C_V = R$

(B) $C_P - C_V = R^{-1}$

(C) $C_P - C_V = R\left[1 + \dfrac{2a}{RTV}\right]$

(D) $C_P - C_V = R\left[1 + \dfrac{2a}{RbV}\right]$

Q.31 **Increasing the pressure, the melting point of ice**

(A) increases
(B) decreases
(C) remains unchanged
(D) increases and becomes maximum

Q.32 **Two blocks of ice when pressed together join to form a single block. This is because**

(A) of heat produced during pressing
(B) of cold produced during pressing
(C) melting point of ice decreases with increase in pressure
(D) melting point of ice increases with increase in pressure

Q.33 **AT normal temperature and pressure, the water boils at $100\,°C$. On top of a mountain, the water will boil at a temperature**

(A) $100\,°C$ (C) less than $100\,°C$
(B) greater than $100\,°C$ (D) will not boil at all

Q.34 **At normal temperature and pressure, the water boils at $100\,°C$. Deep down in a mine, the water will boil at a temperature**

(A) 100 °C   **(C)** less than 100 °C
(B) greater than 100 °C   **(D)** will not boil at all

Q.35 **In a pressure cooker, the cooking process is faster because**

(A) more pressure is available to cook food at 100 °C
(B) more steam is available to cook food at 100 °C
(C) the boiling point of water is lowered by increased pressure
(D) the boiling point of water is raised by increased pressure inside the cooked

Q.36 **Select the wrong one**

(A) in the throttling process, the initial and final enthalpies are equal
(B) enthalpy change in an isobaric process is equal to heat transferred
(C) Gibbs function doesn't change in isothermal and isobaric processes
(D) Gibbs function increases in natural processes

Q.37 **Choose the wrong one**

(A) latent heat cannot become zero
(B) specific heat of saturated water vapour at 100 °C is negative
(C) there is only one triple point of a substance
(D) boiling point of all liquids rises with increase in pressure

Q.38 **An iceberg melts at the base but not at the top because**

(A) the temperature at the base is relatively higher than at the top
(B) due to high pressure, the melting point of ice at the base is lowered
(C) of the warmth of underlying water
(D) cannot say

Q.39 **The melting point of a solid is lowered by increasing pressure on it. When this solid melts, its volume**

(A) increases   **(C)** doesn't change
(B) decreases   **(D)** cannot predict

Q.40 **While cooking rice, it will take the longest time if cooked at**

(A) Mount Everest   **(C)** sea level
(B) Shimla   **(D)** plain delta on earth surface

Q.41 **A closed bottle containing water at 27 °C is taken in a space ship on the surface of moon. If the lid is opened on the moon's surface then**

(A)   the water inside will freeze immediately
(B)   the water will boil and escape
(C)   the water will decompose into H_2 and O_2
(D)   nothing will happen to water

Q.42 Which one of the following is not a thermodynamic relation?

(A) $\left(\dfrac{\partial T}{\partial V}\right)_S = -\left(\dfrac{\partial P}{\partial S}\right)_V$ (C) $\left(\dfrac{\partial S}{\partial V}\right)_T = \left(\dfrac{\partial P}{\partial T}\right)_V$

(B) $\left(\dfrac{\partial T}{\partial P}\right)_S = \left(\dfrac{\partial V}{\partial S}\right)_P$ (D) $\left(\dfrac{\partial S}{\partial P}\right)_T = -\left(\dfrac{\partial T}{\partial V}\right)_P$

Q.43 Specific heat of saturated water vapour at $100\,^{\circ}$C is

(A)   positive   (C)   zero
(B)   negative   (D)   cannot say

Q.44 The Clausius–Clapeyron equation is

(A) $\dfrac{dP}{dT} = \dfrac{L}{T\,(V_2 - V_1)}$ (C) $\dfrac{dP}{dT} = LT\,(V_2 - V_1)$

(B) $\dfrac{dP}{dT} = \dfrac{T}{L\,(V_2 - V_1)}$ (D)   None

Keys and Hints to MCQ Type Questions

Q.1 C	Q.9 A	Q.17 C	Q.25 A	Q.33 C	Q.41 B
Q.2 B	Q.10 B	Q.18 D	Q.26 B	Q.34 B	Q.42 D
Q.3 C	Q.11 C	Q.19 A	Q.27 D	Q.35 D	Q.43 C
Q.4 C	Q.12 D	Q.20 B	Q.28 C	Q.36 D	Q.44 A
Q.5 B	Q.13 A	Q.21 C	Q.29 A	Q.37 A	
Q.6 B	Q.14 A	Q.22 B, D	Q.30 C	Q.38 B	
Q.7 A	Q.15 A	Q.23 B	Q.31 B	Q.39 B	
Q.8 D	Q.16 A	Q.24 D	Q.32 C	Q.40 A	

Hint.7 For a thermodynamic system, the change in the Gibbs free energy is $dG = -SdT + VdP$. Under isothermal and isobaric conditions $dG = 0$. Hence, G is conserved in any isothermal and isobaric process.

Hint.9 For an ideal gas $PV = nRT$. At constant temperature, $PV \propto T$. This implies
$$PdV + VdP = 0 \text{ or } \frac{dP}{P} = -\frac{dV}{V}, \text{ thus giving } -\frac{1}{V}\left(\frac{\partial V}{\partial P}\right)_T = \frac{1}{P} = \kappa_T.$$

Hint.10 For an adiabatic process, $P \propto V^{-\gamma}$, therefore $\dfrac{dP}{P} = -\gamma\dfrac{dV}{V}$. This gives
$$\kappa_S = \frac{1}{\gamma P}.$$

Hint.11 For an ideal gas $V = \dfrac{nRT}{P}$, so that

$$\alpha_P = \left(\frac{1}{V} \frac{\partial V}{\partial T} \right)_P = \frac{1}{V} \frac{nR}{P} = \frac{1}{T}$$

Hint.12 By definition

$$\kappa_T = -\frac{1}{V_m} \left(\frac{\partial V_m}{\partial P} \right)_T$$

$$= -\frac{1}{V_m} \left(-\frac{RT}{P^2} \right) = \frac{V_m - b}{V_m P} = \frac{1}{P} - \frac{b}{V_m P}$$

Hint.19 $dF = -SdT + PdV = 0$ at constant T and V

Hint.22 Assuming $S = S(T, V)$, for small change

$$dS = \left(\frac{\partial S}{\partial T} \right)_V dT + \left(\frac{\partial S}{\partial V} \right)_T dV$$

$$TdS = C_V dT + T \left(\frac{\partial P}{\partial T} \right)_V dV$$

Similarly, $S = S(T, P)$, for a small change

$$dS = \left(\frac{\partial S}{\partial T} \right)_P dT + \left(\frac{\partial S}{\partial P} \right)_T dP$$

Then, we obtain

$$TdS = C_P dT + T \left(\frac{\partial S}{\partial P} \right)_T dP = C_P dT - T \left(\frac{\partial V}{\partial T} \right)_P dP$$

and $dF = SdT - PdV$, $dG = -SdT + PdV$.

Hint.26 Answer is **B**.

$$P = -\left(\frac{\partial E}{\partial V} \right)_N = \frac{2}{3} \left(\frac{N}{V} \right)^{5/3}$$

This allows us to write

$$PV = \frac{2}{3} \frac{N^{5/3}}{V^{2/3}} = \frac{2}{3} E$$

Hint.27 $P = C V^{-2/3}$, and

$$B = -V \frac{\partial P}{\partial V} \propto V^{-5/3}$$

5.13 Exercises

1. What do you understand by thermodynamic potentials? Explain the importance of these functions.

2. Write down Maxwell's thermodynamic equations and prove the relation $\dfrac{E_S}{E_T} = \dfrac{C_P}{C_V}$.

3. Starting from the Maxwell relation, establish Clausius–Clapeyron's equation.

4. Define *Helmholtz free energy* and for an isochoric process and establish the relation

$$U = F - T \left[\frac{\partial F}{\partial T} \right]_V$$

5. Explain why the temperature drops when a gas is subjected to an adiabatic expansion.

6. Write four Maxwell's relations and show that an adiabatic expansion always results in a fall in temperature.

7. Give the physical significance of *Free energy*.

8. Prove the following relation:

$$\left[\frac{\partial S}{\partial V} \right]_T = T \left[\frac{\partial P}{\partial T} \right]_V$$

9. Prove the following thermodynamic relations:

$$T dS = C_V dT + T \left(\frac{\partial P}{\partial T} \right)_V dV$$

$$T dS = C_P dT - T \left(\frac{\partial V}{\partial T} \right)_P dP$$

10. Starting with the definition of enthalpy, obtain the following Maxwell's relation:

$$\left[\frac{\partial T}{\partial P} \right]_S = T \left[\frac{\partial V}{\partial S} \right]_P$$

11. Establish the Clausius–Clapeyron equation

$$\frac{dP}{dT} = \frac{L}{T (v_2 - v_1)}$$

where v_2 and v_1 are specific volumes. From this relation, find out the effect of pressure on the boiling point of a liquid and the melting point of a solid.

12. Prove Maxwell's relation

$$\left[\frac{\partial S}{\partial V}\right]_T = T\left[\frac{\partial P}{\partial T}\right]_V$$

From this relation, establish the Clausius–Clapeyron relation

$$\frac{dP}{dT} = \frac{L}{T\,(v_2 - v_1)}$$

and explain its importance pertaining to the influence of pressure on the boiling point of a liquid and the melting point of a solid.

13. Define the magneto caloric effect.

14. What is adiabatic demagnetization? Deduce an expression for the change in temperature of paramagnetic salt due to adiabatic demagnetization.

15. An elastic rod is stretched by an external force $\mathbb{F}$. Obtain expressions for modified Maxwell's thermodynamic relations for this system.

16. An elastic rod of length L and cross-sectional area A is stretched isothermally by an external force $\mathbb{F}$. Show that entropy change for the rod is given by

$$\left(\frac{\partial S}{\partial L}\right)_T = AE_T\alpha_\mathbb{F}$$

where E_T is isothermal elasticity and $\mathbb{F}$ is isothermal expansivity. Explain qualitatively the cause of the increase in entropy of the rod.

Kinetic Theory of Gases

6

This chapter presents the postulates of the kinetic theory of gases and its mathematical formulation to understand the pressure exerted by various gas molecules. Experimental verification of Maxwell's law is presented in a simpler manner. The chapter also provides a basic discussion on understanding the heat capacities and degree of freedom in a very facile manner. Various transport phenomena, like viscosity, thermal conduction and diffusion, etc., are dealt with in detail in this chapter.

6.1   Postulates of Kinetic Theory of Gases

(i) A gas consists of a large number (N) of molecules (atoms) each of mass m, moving in random directions with a variety of speeds. Experimental data indicate that 1 cm^3 of an ideal gas under standard temperature and pressure (STP) consists of nearly 3×10^{19} molecules.

(ii) On the average, the molecules are far apart from each other. This means their average separation is much higher than the size (diameter) of molecules. Indeed, experiments reveal that the diameter of a gas molecule is around 2–3 Å. The average separation between any two gas molecules at STP is about 30 Å, bigger than the size of gas molecules.

(iii) The gas molecules behave like rigid and perfectly elastic hard spheres. That is, the molecules neither lose energy nor deform in shape when they collide among themselves or with the wall of the container. This means collisions with other gas molecules and with the wall of the vessel are perfectly elastic.

(iv) The gas molecules are in a state of continuous random motion as shown in Fig. 6.1. In principle, they can move in all directions with equal probability. They are assumed to obey the laws of classical mechanics and it is also assumed that they interact with each other only when they collide. Though molecules

© The Author(s) 2022
S. Sharma, *Thermal and Statistical Physics*,
https://doi.org/10.1007/978-3-031-07685-5_6

exert weak attractive (or could be repulsive) forces on each other between collisions, the potential energy associated with these intermolecular attractions is small compared with the kinetic energy of the molecules and is neglected.

(v) The duration of the collision is negligible compared to the time interval between successive collisions. This allows us to ignore the potential energy associated with collisions in comparison to the kinetic energy between collisions.

(vi) The gravitational potential energy does not affect the motion of gas molecules as the magnitude of gravitational forces is of the order of $10^{-43} N$, less than the molecular force whose magnitude is around $10^{-13} N$ for normal separation between the gas molecules.

(vii) All molecules do not move with the same speed. There is a spread of molecular speed ranging from zero to infinity. Experimental evidence for this comes from the finite width of spectral lines.

6.2   Pressure Exerted by an Ideal Gas

Let us consider μ mole of gas confined to a cubical container of side L and volume $V = L^3$ as shown in Fig. 6.1a. Let N be the total number of molecules (each of mass m) so that the number density is n = N/V. In absence of an external field, these molecules are always in random motion and keep on colliding with the walls of the container. That is, they have different velocities and to make the problem mathematically simpler, we divide them into G energy subgroups (1, 2, 3, 4 ..... G), such that each subgroup has n_i (i = 1, 2, 3, ... G) molecules per m^3 and move with average velocity v_i. The molecules are subjected to satisfy the condition

$$N = \sum_{i=1}^{G} n_i = n_1 + n_2 + ...n_G$$

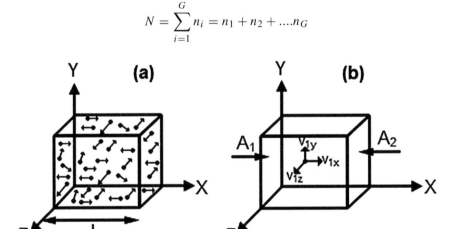

Fig. 6.1 a Schematic illustration of a gas confined to the cube of side L. The arrows inside the cube indicate velocity vectors of randomly moving gas molecules. **b** Resolution of velocity of a molecules into three mutually perpendicular components

Let us first consider group 1 moving with an average velocity v_1. This can be resolved into three mutually perpendicular components v_{1x}, v_{1y} and v_{1z} along X-, Y- and Z-axes, respectively (Fig. 6.1b). Using the properties of vector, one can write

$$v_1^2 = v_{1x}^2 + v_{1y}^2 + v_{1z}^2 \tag{6.1}$$

Instead of dealing with all three directions simultaneously, we will consider the motion of n_1 molecules in group 1 along any one of three axes. The obtained expression for pressure can then be generalized for the other two directions as well as other groups of molecules. Let us consider the motion of one of these molecules (from group 1) along the X-axis, perpendicular to faces A_1 and A_2 of the cube. The initial momentum of this molecule at face A_1 moving along the X-axis will be mv_{1x}. After travelling a distance of L, it reaches face A_2 and makes an elastic collision. As a result, it rebounds (its velocity will be along the negative X-axis, $-v_{1x}$) without any loss of momentum. Its momentum after the collision will be $-mv_{1x}$. Therefore, the change in momentum of the molecule at face A_2 is $mv_{1x} - (-mv_{1x}) = 2mv_{1x}$. After a collision at face A_2, the molecule travels back and strikes again face A_1. Therefore, the molecule covers a distance of 2L before it strikes the same face of the cube.

Let t be the time between any two successive collisions of the molecule with face A_1, so that we can write

$$t = \frac{2L}{v_{1x}}$$

So that along X-axis the momentum imparted per second to face A_1 by a molecule, i.e., the rate of change of momentum of a molecule is

$$\frac{2mv_{1x}}{t} = \frac{mv_{1x}^2}{L}$$

Therefore, the total momentum transferred per second along X-axis by n_1 molecules (in group 1) per second is $\frac{mn_1 v_{1x}^2}{L}$. Since the rate of change of momentum defines the force (Newton's second law of motion), we argue that force exerted along X-direction by n_1 molecules from group 1 on face A_1 is

$$f_{1x} = \frac{mn_1 v_{1x}^2}{L}$$

This force is exerted by n_1 molecules over a surface area L^2. Therefore, the pressure exerted by these n_1 molecules is

$$P_{1x} = \frac{f_{1x}}{L^2} = mn_1 \frac{v_{1x}^2}{L^3} \tag{6.2}$$

Similarly, we can write an expression for pressure exerted by molecules in the remaining groups as follows:

$$P_{2x} = mn_2 \frac{v_{2x}^2}{L^3}$$

$$P_{3x} = mn_3 \frac{v_{3x}^2}{L^3}$$

$$P_{4x} = mn_4 \frac{v_{4x}^2}{L^3}$$

$$\cdot$$
$$\cdot$$

$$P_{Gx} = mn_G \frac{v_{Gx}^2}{L^3}$$

So that the total pressure exerted by all molecules along the X-direction is

$$P_x = \sum_i P_{ix} = \frac{m}{L^3} \sum_{i=1}^{G} n_i v_{ix}^2 \qquad (6.3)$$

Here, we define average value of v_x^2 (due to random motion, $\overline{v_x^2} = \langle v_x \rangle = 0$, but $\langle v_x^2 \rangle \neq 0$) as

$$\overline{v_x^2} = \frac{\sum_{i=1}^{G} n_i v_{ix}^2}{\sum_i^{G} n_i} = \frac{\sum_{i=1}^{G} n_i v_{ix}^2}{N}$$

The above two equations allow us to rewrite P_x as

$$P_x = \frac{mN}{L^3} \overline{v_x^2}$$

As we mentioned in the beginning, the molecules move randomly in all directions, i.e., all directions are equally probable. Physically speaking, the gas molecules do not have a preferred direction of motion. Therefore, we can write

$$\overline{v_x^2} = \overline{v_y^2} = \overline{v_z^2} = \frac{1}{3}\overline{v_x^2 + v_y^2 + v_z^2} = \frac{v^2}{3} \qquad (6.4)$$

Combining Eq. 6.4, and expression for P_x, we obtain

$$P = \frac{1}{3} m \frac{N}{L^3} \overline{v^2} = \frac{1}{3} m \frac{N}{V} \overline{v^2} = \frac{1}{3} mn \overline{v^2} \qquad (6.5)$$

This equation connects macroscopic parameters (P and V) with microscopic properties (mass and mean square velocity) of the system (a system of gas molecules).

$\left\langle v_{rms}^2 \right\rangle = \dfrac{3k_B T}{m}$ is the root mean square velocity and n is the total number of gas molecules confined to volume V, such that the number density of molecules is

$$n = N/V \tag{6.6}$$

Further, in terms of root mean square velocity, the pressure can be written as

$$PV = Nk_B T \tag{6.7}$$

which is the ideal gas equation. One can rewrite it in another form by replacing $N = \mu N_A$ and further using $N_A k_B = R$, where μ is the number of moles of a gas, N_A is Avogadro's constant and R is the universal gas constant. The modified gas equation becomes

$$PV = \mu R k_B T \tag{6.8}$$

The relation $PV = Nk_B T$ expresses an important fact that the pressure of an ideal gas does not depend on the mass m of the molecules. Although more massive molecules transfer greater momentum to the container walls than light molecules, their mean velocity is lower and so they make fewer collisions with the walls. Therefore, the pressure is the same for a gas of light or massive molecules; it depends only on n, the number per unit volume, and the temperature. $v_{rms} = \sqrt{\overline{v^2}}$.

6.2.1 Kinetic Interpretation of Temperature

The pressure exerted by gas molecules is

$$P = \frac{1}{3}mn\overline{v^2} = \frac{1}{3}m\frac{N}{V}\overline{v^2}$$

Rearranging we get

$$\frac{PV}{N} = \frac{2}{3}\left[\frac{1}{2}m\overline{v^2}\right]$$

The quantity $\frac{1}{2}m\overline{v^2}$ is the average kinetic energy $\overline{K.E}$ of the gas molecules in the gas. Further, using the ideal gas law for the molecules, $PV = Nk_B T$ (where N is the total number of gas molecules, not the Avogadro number N_A), we obtain

$$k_B T = \frac{2}{3}\overline{K.E}$$

or

$$\overline{K.E} = \frac{1}{2}m\overline{v^2} = \frac{3}{2}k_B T \qquad (6.9)$$

This equation implies that the *average translational kinetic energy of molecules in random motion in an ideal gas is directly proportional to the absolute temperature of the gas*. This means the higher the temperature, the faster the molecules are moving. Note that $\sqrt{\overline{v^2}}$ is called as root mean square speed (v_{rms}) since we are taking the square root of the mean of the square of the speed.

$$v_{rms} = \overline{v^2} = \sqrt{\frac{3k_B T}{m}} \qquad (6.10)$$

6.2.2   Boyle's Law

From kinetic theory of gases, we obtained the following relation for pressure exerted by the gas molecules.

$$PV = \frac{1}{3}mN\overline{v^2} = \frac{2}{3}N\frac{1}{2}m\overline{v^2} = \frac{2}{3}N\overline{K.E} \qquad (6.11)$$

Since the kinetic energy depends only on temperature, the right-hand side in the above equation will remain constant at a fixed temperature. That is at a constant temperature we can write

$$PV = constant$$

Hence, at a constant temperature, the pressure exerted by a given mass of gas is inversely proportional to its volume. This is Boyle's law (Fig. 6.2).

6.2.3   Gay-Lussac's Law

From Eq. 6.11, we can see that at a fixed volume, the pressure exerted by a given mass of gas is directly proportional to its kinetic energy, i.e., temperature ($P \propto T$). This is known as Gay-Lussac's Law (Fig. 6.3).

6.2.4   Charle's Law

According to Eq. 6.11, when pressure is kept constant, volume of a given mass of gas increases linearly with kinetic energy, i.e., temperature. This is Charle's law (Fig. 6.3).

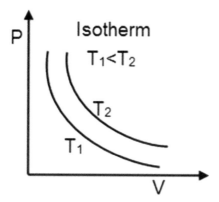

Fig. 6.2 Graphical illustration of Boyle's law

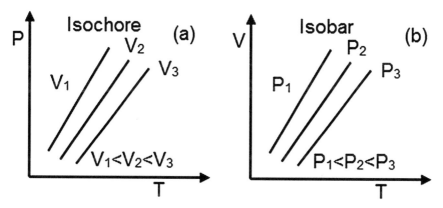

Fig. 6.3 Graphical illustration of **a** Gay-Lussac's Law and **b** Charle's law

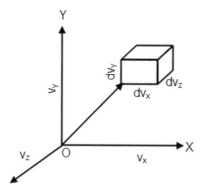

Fig. 6.4 A volume element dv in velocity space

6.2.5 Avogadro's Law

According to this law, at constant temperature and pressure, an equal volume of all gases contains the same number of molecules. Let us consider two different gases at the same temperature and pressure. Then, $p_1 = p_2$ implies

$$\frac{1}{3}m_1 n_1 \overline{v_1^2} = \frac{1}{3}m_2 n_2 \overline{v_2^2}$$

Note that at a constant temperature, the mean kinetic energy of these gases will be equal. So that we can write

$$\frac{1}{2}m_1 \overline{v_1^2} = \frac{1}{2}m_2 \overline{v_2^2}$$

Combining the above relations, we get

$$n_1 = n_2$$

which proves the statement of Avogadro's law. Let us take some examples related to the discussion we had in previous sections.

Ex:6.1 **What is the average translational kinetic energy of a gas molecule at room temperature (300 K). What is the total random translational kinetic energy of 1 mole of this gas?**

Soln: The average kinetic energy per molecule at 300 K is

$$\overline{K.E} = \frac{3}{2}k_B T = \frac{3}{2}(1.38 \times 10^{-23} J K^{-1})(300 K)$$
$$= 6.21 \times 10^{-21} J$$

Further for one mole of gas

$$K_{tr} = \frac{3}{2}N_A k_B T = \frac{3}{2}(6.023 \times 10^{23})(1.38 \times 10^{-23})(300)$$
$$= 3740 J$$

Ex:6.2 **What is the rms speed of an O_2 molecule at room temperature (300 K).**

Soln: For an oxygen molecule, m $(O_2) = 32 \times 1.66 \times 10^{-27}$ kg $= 5.3 \times 10^{-26}$ kg. The rms speed is

$$v_{rms} = \sqrt{\frac{3k_B T}{m}} = \sqrt{\frac{(3)(1.38 \times 10^{-23} J K^{-1})(300K)}{5.3 \times 10^{-26} kg}}$$
$$= 484 \, ms^{-1}$$

6.3   The Maxwell–Boltzmann Law of Distribution of Velocities

6.3.1   The Velocity Distribution

We assume that the following assumptions are satisfied by the gas molecules.
(i) The number of gas molecules is very large so that any region of space has the
same number density and molecules possess the same velocity in all directions.
(ii) The velocities along three perpendicular axes are independent of each other.
(iii) The probability that any randomly selected molecule has a velocity lying between
certain limits is purely a function of velocity and the limits chosen.

Let us consider a randomly moving gas molecule and describe its motion w.r.t an
origin O and OX, OY and OZ as the coordinate axes. Consider the elemental volume
in velocity space defined by dv_x, dv_y and dv_z, shown in Fig. Each velocity vector
can be defined by coordinates v_x, v_y and v_z (along X-, Y- and Z-axes, respectively)
of its end point.

Then we can write the velocity vector satisfying the condition

$$v^2 = v_x^2 + v_y^2 + v_z^2$$

For an elemental volume $dv = dv_x dv_y dv_z$ in velocity space, the number of velocity
vectors that end in elemental volume gives the average number of molecules with
velocities lying in the limit v to v+dv. Assumption (iii) implies that the number
dN_{v_x} of such molecules is proportional to $f(v_x)dv_x$. Therefore, the fraction of the
molecules having velocity components between v_x to $v_x + dv_x$ is

$$\frac{dN_{v_x}}{N} = f(v_x)dv_x \tag{6.12}$$

where dN_{v_x} represents the number of molecules with velocity components in the
range from v_x to $v_x + dv_x$ and N is the total number of gas molecules. The function
$f(v_x)$ is an unknown function to be determined from the constraints imposed on
the system under consideration. Mathematically, the quantity $\dfrac{dN_{v_x}}{N}$ represents the
probability of finding a molecule whose x-component of velocity lies in the range
from v_x to $v_x + dv_x$. Let us denote this quantity by p_x. Since we assumed that the
motion of gas molecules along three different coordinate axes is independent of each
other, we can write an identical expression for probabilities so that molecule can
have velocity components in the range from v_y to $v_y + dv_y$ and v_z to $v_z + dv_z$ along
Y- and Z-axes, respectively, as follows:

$$\frac{dN_{v_y}}{N} = f(v_y)dv_y \tag{6.13}$$

and

$$\frac{dN_{v_z}}{N} = f(v_z)dv_z \tag{6.14}$$

It should be noted that we have considered the same functional dependence in all mutually perpendicular directions, emphasizing that all directions are equally preferred.

Till now we have found expression for the probability that a molecule can have velocity in a certain specified range along particular axes. The probability that a molecule simultaneously have velocity components in the range from v_x to $v_x + dv_x$, v_y to $v_y + dv_y$ and v_z to $v_z + dv_z$ is

$$\frac{d^3 N_{v_x v_y v_z}}{N} = p_x p_y p_z = f(v_x) f(v_y) f(v_z) dv_x dv_y dv_z \qquad (6.15)$$

So that the number of molecules with their velocity components in the range from v_x to $v_x + dv_x$, v_y to $v_y + dv_y$ and v_z to $v_z + dv_z$ is

$$d^3 N_{v_x v_y v_z} = N f(v_x) f(v_y) f(v_z) dv_x dv_y dv_z$$
$$= N f(v_x) f(v_y) f(v_z) dv \qquad (6.16)$$

Note that while writing Eq. 6.15, we have made use of the fact that motion along three different axes is independent of each other so that we can multiply the respective probabilities $p_x p_y p_z$. In Eq. 6.16, $d^3 N_{v_x v_y v_z}$ represents the number of molecules lying in a small volume elements $dv = dv_x dv_y dv_z$. We can denote the velocity vector $\vec{v} = \vec{v}_x + \vec{v}_y + \vec{v}_z$ as a point (say *velocity point*, representing a molecule with velocity components lying in elemental volume dv) in Fig. Hence, the density of such points (in elemental volume dv gives number density of gas molecules) can be expressed as

$$\rho = \frac{d^3 N_{v_x v_y v_z}}{dv_x dv_y dv_z} = N f(v_x) f(v_y) f(v_z) \qquad (6.17)$$

Since molecular motion is isotropic, i.e., velocity space is isotropic, the density of velocity points becomes independent of the angle made by velocity vector $\vec{v}$ to the axes. Therefore, we can write

$$\rho = N f(v_x) f(v_y) f(v_z) = constant$$
$$= N F(v) = N J(v^2)$$

Here, we have introduced two more functions F and J. This equation is valid for a fixed value of velocity $|\vec{v}|$, i.e., it is subjected to the condition that

$$v^2 = v_x^2 + v_y^2 + v_z^2 = constant \qquad (6.18)$$

Physically speaking, the above equation implies that after a certain time when the gas molecules have undergone a large number of collisions, the distribution will become isotropic and for fixed $|\vec{v}|$, $J(v^2)$ becomes constant, and hence its differential, i.e., $dJ(v^2) = 0$. Therefore, we can write

$$\frac{\partial f(v_x)}{\partial v_x} dv_x f(v_y) f(v_z) + \frac{\partial f(v_y)}{\partial v_y} dv_y f(v_x) f(v_z)$$

$$+ \frac{\partial f(v_z)}{\partial v_z} dv_z f(v_x) f(v_y) = 0$$

Dividing both sides by $f(v_x) f(v_y) f(v_z)$, we get

$$\frac{1}{f(v_x)} \frac{\partial f(v_x)}{\partial v_x} dv_x + \frac{1}{f(v_y)} \frac{\partial f(v_y)}{\partial v_y} dv_y + \frac{1}{f(v_z)} \frac{\partial f(v_z)}{\partial v_z} dv_z = 0$$

Equation 6.18 emphasize that v_x, v_y and v_z can vary but v^2 remains fixed. In differential form, we can rewrite it as

$$v_x dv_x + v_y dv_y + v_z dv_z = 0 \tag{6.19}$$

This equation implies that differentials dv_x, dv_y and dv_z are not mutually independent, they can take any value but are subjected to the constraint Eq. 6.19. This constraint can be relaxed by using Lagrange's method of undetermined multipliers. In this method, some new unknown constants are introduced, and then they are determined by imposing the constraints of the system under investigation. In the present case, we multiply Eq. 6.19 by 2B and add it into previous equation to obtain

$$\left[\frac{1}{f(v_x)} \frac{\partial f(v_x)}{\partial v_x} + 2Bv_x \right] dv_x + \left[\frac{1}{f(v_y)} \frac{\partial f(v_y)}{\partial v_y} + 2Bv_y \right] dv_y +$$

$$\left[\frac{1}{f(v_z)} \frac{\partial f(v_z)}{\partial v_z} + 2Bv_z \right] dv_z = 0 \tag{6.20}$$

As the velocity components are independent of each other, Eq. 6.20 can be satisfied only if each term separately vanishes. That is,

$$\left[\frac{1}{f(v_x)} \frac{\partial f(v_x)}{\partial v_x} + 2Bv_x \right] = 0 \tag{6.21}$$

$$\left[\frac{1}{f(v_y)} \frac{\partial f(v_y)}{\partial v_y} + 2Bv_y \right] = 0 \tag{6.22}$$

$$\left[\frac{1}{f(v_z)} \frac{\partial f(v_z)}{\partial v_z} + 2Bv_z \right] = 0 \tag{6.23}$$

We can rewrite Eq. 6.21 as

$$\frac{df(v_x)}{f(v_x)} = -2Bv_x dv_x$$

which after integration reduces to

$$\ln f(v_x) = -\frac{2Bv_x^2}{2} + \ln C$$

where $\ln C$ is the constant of integration. This equation transforms into

$$f(v_x) = Ce^{-Bv_x^2} \tag{6.24}$$

Similarly, we obtain

$$f(v_y) = Ce^{-Bv_y^2} \tag{6.25}$$

and

$$f(v_z) = Ce^{-Bv_z^2} \tag{6.26}$$

Combining Eqs. 6.24 and 6.12, the number of molecules with velocity components in range from v_x to $v_x + dv_x$ are

$$dN_{v_x} = NCe^{-Bv_x^2} dv_x \tag{6.27}$$

Similar analysis gives expressions for number of molecules with velocity range from v_y to $v_y + dv_y$ and v_z to $v_z + dv_z$ as follows:

$$dN_{v_y} = NCe^{-Bv_y^2} dv_y \tag{6.28}$$

$$dN_{v_z} = NCe^{-Bv_z^2} dv_z \tag{6.29}$$

After combining Eq. 6.16 and Eqs. 6.24, 6.25 and 6.26, we obtain the required expression

$$d^3 N_{v_x v_y v_z} = NC^3 e^{-B(v_x^2 + v_y^2 + v_z^2)} dv_x dv_y dv_z$$
$$= NC^3 e^{-Bv^2} dv_x dv_y dv_z \tag{6.30}$$

so that the number density becomes

$$\rho = NC^3 e^{-Bv^2} \tag{6.31}$$

This is required *Maxwell's velocity distribution* function. A qualitative plot between ρ and Cv^2 is shown in Fig. 6.5a. As we see, the molecular number density falls exponentially with velocity and has a maximum value of NC^3 corresponding to v = 0.

6.3.2   Molecular Distribution of Speed

Usually, the molecular distribution function is expressed in other forms from where one can evaluate the number of molecules with speed in the range from v to v+dv. Since it is assumed that all gas molecules are moving randomly in all directions, the distribution is isotropic and we do not expect the gas molecules to show any preferred motion along some direction. Such type of distribution can be easily calculated by considering a spherical shell of radius v and thickness dv in the velocity space (we are talking in terms of velocity space) as shown in Fig. 6.5b. The problem is then to evaluate the number of velocity vectors ending in such a spherical shell and this number is then equal to the number of gas molecules with velocity in the range from v to v+dv.

To evaluate this number, let us consider two spheres of radii v and $v + dv$. The volume of the spherical shell (annulus) is

$$dv = dv_x dv_y dv_z = \frac{4}{3}\pi \left[(v + dv)^3 - v^3 \right] = 4\pi v^2 dv$$

In view of the problem under consideration, it is appropriate to consider the volume element $dv = dv_x dv_y dv_z$ in spherical polar coordinates. An elementary volume in spherical coordinates is given by $dv_x dv_y dv_z = v^2 \sin\theta d\theta d\phi dv$. Where limits of integration for θ range from 0 to π and for ϕ it ranges from 0 to 2π. Therefore, we can write

$$
\begin{aligned}
dN_v = d^3 N_{v_x v_y v_z} &= \int_{\theta=0}^{\pi} \int_{\phi=0}^{2\pi} NC^3 e^{-Bv^2} dv_x dv_y dv_z \\
&= \int_{\theta=0}^{\pi} \int_{\phi=0}^{2\pi} NC^3 e^{-Bv^2} v^2 \sin\theta d\theta d\phi dv \\
&= 4\pi NC^3 v^2 e^{-Bv^2} dv
\end{aligned}
\tag{6.32}
$$

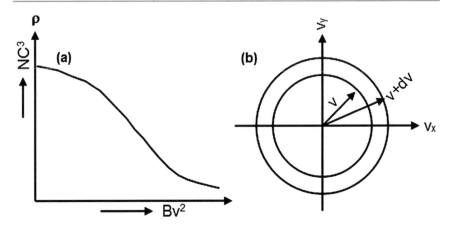

Fig. 6.5 **a** A plot of Maxwell velocity distribution function. **b** A spherical shell of radius v and thickness v+dv

We can rewrite Eq. 6.32 as

$$\frac{dN_v}{N} = 4\pi C^3 v^2 e^{-Bv^2} dv \qquad (6.33)$$

The ratio $\frac{dN_v}{N}$ determines the fraction of molecules with velocity in the range from v to v+dv. This equation contains two unknown constants B and C and they need to be determined.

6.3.3   Evaluating Constants C and B

Total number of molecules can be evaluated using Eq. 6.33.

$$N = \int dN_v = 4\pi N C^3 \int_0^\infty v^2 e^{-Bv^2} dv = 4\pi N C^3 \frac{1}{4}\sqrt{\frac{\pi}{B^3}}$$

Here, we have used the following property of definite integral:

$$\int_{-\infty}^{\infty} x^2 e^{-\alpha x^2} dx = \frac{1}{2}\sqrt{\frac{\pi}{\alpha^3}}$$

The above integral after simplification gives

$$C = \sqrt{\frac{B}{\pi}} \qquad (6.34)$$

Now the problem left with us is to calculate the constant B. Constant B can be determined either calculating the average speed or mean square speed of the molecules.

Fig. 6.6 Maxwell's distribution function for molecular speed at different temperatures

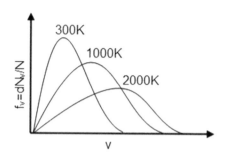

Mean square speed is defined as

$$\overline{v^2} = \frac{\int_0^\infty v^2 dN_v}{\int_0^\infty dN_v} = \frac{\int_0^\infty 4\pi NC^3 e^{-Bv^2} v^2 \times v^2 dv}{\int_0^\infty 4\pi NC^3 e^{-Bv^2} v^2 dv}$$

$$= \frac{\int_0^\infty v^4 e^{-Bv^2} dv}{\int_0^\infty v^2 e^{-Bv^2} dv} = \frac{\frac{3}{8}\sqrt{\frac{\pi}{B^5}}}{\frac{1}{4}\sqrt{\frac{\pi}{B^3}}}$$

$$= \frac{3}{2B} \tag{6.35}$$

Since the mean square speed is

$$\overline{v^2} = \frac{3k_B T}{m}$$

This in combination with Eq. 6.35 gives

$$B = \frac{m}{2k_B T} \tag{6.36}$$

So far we have determined the values of constants B and C. After substituting these values in Eq. 6.33, we obtain

$$\frac{dN_v}{N} = f_v = 4\pi \left[\frac{m}{2\pi k_B T}\right]^{3/2} v^2 e^{-\frac{mv^2}{2k_B T}} dv \tag{6.37}$$

This equation describes the distribution of molecular speeds and is known as Maxwell's distribution law of velocities (Fig. 6.6).

6.4 A Few Important Deductions from Maxwell's Law

Since the gas molecules move randomly in all directions, we cannot characterize a Maxwellian gas with an average velocity, which will be zero (because for every gas molecule moving along positive x-axis, we can always find another molecule moving along negative x-axis). For this reason, we define the average or mean speed $\bar{v}$. As the Maxwell velocity distribution function consists of an increasing quadratic term and a decaying exponential function, it is possible that for a certain value of speed, this function attains a maximum value, and the corresponding speed is known as the most probable speed. Further, the energy of a molecule is defined in terms of root mean square (rms) speed v_{rms}. Therefore, it is imperative to look for expression for these different speeds for a gas obeying Maxwell's distribution. Let us continue with average speed first.

6.4.1   Average or Mean Speed

The average speed is defined as

$$
\begin{aligned}
\bar{v} &= \frac{\int_0^\infty v\, dN_v}{\int_0^\infty dN_v} = \frac{1}{N} \int_0^\infty v\, dN_v \\
&= \frac{1}{N} 4\pi N \left[\frac{m}{2\pi k_B T} \right]^{3/2} \int_0^\infty v^3 e^{-\left(\frac{mv^2}{2k_B T} \right)} dv \\
&= 4\pi \left[\frac{m}{2\pi k_B T} \right]^{3/2} \frac{1}{2 \left(\frac{m}{2k_B T} \right)^2} \\
&= \sqrt{\frac{8 k_B T}{\pi m}}
\end{aligned}
\tag{6.38}
$$

6.4.2   Root Mean Square Speed

In Sect. 6.3.3, we obtained the relationship between constant B $(m/2k_B T)$ and mean square velocity $\overline{v^2}$ as follows:

$$
\overline{v^2} = \frac{3}{2B} = \frac{3 k_B T}{m}
$$

This gives the rms velocity as

$$
v_{rms} = \sqrt{\frac{3 k_B T}{m}}
\tag{6.39}
$$

6.4.3   Most Probable Speed

Most probable speed is defined as the speed at which the Maxwellian distribution function f_v exhibits a maximum. It is denoted by v_p. Since the function f_v exhibits a maximum, its first derivative w.r.t the independent variable must vanish. Therefore, when speed $v = v_p$

$$\frac{df_v}{dv} = 0$$

This means

$$4\pi \left[\frac{m}{2\pi k_B T}\right]^{3/2} \frac{d}{dv}\left[v^2 e^{-\left(\frac{mv^2}{2k_B T}\right)}\right]_{v=v_p} = 0$$

Hence,

$$2v_p e^{-\left(\frac{mv^2}{2k_B T}\right)} + v_p^2 e^{-\left(\frac{mv^2}{2k_B T}\right)}\left(-\frac{2mv_p}{2k_B T}\right) = 0$$

$$or$$

$$2v_p e^{-\left(\frac{mv^2}{2k_B T}\right)}\left[1 - \frac{m}{2k_B T}v_p^2\right] = 0$$

For the non-zero value of v_p, the above equation will hold if

$$1 - \frac{m}{2k_B T}v_p^2 = 0$$

$$or$$

$$v_p^2 = \frac{2k_B T}{m}$$

Therefore, the most probable speed of a gas obeying Maxwellian distribution is

$$v_p = \sqrt{\frac{2k_B T}{m}} \tag{6.40}$$

6.5 Experimental Verification of Maxwell's Law

At present, several tests exist to demonstrate that the velocity distribution in a gas obeys the Maxwell–Boltzmann distribution. We shall briefly discuss a few indirect and direct experimental methods. Let us first continue with one of the indirect methods, namely the Doppler broadening of the spectral lines.

6.5.1 Doppler Broadening of Spectral Lines

The Maxwell–Boltzmann law can be indirectly verified by investigating the width of spectral lines of hot gas atoms. In principle, a monochromatic source of light can be represented by a line of negligible (or no) width. But the Doppler effect causes a finite broadening of otherwise narrow spectral line and therefore set the limit of resolution. As a result, each spectral line has a finite line width due to the fact that each radiating atom is not at complete rest (they are always in random motion due to non-zero ambient temperature). Had these radiating atoms been at rest, the light emitted by each atom can be represented by a single spectral line. But as mentioned in the kinetic theory of gases, these atoms are always in random motion and their velocity distribution is governed by Maxwell's law. Therefore, if an atom (or gas molecule) at rest emits a radiation of frequency ν_0, the frequency emitted by another atom moving towards (away from) reference atom with velocity v will be $\nu_0 \left[1 \pm \dfrac{v}{c} \right]$ (Doppler effect), where c is the velocity of light. Since v can take any value ranging from 0 to ∞, the spectral lines should possess all frequencies or should have infinite spectral width. But, as mentioned in Maxwell's distribution of velocities, the extremely large velocities are very rare (possible only at extremely large temperatures), and as a result, their contribution to intensity is very small.

Let $\lambda_0 \left(= \dfrac{\nu_0}{c} \right)$ be the wavelength of central line and $\lambda_0 \mp x$ of those corresponding to frequencies $\nu_0 \left(1 \pm \dfrac{v}{c} \right)$, then we can write

$$\lambda_0 - x = \frac{c}{\nu_0 \left(1 + \frac{v}{c} \right)} = \frac{c}{\nu_0} \left(1 - \frac{v}{c} \right)$$

$$= \lambda_0 - \frac{v}{\nu_0}$$

After rearranging, this gives

$$x = \frac{v}{\nu_0} = \lambda_0 . \frac{v}{c}$$

The above result clearly indicates that the spread of the spectral line is directly proportional to the molecular velocity v. Let us now evaluate the intensity of spectral lines corresponding to gas molecules with velocity in the range from v to v+dv. The number of particles with velocity in the range from v to v+dv is

$$dN_v = NCe^{-\left(\frac{mv^2}{2k_B T} \right)} dv$$

The corresponding intensity of the spectral line is

$$I = I_0 e^{-\left(\frac{mv^2}{2k_B T} \right)} \tag{6.41}$$

Fig. 6.7 Schematic illustration of Doppler's broadening

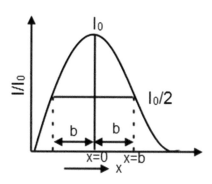

This equation together with result $x = \dfrac{v}{v_0}$ gives

$$I = I_0 e^{-\left(\dfrac{mx^2 v_0^2}{2k_B T}\right)}$$

Here, $I_0 = NC$ denotes the maximum intensity of the radiation emitted by N stationary gas molecules. Let the emission intensity reduce to half of its maximum value when x = b (w.r.t Intensity maxima at x = 0) as shown in Fig. 6.7. Therefore, we can write for intensity at half maxima ($I_h = I_0/2$) as

$$\frac{I_0}{I_h} = 2 = e^{\left(\dfrac{mb^2 v_0^2}{2k_B T}\right)}$$

After taking the natural logarithm, we obtain

$$\ln 2 = \frac{mb^2 v_0^2}{2k_B T}$$

So that the half-width b is

$$b = \frac{1}{v_0}\sqrt{\frac{2k_B T \ln 2}{m}} = \frac{\lambda_0}{c}\sqrt{\frac{2RT \ln 2}{M}}$$

where $M = mN_A$ is the molecular weight and N_A is the Avogadro number. Therefore, we notice that the half-width of spectral lines is inversely proportional to the molecular weight of the gas. This clearly indicates that spectral lines corresponding to cadmium and mercury will be sharp whereas those corresponding to hydrogen will be diffused. This is the reason why heavy nuclei (giving sharp spectral lines) are used for precision work. These points support Maxwell's distribution law.

6.5.2   Zartman and Ko Experiment

Zartman and Ko (1934) gave an elegant experimental proof of the velocity distribution of the gas molecules. This is one of the direct methods to establish the velocity distribution of molecules. The apparatus consists of an oven with an opening A and three slits S_1, S_2 and S_3. It also consists of a drum D which can rotate about an axis passing through O. P is a glass surface over which a beam of silver or bismuth atoms will be deposited. Instead of a glass surface, a photographic film can also be used. The entire apparatus is placed in an evacuated chamber as shown in Fig. 6.8a.

Oven is used for melting the metallic silver. As a result, a beam of silver atoms has emerged through A. The drum is rotated at a speed of approximately six thousand r.p.m. When the drum is stationary, the silver atoms get deposited at the same point on the glass plate fixed inside the drum. When the drum is rotated anticlockwise, a beam of silver atoms enters through opening S_3. Atoms with very high speed reach at plate P first and get deposited on the left end of the plate whereas those moving with low speed reach later and get deposited on the other end (right) of the plate. Continuously moving the cylinder, a sufficient quantity of silver is deposited on the plate P. The relative intensity of silver atoms deposited on the plate can be investigated via a spectro-photometer. The variable density of the deposit across the plate gives a measure of the velocity distribution of gas molecules. A schematic representation of the experimental outcome is illustrated below in Fig. 6.8b. The agreement between experimental data and the theoretically obtained curve indicates the success of the theory.

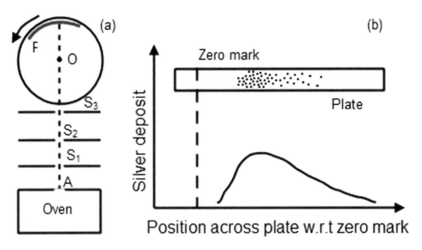

Fig. 6.8 a Schematic diagram of the apparatus used by Zartman and Ko for verifying Maxwell's law for the distribution of velocities. **b** Illustration of the density of deposited silver atoms coming from over with variable velocities and getting deposited over the plate P

Fig. 6.9 Schematic diagram
of the apparatus used by
Estermann, Simpson and
Stern for verifying
Maxwell's law for the
distribution of velocities

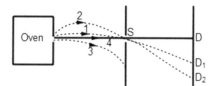

6.5.3   Stern's Experiment

A more precise experiment was designed (1947) by Estermann, Simpson and Stern to
study Maxwell's velocity distribution. The schematic representation of the instrument
is shown in Fig. 6.9. Here, S is a slit and D a hot tungsten wire. This entire apparatus
is enclosed in an evacuated ($\approx 10^{-8}$ mm of Hg) chamber. Cesium atoms emerge from
opening A in the oven. Opening A in the oven and slit S are horizontal. In absence
of a gravitational field, the cesium atoms emerging from opening A will strike the
tungsten wire at point D. But due to the gravitational field, path of cesium atoms
becomes a parabola. The atoms which move along path 3 never reach the wire.
Whereas those following paths 1 and 2 reach at D_1 and D_2, respectively. The atoms
following path 1 move faster than those following path 2.   When a cesium atom hits
the tungsten wire, it gets ionized and evaporates as a positive ion. These positively
charged cesium ions are collected by a negatively charged cylinder surrounding the
wire (not shown in Figure). The magnitude of the current from the collecting cylinder
gives the intensity of atoms striking the wire per unit time. One can move the detector
to different positions on the wire. The measurement of the ion current as a function of
the vertical height of collector gives the details of velocity distribution. This means
a graph between the ionization current along the y-axis and vertical height (speed of
atoms) of the detector along the x-axis gives an idea about the velocity distribution.

Ex:6.3 **Consider a gas obeying the Maxwell–Boltzmann distribution of velocities. Let v_x be the velocity of a gas molecule along X-direction. For what
value of v_x the probability falls to**
(i) 1/e times and
(ii) 1/10 times the maximum value.

Sol: The probability that a gas molecule will have an x-component of velocity
between v_x to $v_x + dv_x$ is

$$P(v_x) = \frac{dN_{v_x}}{N}$$

$$= \left(\frac{m}{2\pi k_B T}\right)^{1/2} e^{-\frac{mv_x^2}{2k_B T}} \qquad (6.42)$$

The probability will be maximum when $v_x = 0$, i.e.,

$$P_{max}(v_x) = \left(\frac{m}{2\pi k_B T}\right)^{1/2} \qquad (6.43)$$

(i) Given that

$$P(v_x) = \frac{1}{e} P_{max}$$

$$= \frac{1}{e} \left(\frac{m}{2\pi k_B T} \right)^{1/2}$$

Substituting this value in Eq. 6.42, we obtain

$$\frac{1}{e} \left(\frac{m}{2\pi k_B T} \right)^{1/2} = \left(\frac{m}{2\pi k_B T} \right)^{1/2} e^{-\frac{mv_x^2}{2k_B T}}$$

or

$$\frac{1}{e} = e^{-\frac{mv_x^2}{2k_B T}}$$

This gives

$$-1 = -\frac{mv_x^2}{2k_B T}$$

or

$$v_x = \sqrt{\frac{2k_B T}{m}}$$

(ii) It is given

$$P(v_x) = \frac{1}{10} P_{max} = \frac{1}{10} \left(\frac{m}{2\pi k_B T} \right)^{1/2}$$

Therefore, as we did in the previous example

$$\frac{1}{10} \left(\frac{m}{2\pi k_B T} \right)^{1/2} = \left(\frac{m}{2\pi k_B T} \right)^{1/2} e^{-\frac{mv_x^2}{2k_B T}}$$

or

$$\frac{1}{10} = e^{-\frac{mv_x^2}{2k_B T}}$$

After taking the logarithm and rearranging, we obtain

$$v_x = \sqrt{\frac{2k_B T}{m} \ln 10}$$

6.6 Heat Capacities of Gases, Classical Approach

In absence of any external field, the energy of an ideal gas is purely kinetic in nature. For one mole of an ideal gas, the total kinetic energy is given by

$$U = N_A < E > = \frac{3}{2} N_A k_B T = \frac{3}{2} RT \tag{6.44}$$

Therefore, the molar heat capacity, i.e., the energy required to raise the temperature of 1 kilo-mole of a substance (ideal gas say) by one kelvin, at a fixed volume, is

$$C_v = \left(\frac{dU}{dT}\right)_V = \frac{3}{2} R = 2.98 \, \text{kcal kmol}^{-1} K^{-1} \tag{6.45}$$

so that the molar heat capacity at constant pressure is given by

$$C_p = C_V + R = \frac{5}{2} R = 4.87 \, \text{kcal kmol}^{-1} K^{-1} \tag{6.46}$$

From here, one can calculate the adiabatic index as

$$\gamma = \frac{C_p}{C_v} = \frac{5}{3} = 1.67 \tag{6.47}$$

These results indicate that *molar heat capacities and their ratios are independent of temperature and same for all gases.* These predictions which are based on the kinetic theory of gases agree well with the experimental data for monoatomic gases but in the case of diatomic and polyatomic gases, discrepancies can be seen [Table 6.1]. The following points can be noted for these polyatomic gases:
(i) The ratio of heat capacities (γ) decreases with an increase in atomicity of a gas, that is, with the rise in the number of atoms in a gas molecule.
(ii) In a few cases, deviations between theoretical and measured values of C_v can be seen.
To further gain an insight into theoretical and experimental data, we will introduce the concept of the degree of freedom of a gas molecule.

Table 6.1 The heat capacities for a few common gases at room temperature

Gas	C_v kcal kmol^{-1}K^{-1}	$(C_p/C_v = \gamma)$
He	2.98	1.66
Ar	2.98	1.67
H_2	4.88	1.41
O_2	5.03	1.4
N_2	4.96	1.4
CO_2	6.8	1.3
NH_3	6.65	1.31

6.6.1   Degrees of Freedom

Degree of freedom (f) of a dynamical system is defined as the number of independent coordinates required to specify its position completely. For instance, consider a gas molecule that is able to move in all directions freely. We need only three independent coordinates to describe its motion accurately. Therefore, a freely moving gas molecule will have three degrees of freedom. In another example, consider the motion of an ant on a straight line (or a stretched string), it has a one degree of freedom. This is true, as long as we are talking about monoatomic gas molecules (He, Ar, etc.). Note that in the case of a freely moving gas molecule, the total energy (due to translational motion) can be written as ($E_{total} = p_x^2/2m + p_y^2/2m + p_z^2/2m$, m is mass of the molecule and p_x, p_y, p_z being momentum along the respective axis) the sum of three independent squared energy terms, one along each axis. *Therefore, the number of degrees of freedom can also be defined as the total number of independent squared energy terms in energy expression of a molecule.*

In the case of diatomic or polyatomic molecules, there is the additional possibility of rotation and vibration and hence more squared energy terms will enter into the total energy. We will learn how to evaluate the degree of freedom for gas molecules with variable atomicity. Before we follow this discussion, let us take a look at an important theorem (*equipartition theorem*) that is connected with average energy associated with a degree of freedom (squared energy term). For this purpose, we will take a particular example with the squared energy term.

6.6.2   Equipartition Theorem

Before giving a general definition of equipartition theorem, let us take an example. Let the energy E of a particular system be given by

$$E = \alpha x^2$$

where α is a constant and x can take any value with equal probability. The probability P(x) of the system having a particular energy αx^2 is

$$P(x) = \frac{e^{-\beta \alpha x^2}}{\int_{-\infty}^{\infty} e^{-\beta \alpha x^2}}$$

and the mean energy is given by

$$< E > = \int_{-\infty}^{\infty} E P(x) dx = \frac{\int_{-\infty}^{\infty} \alpha x^2 e^{-\beta \alpha x^2} dx}{\int_{-\infty}^{\infty} e^{-\beta \alpha x^2} dx} = \frac{1}{2} k_B T$$

It is independent of the constant α and gives a mean energy that is proportional to temperature. The theorem can be extended straightforwardly to the energy being the sum of N independent quadratic terms. In that case, the average energy will be simply

$$U = N\frac{1}{2}k_B T$$

Each quadratic energy dependence of the system is called a mode of the system (or sometimes a degree of freedom (f) of the system).

Mass–Spring System

The mass–spring system, for instance, has two such modes. These correspond to kinetic energy term and potential energy term. For a mass–spring system, the energy is given by

$$E = K.E + P.E = \frac{1}{2}mv^2 + \frac{1}{2}kx^2$$

Therefore, according to the equipartition theorem, the average energy of this system (mass–spring oscillator) will be

$$< E > = \frac{1}{2}k_B T + \frac{1}{2}k_B T = k_B T$$

The result of the example above shows that each mode of the system contributes an amount of energy equal to $1/2k_B T$ to the total mean energy of the system. This result forms the basis of the equipartition theorem, which we state as follows:

Equipartition theorem: There exist various equally valid statements of equipartition theorem. One of them is *If the energy of a classical system is the sum of N quadratic modes (terms), and that system is in contact with a heat reservoir at temperature T, the mean energy of the system is given by*

$$< E > = N \times \frac{1}{2}k_B T$$

Another equally valid definition is
For a system in equilibrium at temperature T, the total energy is equally partitioned among various degrees of freedom, and energy associated with each degree of freedom is $k_B T/2$. This point is schematically illustrated in Fig. 6.12a.

Validity of equipartition theorem: Though the equipartition theorem provides us with an extremely powerful tool to evaluate various thermal systems, it has certain limitations also. Note that the equipartition theorem holds at high temperatures where we can safely ignore the quantum nature of the energy spectrum [see Fig. 6.12b], but should not be so high that we invalidate the approximation of treating (squared energy dependence) the relevant potential wells as perfectly quadratic. At extremely high temperatures, other anharmonic approximations need to be considered and hence equipartition theorem does not hold at extremely high temperatures where energy is not quadratic. Fortunately there exist plenty of room between these two extremes. Below we will consider applications of the equipartition theorem to gases with variable atomicity.

Translational Motion of a Monoatomic Gas Molecule

A monoatomic gas molecule requires only three space coordinates [Fig. 6.10a]. Hence, it has only three degrees of freedom (f = 3). In other words, molecule will have only three translational kinetic energy terms. Its energy is given by

$$E = \frac{1}{2m}\left[p_x^2 + p_y^2 + p_z^2\right] = \frac{1}{2}m\left[v_x^2 + v_y^2 + v_z^2\right]$$

where $v = (v_x, v_y, v_z)$ is the velocity of the atom or particle along x-, y- and z-directions. This energy is the sum of three independent quadratic terms also called modes, and thus the equipartition theorem gives the mean energy as

$$< E > = 3 \times \frac{1}{2}k_B T = \frac{3}{2}k_B T$$

Examples: molecules of rare gases like helium, argon, etc.

Rotational Motion of a Diatomic Gas Molecule

In a diatomic gas molecule (e.g., H_2, O_2, N_2, CO, Cl_2, etc.), an additional possible energy source has to be considered. It is the rotational kinetic energy of the molecule. This gives rise to three more energy terms but only two of them contribute to total energy effectively. The rotational kinetic energy is given by

$$E_{rot} = \frac{L_1^2}{2I_1} + \frac{L_2^2}{2I_2} + \frac{L_3^2}{2I_3}$$

where L_1, L_2 and L_3 are the angular momenta along the three axes shown in Fig. 6.10b and I_1, I_2 and I_3 are the corresponding moments of inertia. We do not need to worry about the direction along the diatomic molecule's bond, the axis labelled 'Z' in the Figure. Because the moment of inertia along this direction is very small, so that the corresponding rotational kinetic energy is very large. *Therefore, rotational modes in this direction cannot be excited at ordinary temperature and are neglected.* Note that such rotational modes are connected with the individual molecular electronic levels and are ignored at ordinary temperature. Hence, only two rotational energy terms (modes) contribute to the total rotational kinetic energy of the molecules. Therefore, the total energy of a diatomic molecule is thus the sum of five terms, three due to translational kinetic energy and two due to rotational kinetic energy.

$$E = \frac{1}{2}m\left[v_x^2 + v_y^2 + v_z^2\right] + \frac{L_1^2}{2I_1} + \frac{L_2^2}{2I_2}$$

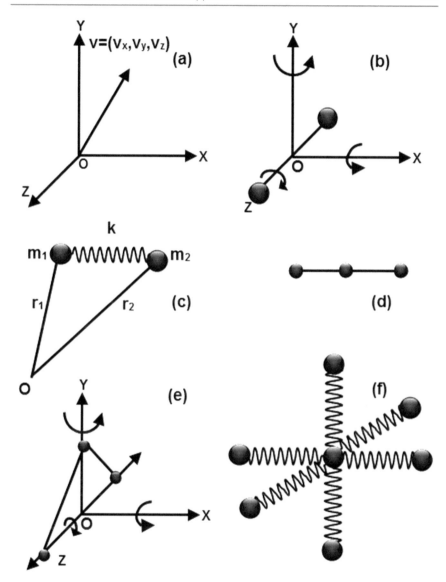

Fig. 6.10 a The velocity of a molecule of a gas in three dimensions. **b** Rotational motion in a diatomic gas molecule. **c** A diatomic molecule modelled as two masses connected by a spring, vibrations in a diatomic molecule. **d** A triatomic linear molecule. **e** A triatomic non-linear molecule

So that

$$< E > = 5 \times \frac{1}{2} k_B T = \frac{5}{2} k_B T$$

Therefore, for a diatomic molecule at ordinary temperature, molar heat capacities C_v, C_p are given by

$$C_v = \left(\frac{dU}{dT} \right)_V = \frac{5}{2} N_A k_B = \frac{5}{2} R = 4.87 \, \text{kcal kmol}^{-1} K^{-1}$$

$$C_p = C_v + R = \frac{7}{2} R = 6.85 \, \text{kcal kmol}^{-1} K^{-1} \tag{6.48}$$

Further, the ratio of molar heat capacities, i.e., $\gamma = C_p/C_v = 7/5 = 1.4$. As can be seen from Table 6.1, these results match well with experimental data for N_2 and O_2.

Vibrational Motion of a Diatomic Gas Molecule

A diatomic molecule can also vibrate as if two atoms are held together by a spring of force constant k. This vibration gives rise to two additional degrees of freedom arising from vibrational kinetic energy and vibrational potential energy. Note that more complicated molecules can vibrate in a variety of ways, for instance by stretching, bending, etc., each mode of vibration results in two degrees of freedom. Note that vibrational mode can only be excited at higher temperature [Fig. 6.12b, as separation between them is quite large, $\Delta E_{trans} < \Delta E_{rot} < \Delta E_{vib}$]; at room temperature, vibrational degrees of freedom do not contribute to the molecule thermal energy. We can say that at room temperature, these vibrational modes are *frozen out*.

In a diatomic molecule, the intramolecular bond can be modelled as a spring as shown in Fig. 6.10c. The two extra energy terms are the kinetic energy due to the relative motion of the two atoms and the potential energy in the bond. Writing the positions of the two atoms as r_1 and r_2 with respect to some fixed origin O, the total energy of the molecule can be written as

$$E = \frac{1}{2} m \left[v_x^2 + v_y^2 + v_z^2 \right] + \frac{L_1^2}{2I_1} + \frac{L_2^2}{2I_2} + \frac{1}{2} \mu \, (\dot{r}_1 - \dot{r}_2)^2 + \frac{1}{2} k \, (r_1 - r_2)^2$$

So that

$$< E > = 7 \times \frac{1}{2} k_B T = \frac{7}{2} k_B T$$

where $\mu = m_1 m_2 / m_1 + m_2$ is the reduced mass of the system. Now, we can evaluate molar heat capacity (for diatomic gas molecules) when vibrational modes are also excited.

$$C_v = \left(\frac{dU}{dT} \right)_V = \frac{7}{2} N_A k_B = \frac{7}{2} R$$

$$C_p = C_v + R = \frac{9}{2} R$$

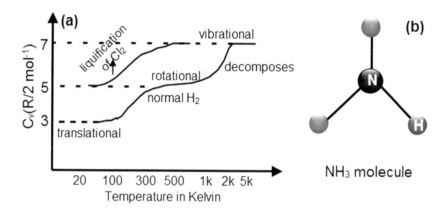

Fig. 6.11 **a** Variation of C_v for H_2 and Cl_2 with temperature. **b** An ammonia molecule

Let us try to understand the point where experimental data and theory encounter each other. A qualitative representation of the data of C_v for H_2 and Cl_2 with temperature is given in Fig. 6.11a. The measured C_v for H_2 at 20 K approaches a value of $\left(\dfrac{3}{2}\right)R$, whereas it increases to $\left(\dfrac{5}{2}\right)R$ at room temperature. With a further rise in temperature, nearly beyond 1000 K, it increases further. We can interpret this observation as follows: at 20 K, in H_2 only translational modes $\left(\dfrac{3}{2}\right)R$ are excited whereas at room temperature rotational modes are also excited and contribute to the heat capacity. Taking a look at the data for Cl_2, the vibrational modes becomes active even at 600 K. Therefore, in this case, C_v approaches $\left(\dfrac{7}{2}\right)R$. We will discuss the case of NH_3 after introducing triatomic linear and non-linear molecules (Fig. 6.12).

Triatomic Molecule (Linear Type)

In the case of triatomic molecules which are linear (CO_2, etc.), the centre of mass lies at the central atom [Fig. 6.10d]. *It, therefore, behaves like a diatomic molecule with three translational degrees of freedom and two rotational degrees of freedom at ordinary temperature.* In total, it has five degrees of freedom (For a linear molecule, however, rotation around its own axis is no rotation because it leaves the molecule unchanged. So there are only two rotational degrees of freedom for any linear molecule).

Triatomic Molecule (Non-Linear Type)

In addition to translation, a triatomic non-linear molecule can rotate, about the three mutually perpendicular axes, as shown in [Fig. 6.10e]. Therefore, it possesses three

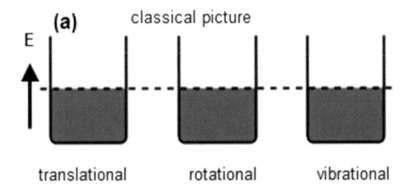

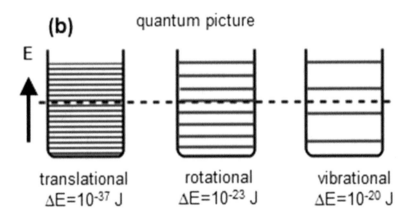

Fig. 6.12 a Schematic illustrating the equipartition theorem and **b** illustration of separation of various energy levels

translational and three rotational degrees of freedom at ordinary temperature. Hence, it has six degrees of freedom. Examples : molecules of H_2O, SO_2, etc.

Polyatomic Molecule

When we use the word polyatomic, it means a molecule with more than two atoms. Let us take an example of polyatomic (non-linear) gas, for instance, SO_2. It has three atoms and three translational and three rotational degrees of freedom at ordinary temperature. The gas molecule will also possess a vibrational degree of freedom, but at a relatively higher temperature (usually greater than 1000 K). Let us denote the vibrational degree of freedom as f_v. Then, the average energy per mole of gas can be written as

$$U = [6 + f_v] \times \frac{1}{2} N_A k_B T$$

$$C_v = \left(\frac{dU}{dT}\right)_V = [6 + f_v] \times \frac{R}{2}$$

Here, one should take care while evaluating the vibrational degree of freedom. As we discussed in the case of a diatomic molecule, vibration (of two atoms) gives rise to two additional degrees of freedom arising from vibrational kinetic energy and vibrational potential energy. Here, in SO_2, we have two pairs (S-O, S-O) of atoms. Hence, in total, two kinetic and two potential energy terms give a total of four vibrational degrees of freedom. If that is the case, then the corresponding average energy and heat capacity are

$$U = [6 + f_v] \times \frac{1}{2} N_A k_B T = [6 + 4] \times \frac{1}{2} N_A k_B T = 5RT$$

$$C_v = [6 + 4] \times \frac{R}{2} = [3 + 2] R = \left[3 + f_{vp}\right] R$$

$$C_p = C_v + R = 6R = \left[4 + f_{vp}\right] R$$

One has to keep in mind that here f_{vp} is the number of vibrational pairs in a molecule (for SO_2, $f_{vp} = 2$, giving a total $f = 4$). C_v and C_p give the general formula for any polyatomic gas molecule. From here, one can also define the adiabatic index as $\gamma = \dfrac{C_p}{C_v} = \dfrac{4 + f_{vp}}{3 + f_{vp}}$.

Let us now apply the concept for NH_3 molecule (three N-H pairs $f_{vp} = 3$, and hence six vibrational degrees of freedom). It has three translational, three rotational and six vibrational degrees of freedom as excited (at a higher temperature). According to the last relation, we can quickly check the following results:

$$C_v = [3 + f_v] R = 6R, \quad C_p = [4 + f_v] R = [4 + 3] R = 7R$$

Whether vibrational modes are excited or not depends upon temperature. Here, we see that for NH_3, $\gamma = 7/6 = 1.17$, which differs significantly from experimental value [Table 6.1]. However, if we assume that vibrational modes are not excited at room temperature (which is the case in general), then $f = 6$ and $C_v = 3R$, $C_p = 4R$, giving $\gamma = 1.33$, which agrees well with the data. This successful interpretation of the heat capacity data of the gases established the kinetic theory of gases.

Heat Capacity of Three-Dimensional Solids

In a solid, the atoms are held rigidly in the lattice and there is no possibility of translational motion. However, the atoms can vibrate about their mean positions. Consider a cubic solid in which each atom is connected by springs (replacing a chemical bond) to six neighbours (one above, one below, one in front, one behind, one to the right and one to the left). In such a situation, each atom inside a solid has three independent vibrational motions, each of which consists of one kinetic energy and one potential energy term. This gives rise to a total of six degrees of freedom for each atom and the average energy per atom is

$$E = 6 \times \frac{1}{2} k_B T = 3 k_B T$$

Hence, the mean thermal energy for a three-dimensional network of N atoms inside a solid is

$$< E > = 3 N k_B T$$

The heat capacity will be

$$C_V = \frac{\partial E}{\partial T} = 3 N k_B$$

If there are Avogadro (N_A) number of atoms in solid, then N $= N_A$ and $C_V = 3 N_A k_B = 3R$ as expected.

NOTE: At ordinary temperature, only translational and rotational modes are excited whereas at sufficiently high temperature (roughly beyond 1000 K) the vibrational modes are also excited and begin to contribute to the total energy of a molecule.

Heat Capacity and Degree of Freedom

In general, for a system with f degree of freedom (i.e., f squared energy terms), the average energy (i.e., energy per particle) is given by

$$\langle E \rangle = f \frac{1}{2} k_B T$$

Therefore, heat capacity per mole, i.e.,

$$C_V = \frac{d \langle E \rangle}{dT} = \frac{f}{2} N_A k_B = \frac{f}{2} R$$

As $C_P - C_V = R$, C_P per mole becomes

$$C_P = \left[1 + \frac{f}{2} \right] R$$

From here, we can write

$$\gamma = \frac{C_P}{C_V} = \frac{\left[1 + \frac{f}{2} \right] R}{\frac{f}{2} R} = \left[1 + \frac{2}{f} \right] \tag{6.49}$$

6.7   Mean Free Path and Transport Phenomena

We consider a gas having a large number of molecules, each of mass m and diameter d. The gas molecules continuously undergo random collisions. Our objective is to look for an expression for the average time between two successive collision events. This time is usually called as *mean collision time* or *scattering time* and denoted as τ. The time τ can be evaluated in two ways:

(i) We consider a particular gas molecule moving in presence of other identical molecules. To make things simple, we assume that the molecule under consideration is travelling at speed v and that the other molecules in the gas are stationary. This we call as **Zeroth-order approximation**. This is clearly a gross oversimplification, which is relaxed in subsequent first-order approximation.

(ii) In the second case, we assume that all molecules are moving with an average velocity. This we call as *first-order approximation*.

Let us continue with the former case, i.e., zeroth-order approximation.

6.7.1   The Mean Collision Time

We assume that only the molecule under consideration moves with speed v and all other molecules are assumed to be at rest. This means the relative speed of this molecule w.r.t any other molecule is v. We also assign a collision cross section σ $(= \pi d^2)$ to each molecule. In a time dt, the volume swept out by this molecule will be $\sigma v dt$. If another molecule lies inside this volume, there will definitely be a collision. Let there be n number of molecules per unit volume. Then the number of molecules contained in the cylinder of volume $\sigma v dt$ will be $n\sigma v dt$. Note that this number is also equal to the number of collisions made by the moving molecule in time dt. Therefore, in time dt, the probability that a collision takes place inside a cylinder of volume $\sigma v dt$ is $n\sigma v dt$ (it is simply the product of the number density of molecules, collision cross-sectional area and elapsed length, vdt).

Let us define P(t) as the probability that a molecule does not collide up to time t. Then from elementary calculus,

$$P(t + dt) = P(t) + \frac{dP}{dt} dt \tag{6.50}$$

Here, P(t+dt) is the probability that a molecule does not collide up to time t multiplied by the probability of the molecule not colliding in subsequent time dt, i.e.,

$$P(t + dt) = P(t) [1 - n\sigma v dt] \tag{6.51}$$

after rearranging the terms, we obtain

$$\frac{1}{P} \frac{dP}{dt} = -n\sigma v \tag{6.52}$$

Further using the fact that P(0)=1, the above equation gives

$$P(t) = e^{-n\sigma vt} \tag{6.53}$$

Further, we will calculate the probability (of a molecule) of surviving without collision up to time t and subsequently colliding in next time dt is $e^{-n\sigma vt}n\sigma vt$. One can check that this is proper probability and satisfy the normalizing condition.

$$\int_0^\infty e^{-n\sigma vt}n\sigma v dt = 1 \tag{6.54}$$

Here, we have made use of the integral $\int_0^\infty e^{-x}dx = 1$. From here, we can calculate the mean scattering time τ as follows:

$$\tau = \int_0^\infty t e^{-n\sigma vt}n\sigma v dt = \frac{1}{n\sigma v}\int_0^\infty (n\sigma vt)e^{-n\sigma vt}d(n\sigma vt)$$

$$= \frac{1}{n\sigma v}\int_0^\infty x e^{-x}dx$$

In the last step, we have made a substitution, x=$n\sigma vt$, and using $\int_0^\infty x e^{-x}dx = 1$, we obtain

$$\tau = \frac{1}{n\sigma v} \tag{6.55}$$

From here, one can also define the collision frequency, i.e., the number of collisions taking place per second as

$$v_c = \frac{1}{\tau} = n\sigma v \tag{6.56}$$

6.7.2   The Collision Cross Section

We consider two spherical molecules with radii, a_1 and a_2, respectively; see Fig. 6.13a. In Fig. 6.13b, a tube of radius $(a_1 + a_2)$ is drawn. A collision will occur if the centre of other molecules either lies on this tube (of radius $(a_1 + a_2)$) or comes inside the tube. Therefore, we can say that the molecule labelled X and Z would collide whereas Y would not. Therefore, the molecule moving with velocity v can be considered to sweep out an imaginary tube of space of cross-sectional area $\pi(a_1 + a_2)^2$. This area or personal space of a molecule is called the collision cross section and denoted by σ, i.e.,

$$\sigma = \pi(a_1 + a_2)^2 \tag{6.57}$$

In a special case, when $a_1 = a_2 = a$,

$$\sigma = \pi d^2 \tag{6.58}$$

where d = $(a_1 + a_2)$ is the diameter of molecule.

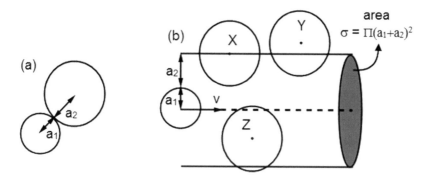

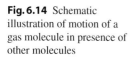

Fig. 6.13 **a** Two spherical molecules with radii a_1 and a_2 in contact with each other and **b** illustrating a molecule sweeping out an imaginary tube of space of cross-sectional area $\sigma = \pi(a_1 + a_2)^2$. If the centre of another molecule lies on or within this tube, a collision will occur

Fig. 6.14 Schematic illustration of motion of a gas molecule in presence of other molecules

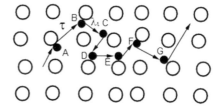

6.7.3 The Mean Free Path, Zeroth-Order Approximation

In the previous section, we just developed the expression for mean collision time

$$\tau = \frac{1}{n\sigma v}$$

Having derived the mean collision time, now we will find an expression for the mean free path. The mean free path is the average distance between two successive collisions and is denoted by λ (Fig. 6.14).

$$\lambda = v\tau = \frac{1}{n\sigma} \tag{6.59}$$

Here, $<v> = v$ under zeroth-order approximation, because all other molecules except the one under consideration are at rest. Recall that σ is the microscopic collision cross section, expressed in m^2. The quantity $n\sigma$ is called as macroscopic collision cross section and is expressed in m^{-1}, i.e., reciprocal of length not area. Equation 6.59 indicates that the mean free path is inversely proportional to the macroscopic collision cross section. This means that the mean free path λ will be smaller for a dense gas. Further, for an ideal gas, $PV = Nk_BT$ or $P = nk_BT$, where $n = N/V$. Using this in Eq. 6.59, we get

$$\lambda = \frac{k_BT}{P\sigma} = \frac{k_BT}{P\pi d^2} \tag{6.60}$$

The expression above predicts that the mean free path is directly proportional to temperature and inversely proportional to pressure. This implies that by reducing pressure (that is creating a vacuum) the λ can be increased. This is indeed done in the case of thin film deposition, where after creating a vacuum in the chamber, a directed beam of atoms can be made to strike on a target substrate resulting in a thin film of the corresponding material.

6.7.4   The Mean Free Path, First-Order Approximation

In Sect. 6.7.3, we made use of the following expression for defining the mean free path under zeroth-order approximation, i.e., when only the molecule under consideration is moving and all others are at rest.

$$\lambda = <v> \tau = \frac{<v>}{n\sigma v} \tag{6.61}$$

In the present case, we will consider the relative motion of other molecules as well. Then what value of v should we take? We should therefore consider the motion of other molecules and take v as the average relative velocity, between two striking molecules, i.e., $<v_r>$ where

$$v_r = v_1 - v_2$$

here, v_1 and v_2 are the velocities of two molecules labelled as 1 and 2, respectively. Now, from here, we can write

$$v_r^2 = (v_1 - v_2)^2 = v_1^2 + v_2^2 - 2v_1.v_2$$

Now we claim that $< v_1.v_2 >= 0$. This is because $< \cos\theta >= 0$, as the molecules are moving randomly in all directions, θ will acquire all possible values ranging from 0 to 2π so that $< \cos\theta >= 0$. Hence, we can write

$$< v_r^2 > = < v_1^2 > + < v_2^2 >= 2 < v^2 > \tag{6.62}$$

where in the last step we used the fact that $< v_1^2 >=< v_2^2 >=< v^2 >$. Now note that we require the value of $< v_r >$ and not $< v_r^2 >$. If the gas molecules follow the Maxwell–Boltzmann probability distribution then to a reasonable approximation, one can write

$$< v_r > \approx \sqrt{< v_r^2 >} \approx \sqrt{2} < 2 >$$

so that we can write

$$\lambda = \frac{<v>}{\sqrt{2}n\sigma <v>} = \frac{1}{\sqrt{2}n\sigma} \tag{6.63}$$

This can be further rewritten using $P = nk_BT$ as

$$\lambda = \frac{k_BT}{\sqrt{2}P\sigma} \qquad (6.64)$$

Ex:6.4 **The diameter of N_2 is 0.37 nm. Calculate the mean free path for N_2 gas at room temperature and atmospheric pressure.**

Sol: The mean free path is

$$\lambda = \frac{1}{\sqrt{2}n\sigma}$$

Here, we have T = 300 K, $P = 1atm = 10^5$ Pa and cross-sectional area $\sigma = \pi d^2 = 4.3 \times 10^{-19}m^2$. This gives the number density

$$n = \frac{P}{k_BT}$$

$$= \frac{10^5}{1.38 \times 10^{-23} \times 300} = 2 \times 10^{25}m^{-3}$$

so that the mean free path is

$$\lambda = \frac{1}{\sqrt{2}n\sigma} = 6.8 \times 10^{-8}m$$

6.8 Transport Phenomenon

In this section, we wish to describe how the gas molecules can transport momentum, energy or particles, from one place to another. We know that a gas molecule has a finite mass and is characterized by random thermal motion. Hence, it possesses momentum as well as kinetic energy. As the gas molecule moves from one place to another, it acts like a potential carrier of physical quantities, such as mass, energy and momentum. For a gas under equilibrium conditions, there shall be no net transport of these physical quantities. However, when we consider non-equilibrium state, so that there is net macroscopic motion in some particular direction, the following different cases may occur.

(i) Different parts of the gas may have different velocities. This will result in a relative motion between different layers of the gas. As a result, the molecules in a faster moving layer will transport greater momentum as compared to those in the slower moving layer. Therefore, across an imaginary plane between these two layers, there will be a net transport of momentum in the preferential direction of motion. This results in the transfer of momentum from faster moving gas

molecules to those moving slowly and equilibrium is achieved. This is characterized by the coefficient of viscosity.

It is important to note that the phenomenon of viscosity in gases and liquids has
an entirely different origin. In the case of the former, it is due to the random
thermal motion of molecules whereas in the latter case frictional force between
any two adjacent layers governs the viscosity.

(ii) Different parts of gases can be at different temperatures. In such a case, the gas
molecules will carry greater thermal energy from regions of higher temperature
to lower temperature. This in turn will lead to the thermal equilibrium between
different parts of the system. This gives rise to the phenomenon of thermal
conduction.

(iv) Different parts of the gas may have different concentrations of the molecules.
In such a case, the molecules migrate from regions of higher concentration to
regions of lower concentration to establish the equilibrium. This is accompanied
by net transport of mass (gas molecules). Such a mass transfer governs the
phenomenon of diffusion.

Therefore, we conclude that viscosity, thermal conduction and diffusion are bulk
properties of the gases and correspond to the transport of momentum, energy and
mass, respectively. Collectively, these phenomena are categorized as transport
phenomena. These processes are of vital importance in studying various physical
properties of solids (e.g., thermal conductivity), liquids (e.g., viscosity) or gases
(e.g., diffusion, etc.). Here, we will limit our discussion only to gases.

6.8.1 Viscosity, Transport of Momentum

Let us consider a fluid sandwiched between two plates, each of surface area A. Let
a shear force be applied to the top plate. The viscosity measures the resistance of a
liquid to the deformation produced by the shear force. For parallel and streamline
flow, the shear stress between the layers is proportional to the velocity gradient in
the direction perpendicular to the layers.

$$\tau_{xz} = \frac{F}{A} = \eta \frac{d<u_x>}{dz}$$

The constant of proportionality, given the symbol η, is called the coefficient of
viscosity.

Consider again Fig. 6.15 in which a fluid is sandwiched between two plates of
area A which each lie in the xy plane. A shear stress $= F/A$ is applied to the fluid

Fig. 6.15 Schematic
illustration of a fluid
sandwiched between two
plates of area A each of
which lies in an xy plane

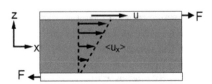

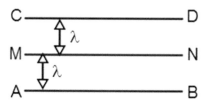

Fig. 6.16 Schematic illustration of three different planes where gas molecules have different velocities, increasing from plane AB to CD. The separation between planes is assumed to be λ. The case when molecules move perpendicular to plane MN

by sliding the top plate over it at speed u and keeping the bottom plate stationary. Applied shear force F gives rise to a velocity gradient $\frac{d \langle u_x \rangle}{dz}$ along z-direction. As a result, $\langle u_x \rangle = 0$ near the bottom plate and $\langle u_x \rangle = u$ near the top plate. This is the case when we are talking about the fluid. In the case of gas, this extra motion in the x-direction is superimposed on the Maxwell–Boltzmann motion in the x-, y- and z-directions (and hence the use of the average $\langle u_x \rangle$, rather than u_x). The shear stress is then given by

$$\tau_{xz} = \eta \frac{d \langle u_x \rangle}{dz} \tag{6.65}$$

Let us consider Fig. 6.15 again and assume that the fluid is a mass of gas moving in parallel layers between two plates each of area A. We consider a plane MN in between two equidistant (*separated by mean free path λ, only then a molecule which has suffered its last collision a distance λ vertically below the plane MN will be able to reach the plane MN. Then molecules which cross a plane of constant z will have travelled on average a distance λ since their last collision, and so they will have travelled a distance λ cos θ parallel to the z-axis since their last collision; see attached Fig. 6.17)* planes AB and CD. For mathematical simplicity, we will follow Fig. 6.16 where planes are separated from each other by a distance of λ. Further, we assume that the velocity of layers of gas molecules increases along the positive z-direction. So that the gas molecules in contact with plane AB have zero velocity and those in contact with top plane CD have some non-zero velocity $\langle u_x \rangle$. This means layers of gas molecules above the plane MN are moving at a faster rate than the layers of gas below it. The faster moving layers above MN exert a tangential force on gas molecules in plane MN, thereby tending to increase its velocity. On the other hand, slowly moving layer AB has the opposite effect on gas molecules in plane MN. As a result, a tangential force is set up across plane MN whose effect is to reduce the difference in velocities of the layers above and below plane MN. This property of gases is called viscosity.

Further, as the gas molecules are in continuous random motion, some of the molecules above MN will move downward and an equal number might be moving upward as there is no mass transport along the vertical direction. As a result, slower moving layers will gain momentum from faster moving layers whereas faster moving

Fig. 6.17 Molecular velocity u for molecules travelling at an angle θ to the z-axis. These will have travelled on average a distance λ since their last collision, and so they will have travelled a distance $\lambda \cos theta$ parallel to the z-axis since their last collision

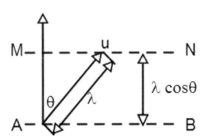

layers above MN will lose momentum. In that sense, viscosity can be considered as a phenomenon of transport of momentum.

Let n be the number of molecules per unit volume and $\langle u \rangle$ be their average velocity. As the motion of molecules is random, one may consider that one-third of them are moving parallel to the coordinate axis. So that on an average, one-sixth (+ve x or −ve x) of the molecules move parallel to any one of the axes. Hence, the number of molecules that cross the plane MN upwards or downwards per unit area per second $= n \langle u_x \rangle /6$.

Let us denote the momentum of each molecule in plane MN by $G = m u_x$ and the momentum gradient (i.e., rate of change of momentum with distance) by

$$\frac{dG}{dz} = m \frac{d \langle u_x \rangle}{dz}$$

in the upward direction perpendicular to plane MN. Note that while defining G we have taken molecular velocity u_x in plane MN. On the other hand, while defining the velocity gradient, it is the average velocity $\langle u_x \rangle$. The latter takes into account the velocity along x-direction due to shear force and the Maxwell–Boltzmann motion in the x-, y- and z-directions. Since we have assumed that planes AB and CD are at a distance λ from plane MN. The momentum of each molecule at plane CD is, therefore,

$$G_{CD} = G + \lambda \frac{dG}{dz}$$

Similarly, the momentum of each molecule at plane AB is, therefore,

$$G_{AB} = G - \lambda \frac{dG}{dz}$$

Now we can evaluate the momentum carried downwards (Π_{zCD}) from plane CD, by the molecules crossing the unit area of plane MN per second

$$\Pi_{zCD} = \frac{1}{6} n \langle u_x \rangle \left[G + \lambda \frac{dG}{dz} \right]$$

Similarly, the momentum carried upwards ($\Pi_{z_{AB}}$) from plane AB, by the molecules crossing the unit area of plane MN per second

$$\Pi_{z_{AB}} = \frac{1}{6}n\langle u_x \rangle \left[G - \lambda \frac{dG}{dz} \right]$$

Therefore, the net momentum transferred downwards per unit area of plane MN per sec is

$$\Pi_{z_{CD}} - \Pi_{z_{AB}} = \frac{1}{6}n\langle u_x \rangle \left[G + \lambda \frac{dG}{dz} \right] - \frac{1}{6}n\langle u_x \rangle \left[G - \lambda \frac{dG}{dz} \right]$$

$$= \frac{1}{3}n\langle u_x \rangle \lambda \frac{dG}{dz} = \frac{1}{3}n\langle u_x \rangle \lambda \frac{dmu_x}{dz}$$

$$= \frac{1}{3}mn\langle u_x \rangle \lambda \frac{du_x}{dz}$$

Here, $G = mu_x$ and u_x is the velocity of the molecules at plane MN. From Eq. 6.65, the coefficient of viscosity η is

$$\eta = \frac{\text{tangential stress}}{\text{velocity gradient}} = \frac{\frac{1}{3}mn\langle u_x \rangle \lambda \dfrac{du_x}{dz}}{\dfrac{du_x}{dz}}$$

$$= \frac{1}{3}mn\langle u_x \rangle \lambda \qquad (6.66)$$

Since, $\lambda = \dfrac{1}{\sqrt{2}n\sigma} = \dfrac{1}{\sqrt{2}n\pi d^2}$, the coefficient of viscosity becomes

$$\eta = \frac{m\langle u_x \rangle}{3\sqrt{2}\sigma} \qquad (6.67)$$

A few important consequences of Eqs. 6.66 and 6.67 are discussed below.

(i) Because $\lambda = \dfrac{1}{\sqrt{2}n\sigma} \propto n^{-1}$, the viscosity is independent of n. Therefore, (at constant temperature) it is independent of pressure. At first sight, it appears to be a weird result. Because, as you increase the pressure, and hence the number density n, one should be better at transmitting the momentum because one will have more molecules to transfer momentum. However, the mean free path reduces correspondingly, so that each molecule becomes less effective at transmitting momentum in such a way as to precisely cancel out the effect of having more of them.

(ii) Since η is independent of n, the only temperature dependence comes from $\langle u_x \rangle \propto \sqrt{T}$, and therefore, $\eta \propto T^{1/2}$.

6.8.2 Thermal Conduction, Transport of Energy

The thermal conductivity of a gas is treated in the same way as the viscosity. Before we develop an expression for the thermal conductivity of the gases, let us recall a few concepts that we have learned earlier in thermal physics. The heat has been defined as energy in transit whenever a thermal gradient exists. The amount of heat flowing along a temperature gradient depends upon the thermal conductivity of the material which we will now define. The process of thermal conductivity can be understood in one dimension using the diagram shown in Fig. 6.18. Heat flow is observed from the hot to cold end of a material and therefore flows against the temperature gradient. The heat flow can be described by a quantity known as heat flux vector J, whose direction lies along the direction of the flow of heat. Its magnitude is equal to the heat energy flowing per unit time per second and is measured in $Js^{-1}m^{-2} = Wm^{-2}$. The heat flux J_z along z-direction is given by

$$J_z = -k\frac{dT}{dz}$$

where the negative sign merely indicates that heat flows in a direction opposite to the thermal gradient. The constant k is known as the thermal conductivity of the gas.

Let us try to understand that how do molecules in a gas carry heat? The mean translational kinetic energy of the gas molecule is $\frac{1}{2}mu^2 = \frac{3}{2}k_BT$, and it depends on the temperature. Therefore, to increase the temperature of a gas by 1 K, one has to increase the mean kinetic energy per molecule by $\frac{3}{2}k_B$. The heat capacity C of the gas is defined as the heat required to increase the temperature of the gas by 1 K. The heat capacity c of a molecule of a gas is, therefore, equal to $\frac{3}{2}k_B$ (though we will discuss later that it can be larger than this if the molecule can store energy in forms other than translational kinetic energy). The derivation of the thermal conductivity of a gas is very similar to that of viscosity.

As argued earlier, for n molecules per unit volume with $\langle u \rangle$ being their average velocity, the number of molecules that cross the plane MN upwards or downwards per unit area per second $= n \langle u \rangle /6$.

Fig. 6.18 Schematic illustration of heat flow. Heat flows in the opposite direction to the temperature gradient

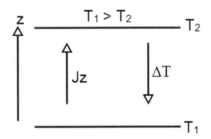

Let E be the energy of each molecule in plane MN and $\dfrac{dE}{dz}$ be the energy gradient in the upward direction perpendicular to plane MN. Further, we also assume that the planes AB and CD are at a distance λ from plane MN.

The energy of each molecule at plane CD is, therefore,

$$E_{CD} = E + \lambda \frac{dE}{dz}$$

Similarly, the energy of each molecule at plane AB is, therefore,

$$E_{AB} = E - \lambda \frac{dE}{dz}$$

Now we can evaluate the energy carried downwards (E_{zCD}) from plane CD, by the molecules crossing the unit area of plane MN per second

$$E_{zCD} = \frac{1}{6} n \langle u \rangle \left[E + \lambda \frac{dE}{dz} \right]$$

Similarly, the energy carried upwards (E_{zAB}) from plane AB, by the molecules crossing the unit area of plane MN per second

$$E_{zAB} = \frac{1}{6} n \langle u \rangle \left[E - \lambda \frac{dE}{dz} \right]$$

Therefore, the net momentum transferred downwards per unit area of plane MN per sec is

$$E_{zCD} - E_{zAB} = \frac{1}{6} n \langle u \rangle \left[E + \lambda \frac{dE}{dz} \right] - \frac{1}{6} n \langle u \rangle \left[E - \lambda \frac{dE}{dz} \right]$$

$$= \frac{1}{3} n \langle u \rangle \lambda \frac{dE}{dz}$$

Now the energy carried by one molecule of a gas is $E = \dfrac{3}{2} k_B T = m c_v T = CT$, where m is the mass of a molecule and c_v is the specific heat of a molecule and $m c_v = C$ given the heat capacity of a molecule at constant volume and T is the absolute temperature. Therefore, net energy transported downwards across the unit area of plane MN per second is

$$E_{\text{downwards}} = \frac{1}{3} n \langle u \rangle \lambda \frac{d m c_v T}{dz} = \frac{1}{3} m n c_v \langle u \rangle \lambda \frac{dT}{dz}$$

The coefficient of thermal conductivity of a gas is defined as follows:

$$k = \frac{\text{Heat flux}}{\text{Temperature gradient}} = \frac{\frac{1}{3}mnc_v \langle u \rangle \lambda \frac{dT}{dz}}{\frac{dT}{dz}}$$

$$= \frac{1}{3}mnc_v \langle u \rangle \lambda = \frac{1}{3}C_v \langle u \rangle \lambda \tag{6.68}$$

Here, we have used $C_v = mnc_v$ as the heat capacity per unit volume. Since mean free path is

$$\lambda = \frac{1}{\sqrt{2}n\sigma}$$

where sigma is the collision cross-sectional area $\sigma = \pi d^2$. The thermal conductivity can also be written as

$$k = \frac{1}{3\sqrt{2}} \frac{mc_v \langle u \rangle}{\sigma} \tag{6.69}$$

A few important consequences of this equation are as follows:
(i) k is independent of pressure.
(ii) $k \propto T^{1/2}$, the temperature dependence is from $\langle u \rangle \propto T^{1/2}$.

6.8.3   Diffusion, Transport of Mass

Consider a distribution of similar molecules or particles with variable number density n(z). Because of the concentration gradient along the z-axis, the molecules will move from higher concentration region to lower concentration region. The flux ϕ_z (measured in $m^{-2}s^{-1}$) of the molecules parallel to the z-direction and say perpendicular to plane MN is

$$\phi_z = -D\left(\frac{\partial n}{\partial z}\right)$$

Here, D is the coefficient of self-diffusion.

Let the gas molecules concentration at plane MN be n and the concentration gradient along the positive z-direction is represented by $\frac{dn}{dz}$. We also assume that planes AB and CD are at distance λ from MN. Recall that λ is the mean free path. The concentration of gas molecules at plane CD is, therefore,

$$n_{CD} = n + \lambda\frac{dn}{dz}$$

Similarly at plane AB

$$n_{AB} = n - \lambda \frac{dn}{dz}$$

The number of molecules moving from a higher concentration region, i.e., from plane CD crossing the plane MN per unit area per second:

$$n_{down} = \left[n + \lambda \frac{dn}{dz} \right] \frac{\langle u \rangle}{6}$$

The number of molecules moving from the lower concentration region, i.e., from plane AB crossing the plane MN per unit area per second:

$$n_{up} = \left[n - \lambda \frac{dn}{dz} \right] \frac{\langle u \rangle}{6}$$

Net number of gas molecules that cross the unit area of the plane MN downwards per second are therefore

$$n_{down} - n_{up} = \left[n - \lambda \frac{dn}{dz} \right] \frac{\langle u \rangle}{6} - \left[n - \lambda \frac{dn}{dz} \right] \frac{\langle u \rangle}{6}$$

$$= \frac{1}{3} \langle u \rangle \lambda \frac{dn}{dz}$$

From Eq. 6.70, the diffusion coefficient is defined as

$$D = \frac{\text{Diffusion flux}}{\text{concentration gradient}} = \frac{\frac{1}{3} \langle u \rangle \lambda \frac{dn}{dz}}{\frac{dn}{dz}}$$

$$= \frac{1}{3} \langle u \rangle \lambda \tag{6.70}$$

Equation 6.70 can be further written as

$$D = \frac{1}{3} \frac{\rho \langle u \rangle \lambda}{\rho} = \frac{\eta}{\rho} \tag{6.71}$$

where $\rho = mn = \frac{mN}{V}$ and $\eta = \frac{1}{3} \rho \langle u \rangle \lambda$, the coefficient of viscosity. From here, we draw the following conclusions.

As $\lambda \propto \frac{1}{n} \propto \frac{T}{P}$ and $\langle u \rangle \propto T^{1/2}$. Thus giving

$$D \propto P^{-1}$$
$$D \propto T^{3/2}$$

6.9   Solved Problems

Q.1 **A gas of molecules each having mass m is in thermal equilibrium at a temperature T. Let v_x, v_y, v_z be the Cartesian components of velocity, $\vec{v}$ of a molecules. Then obtain the mean value of the following**

(i) $\langle v_x \rangle$ **(iv)** $\langle v_x v_z \rangle$

(ii) $\langle v_x^2 \rangle$ **(v)** $\left\langle \left(v_x + \alpha v_y \right)^2 \right\rangle$

(iii) $\langle v_x v_y \rangle$

where α and β are constants.

Sol: (i) Since gas molecules have random motion, the velocity can take value from $-\infty$ to ∞. The mean of $\langle v_x \rangle$ is

$$\langle v_x \rangle = \int_{-\infty}^{\infty} v_x P(v_x) dv_x$$

$$= \int_{-\infty}^{\infty} v_x \left(\frac{m}{2\pi k_B T} \right)^{1/2} e^{-mv_x^2/2k_B T} dv_x$$

$$= \left(\frac{m}{2\pi k_B T} \right)^{1/2} \int_{-\infty}^{\infty} v_x e^{-mv_x^2/2k_B T} dv_x = 0$$

Here, we have used $\int_{-\infty}^{\infty} x e^{-\alpha x^2} dx = 0$. Similarly, we can prove that

$$\langle v_x \rangle = \langle v_y \rangle = 0$$

(ii) The mean of $\langle v_x^2 \rangle$ is

$$\langle v_x^2 \rangle = \int_{-\infty}^{\infty} v_x^2 P(v_x) dv_x$$

$$= \int_{-\infty}^{\infty} v_x^2 \left(\frac{m}{2\pi k_B T} \right)^{1/2} e^{-mv_x^2/2k_B T} dv_x$$

$$= \left(\frac{m}{2\pi k_B T} \right)^{1/2} \int_{-\infty}^{\infty} v_x^2 e^{-mv_x^2/2k_B T} dv_x$$

$$= \left(\frac{m}{2\pi k_B T} \right)^{1/2} \frac{1}{2} \sqrt{\frac{\pi}{(m/2k_B T)^3}} = \frac{k_B T}{m}$$

Here, we have used $\int_{-\infty}^{\infty} x^2 e^{-\alpha x^2} dx = \frac{1}{2} \sqrt{\frac{\pi}{\alpha^3}}$. Similarly, we can prove that

$$\langle v_y^2 \rangle = \langle v_y^2 \rangle = \frac{k_B T}{m}$$

(iii) The mean value of

$$\langle v_x v_y \rangle = \int_{-\infty}^{\infty} \int_{-\infty}^{\infty} v_x v_y P(v_x, v_y) dv_x dv_y$$

$$= \int_{-\infty}^{\infty} \int_{-\infty}^{\infty} v_x v_y P(v_x) P(v_y) dv_x dv_y$$

$$= \int_{-\infty}^{\infty} v_x P(v_x) dv_x \int_{-\infty}^{\infty} v_y P(v_y) dv_y$$

$$= \int_{-\infty}^{\infty} v_x \left(\frac{m}{2\pi k_B T} \right)^{1/2} e^{-\frac{mv_x^2}{2k_B T}} dv_x$$

$$\int_{-\infty}^{\infty} v_y \left(\frac{m}{2\pi k_B T} \right)^{1/2} e^{-\frac{mv_y^2}{2k_B T}} dv_y$$

$$= \left(\frac{m}{2\pi k_B T} \right) \int_{-\infty}^{\infty} v_x e^{-\frac{mv_x^2}{2k_B T}} dv_x \int_{-\infty}^{\infty} v_y e^{-\frac{mv_y^2}{2k_B T}} dv_y = 0$$

Similarly,

$$\langle v_x v_z \rangle = \langle v_y v_z \rangle = 0$$

(iv) Similar to (iii) above, we can prove that

$$\langle v_x v_z \rangle = 0$$

(v) Consider

$$\langle (v_x + \alpha v_y)^2 \rangle = \langle \left(v_x^2 + \alpha^2 v_y^2 + 2\alpha v_x v_y \right) \rangle$$

$$= \langle v_x^2 \rangle + \langle \alpha^2 v_y^2 \rangle + \langle 2\alpha v_x v_y \rangle$$

$$= [1 + \alpha^2] \frac{k_B T}{m}$$

Q.2 Given Maxwellian's distribution formula

$$N_v dv = 4\pi N \left[\frac{m}{2\pi k_B T} \right]^{3/2} e^{-\frac{mv^2}{2k_B T}} v^2 dv$$

Using this distribution function
(i) Show that number of molecules in energy range ϵ to $\epsilon + d\epsilon$ is given by

$$N_\epsilon d\epsilon = 2\pi N \left(\frac{1}{\pi k_B T} \right)^{3/2} \epsilon^{1/2} e^{-\epsilon/k_B T} d\epsilon$$

(ii) Show that probability of molecules having energy between ϵ to $\epsilon + d\epsilon$ is given by

$$P_\epsilon d\epsilon = \frac{N_\epsilon d\epsilon}{N}$$

$$= 2\pi N \left(\frac{1}{\pi k_B T}\right)^{3/2} \epsilon^{1/2} e^{-\epsilon/k_B T} d\epsilon$$

(iii) Show that most probable energy

$$\epsilon_{mp} = \frac{1}{2} k_B T$$

(iv) Show that number of molecules having most probable energy is

$$N_{\epsilon_{mp}} = \sqrt{\frac{2}{\pi e}} \frac{1}{k_B T} N$$

(v) Show that the probability at ϵ_{mp} is

$$P_{\epsilon_{mp}} = \sqrt{\frac{2}{\pi e}} \frac{1}{k_B T}$$

(vi) Show that the mean energy is

$$\langle E \rangle = \frac{3}{2} k_B T$$

(vii) Give a typical plot for this energy distribution at two different temperatures.

Sol: (i) Given Maxwell's velocity distribution

$$N_v dv = 4\pi N \left[\frac{m}{2\pi k_B T}\right]^{3/2} e^{-\frac{mv^2}{2k_B T}} v^2 dv$$

The kinetic energy

$$\epsilon = \frac{1}{2} m v^2$$

$$v = \left(\frac{2\epsilon}{m}\right)^{1/2}$$

So that,

$$dv = \frac{1}{2} \left(\frac{2\epsilon}{m}\right)^{-1/2} \left(\frac{2d\epsilon}{m}\right)$$

Here, dv is the range of speed that corresponds to the energy range $d\epsilon$. Let $N_\epsilon d\epsilon$ represents the number of molecules in the energy range from ϵ to $\epsilon + d\epsilon$. Therefore, following equality must hold.

$$N_\epsilon d\epsilon = N_v dv$$

$$= 4\pi N \left[\frac{m}{2\pi k_B T}\right]^{3/2} e^{-\frac{m\left(\frac{2\epsilon}{m}\right)}{2k_B T}} \frac{2\epsilon}{m}$$

$$\frac{1}{2}\left(\frac{2\epsilon}{m}\right)^{-1/2}\left(\frac{2d\epsilon}{m}\right)$$

$$= 2\pi N \left[\frac{1}{\pi k_B T}\right]^{3/2} \epsilon^{1/2} e^{-\frac{\epsilon}{k_B T}} d\epsilon$$

(ii) From the previous result in (i), one can easily calculate the probability ($P_\epsilon d\epsilon$) of a molecule having an energy between ϵ and $\epsilon + d\epsilon$.

$$P_\epsilon d\epsilon = \frac{N_\epsilon d\epsilon}{N}$$

$$= 2\pi \left[\frac{1}{\pi k_B T}\right]^{3/2} \epsilon^{1/2} e^{-\frac{\epsilon}{k_B T}} d\epsilon$$

(iii) The most probable energy (ϵ_{mp}) corresponds to maximum value of P_ϵ. Therefore, one can obtain ϵ_{mp} for which $\dfrac{dP_\epsilon}{d\epsilon} = 0$. Therefore,

$$2\pi \left(\frac{1}{\pi k_B T}\right)^{\frac{1}{2}} \left[\frac{1}{2}\epsilon^{-\frac{1}{2}} e^{-\frac{\epsilon}{k_B T}} + \epsilon^{\frac{1}{2}}\left(-\frac{1}{k_B T}\right) e^{-\frac{\epsilon}{k_B T}}\right] = 0$$

After rearranging, we get

$$\epsilon_{mp} = \frac{1}{2}k_B T$$

(iv) The number of molecules with energy equal to most probable energy is given by putting $\epsilon_{mp} = \dfrac{1}{2}k_B T$ in expression obtained in (i). That is required number is

$$N_{\epsilon_{mp}} = 2\pi N \left[\frac{1}{2\pi k_B T}\right]^{3/2} \left[\frac{1}{2}k_B T\right] e^{-\frac{1}{2}}$$

$$= \sqrt{\frac{2}{\pi e}} \left[\frac{N}{k_B T}\right]$$

(v) The probability at energy ϵ_{mp} is

$$P_{\epsilon_{mp}} = \frac{N_{\epsilon_{mp}}}{N}$$

$$= \sqrt{\frac{2}{\pi e}} \left[\frac{1}{k_B T} \right]$$

(vi) The mean energy is given by

$$\langle E \rangle = \frac{\int_0^\infty \epsilon N_\epsilon d\epsilon}{N}$$

$$= \int_0^\infty \epsilon P_\epsilon d\epsilon$$

$$= 2\pi \left[\frac{1}{\pi k_B T} \right]^{3/2} \int_0^\infty \epsilon^{3/2} e^{-\frac{\epsilon}{k_B T}} d\epsilon$$

Let $\epsilon^{1/2} = x$, so that the above integral reduces to

$$\langle E \rangle = 2\pi \left[\frac{1}{\pi k_B T} \right]^{3/2} \int_0^\infty x^3 e^{-\frac{x^2}{k_B T}} 2x dx$$

$$= 4\pi \left[\frac{1}{\pi k_B T} \right]^{3/2} \int_0^\infty x^4 e^{-\frac{x^2}{k_B T}} dx$$

$$= 4\pi \left[\frac{1}{\pi k_B T} \right]^{3/2} \left[\frac{3\sqrt{\pi}}{8} (k_B T)^{5/2} \right]$$

$$= \frac{3}{2} k_B T$$

here, we have made use of the following integral:

$$\int_0^\infty x^4 e^{-\alpha x^2} dx = \frac{3}{8} \sqrt{\frac{\pi}{\alpha^5}}$$

Q.3 **A gas of molecules each having mass m is in thermal equilibrium at a temperature T. Let v_x, v_y, v_z be the Cartesian components of velocity, $\vec{v}$ of a molecules. The mean value $(v_x - \alpha v_y + \beta v_z)^2$ is**

[JAM-2010]

(A) $\left[1 + \alpha^2 + \beta^2 \right] \dfrac{k_B T}{m}$ (C) $\left[\beta^2 - \alpha^2 \right] \dfrac{k_B T}{m}$

(B) $\left[1 - \alpha^2 + \beta^2 \right] \dfrac{k_B T}{m}$ (D) $\left[\alpha^2 + \beta^2 \right] \dfrac{k_B T}{m}$

Sol: We can write

$$(v_x - \alpha v_y + \beta v_z)^2 = v_x^2 + \alpha^2 v_y^2 + \beta^2 v_z^2 - 2\alpha v_x v_y - 2\alpha\beta v_y v_z + 2\beta v_x v_z$$

so that

$$\langle (v_x - \alpha v_y + \beta v_z)^2 \rangle = \langle v_x^2 \rangle + \alpha^2 \langle v_y^2 \rangle + \beta^2 \langle v_z^2 \rangle - 2\alpha \langle v_x v_y \rangle$$
$$- 2\alpha\beta \langle v_y v_z \rangle + 2\beta \langle v_x v_z \rangle$$

Using the fact that

$$\langle v_x \rangle = \langle v_y \rangle = \langle v_z \rangle = 0$$
$$\langle v_x^2 \rangle = \langle v_y^2 \rangle = \langle v_z^2 \rangle = \frac{k_B T}{m}$$
$$\langle v_x v_y \rangle = \langle v_y v_z \rangle = \langle v_z v_x \rangle = 0$$

we obtain

$$\langle (v_x - \alpha v_y + \beta v_z)^2 \rangle = [1 + \alpha^2 + \beta^2] \frac{k_B T}{m}$$

Therefore, **A** is correct.

Q.4 **A gas of molecular mass m is at temperature T. If the gas obeys the Maxwell–Boltzmann velocity distribution, the average speed of molecules is given by**

[JAM-2011]

(A) $\sqrt{\dfrac{k_B T}{m}}$ (C) $\sqrt{\dfrac{2k_B T}{\pi m}}$

(B) $\sqrt{\dfrac{2k_B T}{m}}$ (D) $\sqrt{\dfrac{8k_B T}{\pi m}}$

Sol: The average speed is defined as

$$\bar{v} = \frac{\int_0^\infty v \, dN_v}{\int_0^\infty dN_v} = \frac{1}{N} \int_0^\infty v \, dN_v$$
$$= \frac{1}{N} 4\pi N \left[\frac{m}{2\pi k_B T} \right]^{3/2} \int_0^\infty v^3 e^{-\left(\frac{mv^2}{2k_B T}\right)} dv$$
$$= 4\pi \left[\frac{m}{2\pi k_B T} \right]^{3/2} \frac{1}{2\left(\frac{m}{2k_B T} \right)^2}$$
$$= \sqrt{\frac{8k_B T}{\pi m}}$$

Therefore, **D** is correct.

Q.5 **A tiny dust particle of mass** 1.4×10^{-11} **kg is floating in air at 300K. Ignoring gravity, its rms speed (in** μ ms^{-1} **) due to random collisions with air molecules will be closest to**

[JAM-2012]

(A) 0.3   (C) 30
(B) 3   (D) 300

Sol: The rms speed is defined as

$$v_{rms} = \sqrt{\frac{3k_B T}{m}} = \sqrt{\frac{3 \times 1.38 \times 10^{-23} \times 300}{1.4 \times 10^{-11}}} = 30 \times 10^{-6} \, \text{ms}^{-1}$$

Q.6 **A spherical closed container with a smooth inner wall contains a mono-atomic ideal gas. If the collisions between the wall and the atoms are elastic, then the Maxwell speed distribution function** $\left(\dfrac{\partial n_v}{\partial v}\right)$ **for the atoms is best represented by**

[JAM-2016]

Sol: From Maxwell's distribution

$$\frac{dN_v}{dv} \propto v^2 e^{-\frac{mv^2}{2k_B T}} \tag{6.72}$$

Therefore, the correct option is **C**.

Q.7 **One mole of an ideal gas with average molecular speed** v_0 **is kept in a container of fixed volume. If the temperature of the gas is increased such that the average speed gets doubled, then**

[JAM-2016]

(A)   the mean free path of the gas molecule will increase.
(B)   the mean free path of the gas molecule will not change.
(C)   the mean free path of the gas molecule will decrease.
(D)   the collision frequency of the gas molecule with the wall of the container remains unchanged.

Sol: For fixed volume if the temperature is increased, then pressure is also increased by the same amount so mean free path will not change.

$$\lambda = \frac{k_B T}{\sqrt{2}\pi d^2 P}$$

Q.8 **The mean free path of a gas molecule at pressure P and temperature T is 2×10^{-7} m. If the temperature is doubled, the mean free path will be**

(A) 10^{-7} m   (C) 3×10^{-7} m

(B) 2×10^{-7} m   (D) 4×10^{-7} m

Sol: The mean free path is given by

$$\lambda = \frac{k_B T}{\sqrt{2}\pi d^2 P} \propto T$$

Therefore, when T is doubled, the mean free path is doubled. Hence, **D** is correct.

Q.9 **Electrons of mass m in a thin, long wire at a temperature T follow a one-dimensional Maxwellian velocity distribution. The most probable speed of these electrons is**

[JEST-2015]

(A) $\sqrt{\dfrac{k_B T}{2\pi m}}$ (C)   0

(B) $\sqrt{\dfrac{2k_B T}{m}}$ (D) $\sqrt{\dfrac{8k_B T}{\pi m}}$

Sol: For a one-dimensional Maxwellian distribution

$$P(v_x) = \frac{dN_{v_x}}{N} = \sqrt{\frac{m}{2\pi k_B T}}\, e^{-\frac{mv_x^2}{2k_B T}}$$

When $v = v_{mp}$, the most probable speed, this function acquires a maximum value and its derivative w.r.t v must vanish. Therefore,

$$\frac{d}{dv}\left(\frac{dN_{v_x}}{N}\right) = 0$$

$$\sqrt{\frac{m}{2\pi k_B T}}\,\frac{d}{dv}e^{-\frac{mv^2}{2k_B T}} = 0$$

$$\sqrt{\frac{m}{2\pi k_B T}}\,e^{-\frac{mv^2}{2k_B T}}\left[-\frac{m}{2k_B T}(2v)\right] = 0$$

This is zero only when $v = v_{mp} = 0$. Therefore, **C** is correct.

Q.10 **A classical gas of molecules, each of mass m, is in thermal equilibrium at the absolute temperature T. The velocity components of the molecules along the Cartesian axes are v_x, v_y and v_z. The mean value of $(v_x + v_y)^2$ is**

[GATE-2012]

(A) $\dfrac{k_B T}{m}$

(B) $\dfrac{3}{2}\dfrac{k_B T}{m}$

(C) $\dfrac{k_B T}{2m}$

(D) $\dfrac{2k_B T}{m}$

Sol: We can write

$$(v_x + v_y)^2 = v_x^2 + v_y^2 + 2v_x v_y$$

Therefore,

$$\langle (v_x + v_y)^2 \rangle = \langle v_x^2 \rangle + \langle v_y^2 \rangle + 2\langle v_x v_y \rangle$$

Since,

$$\langle v_x^2 \rangle = \langle v_y^2 \rangle = \langle v_z^2 \rangle = \frac{k_B T}{m}$$
$$\langle v_x v_y \rangle = \langle v_y v_z \rangle = \langle v_z v_x \rangle = 0$$

$$\langle (v_x + v_y)^2 \rangle = \frac{2k_B T}{m}$$

Therefore, **D** is correct.

Q.11 **Consider a gas of atoms obeying the Maxwell–Boltzmann statistics. The average value of $e^{\vec{p}.\vec{a}}$ over all the moments $\vec{p}$ of each of the particles (where $\vec{a}$ is a constant vector and a is the magnitude, m is the mass of each atom, T is temperature and k is Boltzmann's constant) is**

[GATE-2013]

(A) 1

(B) 0

(C) $e^{-\frac{a^2}{2}mk_B T}$

(D) $e^{-\frac{3a^2}{2}mk_B T}$

Sol: The mean value of

$$\langle e^{\vec{p}.\vec{a}} \rangle = \int \int \int_{-\infty}^{\infty} f(p_x p_y p_z)dp_x dp_y dp_z$$

here $f(p_x p_y p_z)$ is the Maxwell probability distribution function at temperature T.

$$\left\langle e^{\vec{p}.\vec{a}} \right\rangle = \int_{-\infty}^{\infty} C_x e^{-\frac{p_x^2}{2mk_BT}} e^{p_x a_x} dp_x \int_{-\infty}^{\infty} C_y e^{-\frac{p_y^2}{2mk_BT}} e^{p_y a_y} dp_y$$

$$\int_{-\infty}^{\infty} C_z e^{-\frac{p_z^2}{2mk_BT}} e^{p_z a_z} dp_z$$

$$= e^{-\frac{(a_x^2+a_y^2+a_z^2)mk_BT}{2}} \int_{-\infty}^{\infty} C_x e^{-\frac{(p_x-mk_BTa_x)}{2mk_BT}}$$

$$\int_{-\infty}^{\infty} C_y e^{-\frac{(p_y-mk_BTa_y)}{2mk_BT}} \int_{-\infty}^{\infty} C_z e^{-\frac{(p_z-mk_BTa_z)}{2mk_BT}}$$

$$= e^{-\frac{(a_x^2+a_y^2+a_z^2)mk_BT}{2}} 1.1.1 = e^{-\frac{a^2}{2}mk_BT}$$

Hence, **C** is correct.

Q.12 **Consider a Maxwellian distribution of the velocity of the molecules of an ideal gas. Let V_{mp} and V_{rms} denote the most probable velocity and the root mean square velocity, respectively. The magnitude of the ratio V_{mp}/V_{rms} is**

[NET-JRF Dec-2011]

(A) 1   (C) $\sqrt{2/3}$

(B) 2/3   (D) 3/2

Sol: For Maxwellian's distribution, the most probable and rms speeds are given by

$$v_{mp} = \sqrt{\frac{2k_BT}{m}} \text{ and } v_{rms} = \sqrt{\frac{3k_BT}{m}}$$

$$\frac{v_{mp}}{v_{rms}} = \sqrt{2/3}$$

Q.13 **The speed v of the molecules of mass m of an ideal gas obeys Maxwell's velocity distribution law at an equilibrium temperature T. Let (v_x, v_y, v_z) denote the components of the velocity and k_B be the Boltzmann constant. The average value of $(\alpha v_x - \beta v_y)^2$, where α and β are constants is**

[NET-JRF Dec-2013]

(A) $(\alpha^2 - \beta^2)\dfrac{k_BT}{m}$ (C) $(\alpha + \beta)^2\dfrac{k_BT}{m}$

(B) $(\alpha^2 + \beta^2)\dfrac{k_BT}{m}$ (D) $(\alpha - \beta)^2\dfrac{k_BT}{m}$

Sol: The average value is

$$\langle (\alpha v_x - \beta v_y)^2 \rangle = \alpha^2 \langle v_x^2 \rangle + \beta^2 \langle v_y^2 \rangle - 2\alpha\beta \langle v_x.v_y \rangle$$

Now using the fact that

$$\langle v_x^2 \rangle = \langle v_y^2 \rangle = \langle v_z^2 \rangle = \frac{k_B T}{m}$$
$$\langle v_x v_y \rangle = \langle v_y v_z \rangle = \langle v_z v_x \rangle = 0$$

we obtain

$$\langle (\alpha v_x - \beta v_y)^2 \rangle = [\alpha^2 + \beta^2] \frac{k_B T}{m}$$

Q.14 In one model of a solid, the material is assumed to consist of a regular array of atoms in which each atom has a fixed equilibrium position and is connected by springs to its neighbours. Each atom can vibrate in the x-, y- and z-directions. The total energy of an atom in this model is

$$E = \frac{1}{2} m \left[v_x^2 + v_y^2 + v_z^2 \right] + \frac{1}{2} k_B \left[x^2 + y^2 + z^2 \right]$$

What is the average energy of an atom in the solid when the temperature is T? What is the total energy of one mole of such a solid?

Sol: The energy expression contains six ($f = 6$) squared energy terms. According to the equipartition theorem, energy per atom is

$$< E > = f \times \frac{1}{2} k_B T = \frac{6}{2} k_B T = 3 k_B T$$

For one mole of a solid

$$U = N_A < E >$$
$$= 3 N_A k_B T = 3RT$$

where N_A is Avogadro's number and we have used the fact that $N_A k_B = R$.

Q.15 Consider a non-linear triatomic molecule in three dimensions. The heat capacity of this molecule at high temperature (temperature much higher than the vibrational and rotational energy scales of the molecule but lower than its bond dissociation energies) is

GATE-2017

Soln: At low temperature, the molecules will have only six degrees of freedom (three translational and three rotational). The mean energy per molecule is, therefore,

$$< E > = \frac{1}{2} f k_B T = \frac{6}{2} k_B T$$

However, at high temperatures, vibration modes will also be active so there will be four (2 kinetic energy and two potential energy terms) extra degrees of freedom due to the vibration of atoms. Hence, the average energy per molecule becomes

$$< E > = \frac{10}{2} k_B T$$

Therefore, the specific heat is

$$C_V = \frac{10}{2} k_B$$

Q.16 A system of N non-interacting classical point particles is constrained to move on the two-dimensional surface of a sphere. The internal energy of the system is

[GATE-2010]

(A) $\frac{3}{2} N k_B T$ **(C)** $N k_B T$

(B) $\frac{1}{2} N k_B T$ **(D)** $\frac{5}{2} N k_B T$

Sol: As the N non-interacting classical point particles are constrained to move on the two-dimensional surface of a sphere. There are 2N degrees of freedom. Therefore, the average energy will be

$$< E > = \frac{1}{2} N k_B T + \frac{1}{2} N k_B T = N k_B T$$

Q.17 In two dimensions, an ensemble of N classical particles has the energy of the form

$$E = \frac{p_x^2 + p_y^2}{2m} + \frac{k}{2}(x^2 + y^2)$$

The average internal energy of the system at temperature T is

(A) $\dfrac{3}{2}Nk_BT$ **(C)** $3Nk_BT$

(B) $2Nk_BT$ **(D)** Nk_BT

Sol: The Hamiltonian contains four squared energy terms. According to the equipartition theorem, each square term in energy contributes $(1/2)k_BT$ to the total energy of the system. Therefore, for a system of N two-dimensional oscillators,

$$< E > = N\left[\frac{1}{2}k_BT + \frac{1}{2}k_BT + \frac{1}{2}k_BT + \frac{1}{2}k_BT\right] = 2Nk_BT$$

Hence, B is correct.

Q.18 **The specific heat per molecule of a gas of diatomic molecules at high temperatures is**

[NET-JRF June2016]

(A) $8k_B$ **(C)** $4.5k_B$

(B) $3.5k_B$ **(D)** $3k_B$

Sol: At sufficiently high temperature, translational, rotational as well as vibrational degrees of freedom will be excited. A diatomic molecule has seven (three translational, two rotational and two vibrational) degrees of freedom. Hence,

$$< E > = \frac{7}{2}k_BT$$

$$C_V = \frac{7}{2}k_B = 3.5k_B$$

Therefore, B is correct.

Q.19 **In low-density oxygen gas at low temperature, only the translational and rotational modes of the molecules are excited. The specific heat per molecule of the gas is**

[NET-JRF Dec-2014]

(A) $\dfrac{1}{2}k_B$ **(C)** $\dfrac{3}{2}k_B$

(B) k_B **(D)** $\dfrac{5}{2}k_B$

Sol: For given conditions, the oxygen molecule will have three translational and two rotational degrees of freedom. Hence,

$$< E > = \frac{5}{2} k_B T$$

$$C_V = \frac{5}{2} k_B = 2.5 k_B$$

Therefore, D is correct.

Q.20 The Hamiltonian of a classical non-linear one-dimensional oscillator is

$$H = \frac{p^2}{2m} + \lambda x^4$$

where $\lambda > 0$ is a constant. The specific heat of a collection of a collection of N independent such oscillators is

[NET-JRF June-2019]

(A) $\frac{3}{2} N k_B$

(B) $\frac{3}{4} N k_B$

(C) $N k_B$

(D) $\frac{1}{2} N k_B$

Sol: We can make direct use of the equipartition theorem here. For non-linear term with exponent n=4, the average energy will be $\frac{k_B T}{n} = \frac{k_B T}{4}$. Therefore,

$$< E > = \frac{1}{2} k_B T + \frac{1}{4} k_B T = \frac{3}{4} N k_B T$$

$$C_V = \frac{3}{4} N k_B$$

Therefore, B is correct.

Q.21 At a given temperature T, the average energy per particle of a non-interacting gas of two-dimensional classical harmonic oscillators is ....$k_B T$?

[GATE-2014]

Sol: Let us consider a single two-dimensional classical harmonic oscillator. Its energy will be

$$E = \frac{1}{2m} \left[p_x^2 + p_y^2 \right] + \frac{1}{2} k_B \left[x^2 + y^2 \right]$$

Therefore, the average energy per oscillator is

$$< E > = \frac{1}{2} k_B T + \frac{1}{2} k_B T + \frac{1}{2} k_B T + \frac{1}{2} k_B T = 2 k_B T$$

Q.22 **A solid consists of N atoms. Assuming N atoms to be independent oscilla-tors, all moving with the same frequency ω. The average energy of such a solid at room temperature T is?**

Sol: Each atom can be considered as three linearly independent oscillators along x-, y- and z-axes. The Hamiltonian for such a single three-dimensional oscillator is

$$H = \frac{1}{2m}\left[p_x^2 + p_y^2 + p_z^2\right] + \frac{1}{2}m\omega^2\left[x^2 + y^2 + z^2\right]$$

Equivalently for N three-dimensional oscillators, the Hamiltonian is

$$H = \sum_{i=1}^{3N}\left[\frac{1}{2m}p_i^2 + \frac{1}{2}m\omega^2 x_i^2\right]$$

For a single three-dimensional oscillator, the Hamiltonian consists of six quadratic energy terms. Therefore, the average energy for a single oscilla-tor will be

$$< E > = 6\frac{1}{2}k_B T = 3k_B T$$

For N such three-dimensional oscillators

$$< U > = 6N\frac{1}{2}k_B T = 3Nk_B T$$

Q.23 **The molar specific heat of a gas as given from the kinetic theory is 5/2R. If it is not specified whether it is C_P or C_V, one could conclude that the molecules of the gas**

(A) are definitely monoatomic.
(B) are definitely rigid diatomic.
(C) are definitely non-rigid diatomic.
(D) can be monoatomic or rigid diatomic.

Sol: For a monoatomic molecule (three translational degree of freedom only), $< E >= 3/2k_B T$, $C_V = 3/2k_B$, therefore, $C_P = C_V + R = 3/2R + R = 5/2R$ and for a rigid diatomic (three translational + two rotational degree of freedom) molecule $C_V = 5/2R$. Hence, D is correct.

Q.24 **An ideal gas consists of three-dimensional polyatomic molecules. The temperature is such that only one vibrational mode is excited. If R denotes the gas constant, then the specific heat at a constant volume of one mole of the gas at this temperature is**

[JAM-2018]

(A)   3R   **(C)**   4R

(B) $\frac{7}{2}R$ **(D)** $\frac{9}{2}R$

Sol: A polyatomic gas will have three translational and three rotational degrees of freedom. Since one vibrational mode is excited, it means (kinetic and potential energy terms) two more squared energy terms will be there in the total energy of the molecule. Therefore, average energy and corresponding heat capacities for one mole of gas are

$$U = 8 \times \frac{1}{2}RT, \quad C_v = 4R$$

The correct option is (C).

Q.25 **A rigid triangular molecule consists of three non-collinear atoms joined by rigid rods. The constant pressure molar specific heat C_P of an ideal gas consisting of such molecules is**

[JAM-2015]

(A)   6R   **(C)**   4R

(B)   5R   **(D)**   3R

Sol: For triangular molecules at ordinary temperature, degrees of freedom (three translational and three rotational) are $= 6$. Thus, for a mole of gas, $U = \frac{1}{2}6RT$. Giving

$$C_V = \frac{\partial U}{\partial T} = 3R, \quad C_P = C_V + R = 4R$$

Q.26 **In one dimension an ensemble of N classical particles has an energy of the form**

$$E = \frac{p_x^2}{2m} + \frac{1}{2}kx^2$$

The average internal energy of the system at temperature T is

[JAM-2014]

(A) $\dfrac{3}{2}Nk_BT$ **(C)** $3Nk_BT$

(B) $\dfrac{1}{2}Nk_BT$ **(D)** Nk_BT

Sol: The energy corresponding to the individual particle is

$$E = \frac{p_x^2}{2m} + \frac{1}{2}kx^2$$

There are two squared energy terms; therefore, by equipartition theorem

$$< E > = \frac{1}{2m} < p_x^2 > + \frac{1}{2}k_Bx^2 = \frac{1}{2}k_BT + \frac{1}{2}k_BT = k_BT$$

For N particles $< E >= Nk_BT$. Answer is **D**.

Q.27 **Consider a grand canonical ensemble of a system of one-dimensional non-interacting classical harmonic oscillators (each of frequency ω). Which one of the following equations is correct? Here, the angular bracket < ... > indicates the ensemble average. N, E and T represent the number of particles, energy and temperature, respectively. k_B is the Boltzmann constant.**

[JEST-2019]

(A) $< E >= N\frac{k_BT}{2}$ **(C)** $< E >= Nk_BT$

(B) $< E >=< N > \frac{k_BT}{2}$ **(D)** $< E >=< N > k_BT$

Sol: For a single one-dimensional oscillator

$$E = \frac{p^2}{2m} + \frac{1}{2}kx^2$$

Since there are two quadratic energy terms; therefore, by equipartition theorem, the average energy per oscillator is

$$< E > = \frac{1}{2}k_BT + \frac{1}{2}k_BT = k_BT$$

In a grand canonical ensemble, both N and E are variables. Therefore, the ensemble average for a given system of the oscillator at fixed T will be

$$< U > = < N > k_BT$$

Q.28 **A vessel contains four moles of oxygen and two moles of argon at absolute temperature T. The total internal energy of the given gas mixture is**

(A) 6RT
(B) 9RT

(C) 11RT
(D) 13RT

Sol: The average energy for f degree of freedom (and n=1mole=N_A, Avogadro number of atoms) is

$$< E > = f\frac{1}{2}N_A k_B T = f\frac{1}{2}RT$$

This is the average energy per mole for a gas with f degree of freedom per gas molecule. Now, for n moles, the average energy becomes

$$U = n \times f\frac{1}{2}RT$$

For oxygen (O_2), a diatomic gas, there are five (three translational and two rotational) degrees of freedom. On the other hand, Argon being monoatomic gas has only three. Therefore,

$$U_{total} = 4 \times 5\frac{1}{2}RT + 2 \times 3\frac{1}{2}RT = 13RT$$

Q.29 **Two perfect gases at absolute temperatures T_1 and T_2 are mixed and there is no loss of energy during this process. If n_1 and n_2 are respective number of molecules of the gases, the temperature of the mixture will be**

(A) $\dfrac{n_1 T_1 + n_2 T_2}{n_1 + n_2}$

(B) $\dfrac{n_2 T_1 + n_1 T_2}{n_1 + n_2}$

(C) $T_1 + \dfrac{n_2}{n_1}T_1$

(D) $T_2 + \dfrac{n_1}{n_2}T_1$

Sol: The average kinetic energy (K.E) per molecule of a perfect gas is $\frac{3}{2}k_B T$. Therefore, for the first gas, the average K.E for n_1 molecules will be

$$E_1 = \frac{3}{2}n_1 k_B T_1 \text{ For second gas } E_2 = \frac{3}{2}n_2 k_B T_2$$

The total energy of gases before mixing

$$E = E_1 + E_2 = \frac{3}{2}n_1 k_B T_1 + \frac{3}{2}n_2 k_B T_2 = \frac{3}{2}[n_1 T_1 + n_2 T_2]k_B$$

Let T be the final temperature of the mixture. So that K.E of the molecules $(n_1 + n_2)$ in the mixture will be

$$E_{mix} = \frac{3}{2}[n_1 + n_2]k_B T$$

Since there is no loss of energy during mixing, therefore

$$\frac{3}{2} [n_1 T_1 + n_2 T_2] k_B = \frac{3}{2} [n_1 + n_2] k_B T$$

$$T = \frac{n_1 T_1 + n_2 T_2}{n_1 + n_2}$$

Hence, A is correct.

Q.30 **Two thermally insulated vessels 1 and 2 are filled with air at temperature (T_1, T_2), pressure (P_1, P_2) and volume (V_1, V_2), respectively. If the valve joining the two vessels is opened, the temperature inside the vessel at equilibrium will be**

(A) $T_1 + T_2$

(B) $\dfrac{T_1 + T_2}{2}$

(C) $T_1 T_2 \left[\dfrac{P_1 V_1 + P_2 V_2}{P_1 V_1 T_2 + P_2 V_2 T_1} \right]$

(D) $T_1 T_2 \left[\dfrac{P_1 V_1 + P_2 V_2}{P_1 V_1 T_1 + P_2 V_2 T_2} \right]$

Sol: Similar to the question above, the final temperature after mixing will be

$$\frac{3}{2} [n_1 T_1 + n_2 T_2] k_B = \frac{3}{2} [n_1 + n_2] k_B T$$

$$or$$

$$T = \frac{n_1 T_1 + n_2 T_2}{n_1 + n_2}$$

Now assuming the gases to be ideal, they will satisfy the ideal gas equation, i.e.,

$$P_1 V_1 = n_1 R T_1, \quad n_1 = \frac{P_1 V_1}{R T_1}, \quad n_2 = \frac{P_2 V_2}{R T_2}$$

Using the value of n_1 and n_2 in T, we obtain

$$T = T_1 T_2 \left[\frac{P_1 V_1 + P_2 V_2}{P_1 V_1 T_2 + P_2 V_2 T_1} \right]$$

Hence, C is correct.

6.10 Multiple Choice Type Questions

Q.1 **For which gas the ratio of specific heats (C_P/C_V) will be the largest?**
[JEST-2014]

(A) monoatomic
(B) diatomic
(C) triatomic
(D) hexa-atomic

Q.2 **The degrees of freedom for a monoatomic gas molecule at room temperature are**

(A) 1
(B) 2
(C) 3
(D) 5

Q.3 **The temperature of a gas is doubled, its root mean square speed becomes**

(A) half
(B) double
(C) $\sqrt{2}$ times
(D) $\dfrac{1}{\sqrt{2}}$ times

Q.4 **Diffusion arises due to**

(A) concentration gradient
(B) Pressure difference
(C) Joule–Kelvin effect
(D) Joule effect

Q.5 **For a gas with molecular density n and molecular diameter d, the mean free path of the molecule is**

(A) $\dfrac{\pi}{nd^2}$
(B) $\dfrac{1}{\pi nd}$
(C) $\dfrac{1}{\sqrt{2}\pi nd^2}$
(D) $\dfrac{\pi nd^2}{\sqrt{2}}$

Q.6 **The average velocity of the gas molecules in equilibrium is**

(A) proportional to $\sqrt{T}$
(B) proportional to T
(C) proportional to T^2
(D) equal to zero

Q.7 **The average speed of the gas molecules in equilibrium is**

(A) proportional to $\sqrt{T}$
(B) proportional to T
(C) proportional to T^2
(D) equal to zero

Q.8 **The mean square speed of gas molecules in equilibrium is**

    (A) proportional to $\sqrt{T}$ **(C)** proportional to T^2

    (B) proportional to T   **(D)** equal to zero

Q.9 **The mean free path for a gas molecule is inversely proportional to**

    (A) square of the diameter of the molecule.

    (B) square root of diameter of the molecule.

    (C) molecular diameter.

    (D) fourth power of the molecular diameter.

Q.10 **The mean free path of a gas varies with absolute temperature as**

    (A) T   **(C)** T^{-1}

    (B) T^2 **(D)** $T^{3/2}$

Q.11 **The mean free path of a gas varies with pressure as**

    (A) P   **(C)** P^{-1}

    (B) P^2 **(D)** P^{-2}

Q.12 **In a certain gas the molecular density and diameter of gas molecules are, n and d, respectively. Then mean free path λ of molecules is**

    (A) $\dfrac{\pi}{nd^2}$ **(C)** $\dfrac{1}{\sqrt{2}n\pi d^2}$

    (B) $\dfrac{1}{\pi nd}$ **(D)** $\dfrac{1}{3\sqrt{2}n\pi d^3}$

Keys and Hints to MCQ Type Questions

Q.1 A   Q.3 C   Q.5 C   Q.7 A   Q.9 A   Q.11 C

Q.2 C   Q.4 A   Q.6 D   Q.8 B   Q.10 A   Q.12 C

6.11 Exercises

1. Explain what do you understand by the mean free path.

2. Explain what do you understand by degrees of freedom. How many degrees of freedom a diatomic molecule will have at ordinary temperature?

3. State the equipartition theorem. Under what conditions it is valid?

4. What do you understand by degrees of freedom? State principle of equipartition of energy.

5. Give a kinetic interpretation of temperature.

6. Give an interpretation of temperature on the basis of the kinetic theory of gases.

7. Draw the Maxwell–Boltzmann velocity distribution curve in the following cases
   (i) At temperature T and 16T for an ideal gas.
   (ii) At same temperature for H_2 and O_2.

8. Give an expression for the mean free path under zeroth-order approximation. Explain each term.

9. Discuss the effect of temperature and pressure on the mean free path of a gas molecule.

10. Discuss the effect of temperature and pressure on the viscosity of a gas.

11. Show that temperature is a measure of the average kinetic energy of molecules.

12. Give an expression for the thermal conductivity of a gas. Explain the parameters involved.

13. Show that the mean free path for a molecule in an ideal gas at temperature T and pressure P is given by

$$\lambda = \frac{kT}{\sqrt{2}P\pi d^2}$$

14. Show that the pressure exerted by a perfect gas is 2/3 of the kinetic energy of the gas molecules in a unit volume.

15. Derive an expression for pressure exerted by gas on the basis of kinetic theory.

16. What is the kinetic model of a gas? How is the pressure of a gas explained in this model? Deduce an expression for the pressure.

17. Considering various postulates underlying the kinetic theory of gases, show that the mean kinetic energy of translation of a molecule is $\frac{3}{2}kT$.

18. On the basis of the kinetic theory of gases, derive an expression for the thermal conductivity and viscosity of the gas. Hence, obtain the relation between them.

19. How do you interpret (i) pressure and (ii) temperature on the basis of the kinetic theory of gases?

20. What is the meaning of the mean free path of the molecules of a gas whose diameter is d? Show that it is equal to $\frac{1}{n\sigma}$, where $n = N/V$ the number density and $\sigma = \pi d^2$.

21. Express Maxwell's law of distribution of speeds in terms of the kinetic energy of the molecules. Hence, find the most probable and the average energy of the molecules.

22. Develop an expression for the viscosity of a gas on the basis of kinetic theory and discuss its dependence on temperature and pressure.

23. What do you understand by degrees of freedom of a molecule in a thermal system? State the law of equipartition of energy.

24. Write a short note on the following:
    (i) Mean free path.
    (ii) Degrees of freedom.
    (iii) Maxwell's law of distribution of velocity.

25. Starting with assumptions of the kinetic theory of gases, show that the pressure exerted by a gas of molecules each having mass m is

$$P = \frac{1}{3}mn\overline{v^2}$$

where $\overline{v^2}$ is the mean square velocity and n=N/V is the number density of molecules.

26. Explain Maxwell's distribution law of velocities for gas molecules. Give a brief description of Stern's method verifying Maxwell's law.

27. Show that probability of a molecule with x-component of its velocity in the range from v_x to $v_x + dv_x$ is given by

$$P(v_x)dv_x = \left(\frac{m}{2\pi kT}\right)^{1/2} e^{-\frac{mv_x^2}{2kT}} dv_x$$

Draw a graph between $P(v_x)$ and v_x. Further show that
(i) P is symmetrically distributed about $v_x = 0$.
(ii) The maximum value of P is $\left[\frac{m}{2\pi kT}\right]^{1/2}$.

28. Derive Maxwell's speed distribution law for gas molecules. Draw a graph between the number of particles and speed. Show that the most probable speed v_{mp} and root mean square speed v_{rms} are given by

$$v_{mp} = \sqrt{\frac{2kT}{m}}$$

$$v_{rms} = \sqrt{\frac{3kT}{m}}$$

29. Given Maxwellian's distribution formula

$$N_v dv = 4\pi N \left[\frac{m}{2\pi kT}\right]^{3/2} e^{-\frac{mv^2}{2kT}} v^2 dv$$

Using this distribution function, deduce an expression for
(i) the number of molecules in energy range from E to E+dE.
(ii) the most probable energy.
(iii) the mean energy.

Real Gases

7

In this book, most of the discussion has been focused on ideal gases which obey the PV = nRT equation. It should be noted that real gas behaviour deviates from this equation at high pressure and low volume. For instance, if somehow we can make a real gas cold enough it will liquefy, and this is a particular case that the ideal gas equation does not predict or describe. In liquids, the intermolecular attractions becomes significant, and so far we have not talked about them. In fact, even before the liquefaction of gases starts, departures from ideal gas behaviour appears and needs to be addressed. This chapter focuses on these issues by introducing various extensions to the ideal gas equation, including those given by Van der Waals and Virial expansion.

7.1 Behaviour of Real Gases

In gases, the intermolecular separations are much larger compared to solid or liquid. The molecules in a gas are free to move in all directions. At very low pressure, the forces of intermolecular interactions (attractions) are negligible, and a real gas behaves like an ideal gas obeying the equation

$$PV = RT \tag{7.1}$$

In other words, one can define an ideal gas as a real gas at low pressure. For a real gas, the internal energy (U) depends upon both, temperature and pressure. For an ideal gas, satisfying PV = RT,

$$\left(\frac{dU}{dP}\right)_T = 0$$

© The Author(s) 2022
S. Sharma, *Thermal and Statistical Physics*,
https://doi.org/10.1007/978-3-031-07685-5_7

Here, dU represents a small change in internal energy of an ideal gas with a small change in pressure dP at constant temperature T. Now, one can write

$$\left(\frac{dU}{dV}\right)_T = \left(\frac{dU}{dP}\right)_T \left(\frac{dP}{dV}\right)_T$$

From Eq. 7.1,

$$\frac{dP}{dV} = -\frac{RT}{V^2}$$
$$= -\frac{RT}{V}\frac{1}{V}$$
$$= -\frac{P}{V}$$

Therefore,

$$\left(\frac{dU}{dV}\right)_T = \left(\frac{dU}{dP}\right)_T \left(-\frac{P}{V}\right)_T$$

Since $-P/V$ is not equal to zero, this implies

$$\left(\frac{dU}{dP}\right)_T = 0, \quad \text{so that} \quad \left(\frac{dU}{dV}\right)_T = 0$$

Therefore, for an ideal gas $\left(\frac{dU}{dP}\right)_T = \left(\frac{dU}{dV}\right)_T = 0$, the internal energy for an ideal gas becomes a function of T only, i.e., $U = f(T)$ only. In most of the cases, a real gas can be approximated as an ideal gas within a certain error tolerable in calculations. In general for all practical purposes, real gases around atmospheric pressure can be approximated as ideal gases.

In principle, an ideal or perfect gas strictly obeys the gas laws, namely Boyle's law, Charle's Law as well as the ideal gas equation, i.e., PV = RT under all conditions. This equation of state governs the isotherms as plotted in Fig. 7.1.

Fig. 7.1 Isotherms for an ideal gas at different temperatures

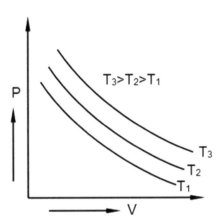

7.2   Deviation from Ideal Gas Equation

In the previous section, we learnt that a perfect or an ideal gas always obeys the equation PV = RT under all conditions. This equation leads to the isotherms (a graph between P and V at constant temperature) as shown in Fig. 7.1. However, at high pressure and low temperature, the real gases don't behave like this and deviations in ideal gas behaviour appear. To describe these deviations from an ideal gas behaviour, various gas equations have been proposed. A few of them are purely empirical while others are phenomenological. If a real gas is cooled, it will liquefy, and this is something that the ideal gas equation is not capable of describing very well. In a liquid, the forces of intermolecular attractions are really significant and have been ignored in the ideal gas model. Even before the start of the liquefaction of a gas, departures from ideal gas behaviour can be seen. The Van der Waals equation took into account the intermolecular forces and finite size of gas molecules and successfully explained the process of liquefaction in real gases. Below in Sect. 7.7, we will describe the Van der Waals model to account for these interactions.

7.3   Boyle Temperature

As we mentioned in the beginning of this chapter, a perfect gas obeys the ideal gas equation (PV = RT) at all temperatures and pressures. This relation also indicates that for a given quantity of a gas at a given temperature T, the product PV determined over a wide range of pressure should be constant, i.e., PV = constant at fixed T (Boyle's law). However, as found experimentally, these laws are not obeyed strictly by any real gas. Various scientists have investigated the behaviour of real gases over a wide range of temperatures and pressures. For a real gas, at a fixed temperature, PV = constant (Boyle's law) should be obeyed. In contrast, Amagat's experiments on CO_2 gave interesting results. These results are qualitatively shown in Fig. 7.2. Some of the important observations are as follows:

(i) For temperatures lower than critical temperature ($T < T_c$), the isotherms consist of a straight line parallel to the PV-axis. This implies that below the critical temperature, a decrease in volume occurs at constant pressure. This corresponds to the process of condensation as in Fig. 7.5b.

(ii) As the temperature increases, the curvature of the isotherms reduces and the straight part of the isotherms also disappears.

(iii) Each isotherm consists of a minima that shifts away from the origin towards the right side when temperature increases. After a particular temperature (T_c), the minima shifts towards the left. The locus of these minima constitutes a curve similar to a parabola.

(iv) At certain temperature ($T > T_c$), this parabola cuts the P = 0 axis. This temperature is known as the Boyle temperature, T_B.

(v) For any temperature T, larger than T_B, the value of PV steadily increases.

Therefore, experimental findings were totally different from those expected on the basis of the ideal gas equation. For a theoretical physicist, it was a challenging task

Fig. 7.2 Illustration of PV
isotherms for CO_2 obtained
by Amagat

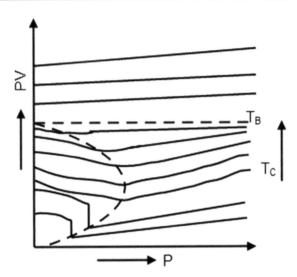

to explain these experimental findings. To address these issues, various equations of
state were proposed, but none of them was capable to explain the results over a wide
range of temperatures and pressures.

Therefore, attempts were made to modify Boyle's law. Kamerlingh Onnes proposed an empirical equation of the type:

$$PV = A + BP + CP^2 + DP^3 + ...$$ (7.2)

where A, B, C and D are constants at a fixed temperature and depend upon the
nature of the gas. These are known as *Virial coefficients*. A is called the first Virial
coefficient, B the second Virial coefficient and so on. These coefficients follow the
order, $A > B > C > D$. Since this equation is expected to reproduce the ideal gas
equation, so as $P \rightarrow 0$, the first Viral coefficient (A) will be equal to RT.

The second Virial coefficient B has special importance. For all gases, it varies
in a similar way. It is negative at low temperatures. With a rise in temperature, its
value becomes less negative, crosses zero and become positive with a further rise in
temperature. The temperature at which second Virial coefficient B is zero is known
as the *Boyle temperature*. It is denoted by T_B. AT this temperature, Boyle's law
holds over a wide range of pressure provided C and D are negligible. Under these
conditions, Eq. 7.2 (at $T = T_B$) implies

$$PV = A$$

so that at Boyle' temperature

$$B = \frac{d}{dP}(PV)$$

Note that constant C is used to be very small and positive. It is to be noted that
the equation proposed by Onnes lacked physical backing. In order to explain the

observed temperature-dependent properties of real gases, Van der Waals proposed another equation by incorporating modifications in existing kinetic theory. We will discuss this in detail.

7.4 Van der Waals Equation of State for Real Gases

While developing Van der Waals equation of state for a real gas, the following simplifying assumptions are made. These assumptions are as follows:
(i) Gas molecules have finite size (volume) and cannot be regarded as point masses.
(ii) The gas molecules attract one another with a weak force of attraction that depends upon the distance between them.

7.4.1 Correction for Volume, Finite Size of Gas Molecule

The finite size of the gas molecules limits the actual space available for the movement of gas molecules. It is less than the volume of the vessel. Each molecule has a sphere of influence around them whose size is twice the radius of each molecule. Other gas molecules cannot penetrate within this sphere of influence. Assume that the radius of each molecule is r. When another gas molecule comes close to this gas molecule, the centre-to-centre distance between these two colliding molecules will be $d = 2r$. This means that around any random molecule, a spherical volume ($V_s = 4/3\pi(2r)^3$) will be denied to any other gas molecule. This volume of exclusion is called as the sphere of exclusion, and its volume ($V_m = 4/3\pi r^3$) is eight times the volume of a gas molecule, i.e., we can write $V_s = 8V_m$ (Fig. 7.3).

Let us consider a space of volume V to be filled by n molecules whose radius is r. Then, the volume available
for the first molecule $= V$
and for second molecule $= V - 8V_m = V - V_s$.
Similarly, the volume available
for nth molecule $= [V - (n-1)V_s]$.

Fig. 7.3 Schematic illustration of sphere of influence

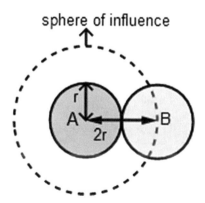

Hence, the average space available for each molecule is

$$V_a = \frac{V + (V - V_s) + (V - 2V_s) + \ldots + [V - (n-1)V_s]}{n}$$

$$= \frac{nV}{n} - \frac{V_s}{n}[1 + 2 + 3 + \ldots + (n-1)]$$

$$= V - \frac{V_s}{n}\frac{n(n-1)}{2} = V - \frac{nV_s}{2} + \frac{V_s}{2}$$

Since the number of molecules are very large, the term $V_s/2$ can be neglected. Therefore, the average space available for each gas molecule is

$$V_a = V - \frac{nV_s}{2} = V - \frac{n8V_m}{2}$$

$$= V - 4nV_m = V - b$$

where $b = 4nV_m$, four times the actual volume of the gas molecule. Therefore, the volume V in the Van der Waals equation of state should be replaced by $V_{avg} = V - b$.

7.4.2   Correction for Pressure, Intermolecular Forces

Let us consider a gas molecule (labelled as A) well inside the vessel as shown in Fig. 7.4. This molecule is influenced (attracted) by other molecules equivalently in all directions. As a result, net force acting on it is zero. The moment it strikes the wall of the vessel, it is pulled back by other molecules. Therefore, its velocity and hence momentum with which it strikes the wall would be less than the momentum with which it will strike in the absence of the intermolecular force of attraction. However, this is not true for the molecule close to the vessel (marked as B). There will be a net inward force acting on this molecule. Whenever a molecule strikes the walls of the vessel, it has to overcome this inward force (arising from intermolecular attraction).

Fig. 7.4 Illustration of forces acting on a gas molecule at different places inside a closed vessel

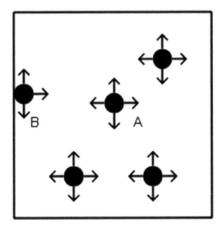

This results in a loss of kinetic energy of the molecule striking the wall. Therefore, the momentum transferred to the wall will be lower than in the case of an ideal gas (where intermolecular interactions are ignored). Therefore, molecular interactions cause a decrease in pressure. This decrease in pressure is known as *cohesive pressure*. If we denote this drop by ΔP, the modified ideal gas equation becomes

$$(P + \Delta P)(V - b) = RT$$

It is important to note that cohesive pressure ΔP depends upon two factors.
(i) number of molecules per unit volume (N/V) in the layer close to the wall (on which an inward force act);
(ii) number of molecules per unit volume (N/V) in the layer just below the surface layer, i.e., those molecules which are pulling the surface molecules inward. Therefore,

$$\Delta P \propto \left(\frac{N}{V}\right)^2 = \frac{a'N^2}{V^2} = \frac{a}{V^2} \tag{7.3}$$

where we have put $a'N^2 = a$. Equations 7.3 and 7.3 indicate that intermolecular interactions and finite volume, respectively, modify the pressure exerted on the walls of the container and the volume available for the molecules. In other words, we must replace pressure P by the sum of observed pressure for real gas and the corresponding drop ΔP caused by intermolecular interactions, i.e., $\left(P + \frac{a}{V^2}\right)$. Similarly, volume correction requires volume V to be replaced by $(V - b)$. Therefore, with these corrections, the equation of state for an ideal gas changes to

$$\left(P + \frac{a}{V^2}\right)(V - b) = RT \tag{7.4}$$

Equation 7.4 is *known as Van der Waals equation of state*. Here, a and b are known as Van der Waals constant which is assumed to be the same for a gas at all temperatures and different for different gases.

7.5   Andrew's Experiment on CO_2 Gas, Isotherms of a Real Gas

Andrew studied the isotherms of CO_2 in 1869. Figure 7.5a displays the apparatus used for this purpose. It consists of two identical glass tubes marked as 'A' and 'B' having two capillary tubes at the top along with a valve (shown in dark colour) in the middle. In the first step, pure dry air is passed through tube A for quite a long time and then it is sealed at both ends. Similarly, in tube 'B' CO_2 is passed for quite a long time (approximately 24 h) and then it is sealed at both ends. These tubes are fixed in the H-shaped copper vessel having two stoppers (screws) S and S' at the bottom. This vessel is filled with water. One can adjust the pressure on tubes (A and B) by moving the screws in or out. This way, pressure of the order of 400 atmospheres can

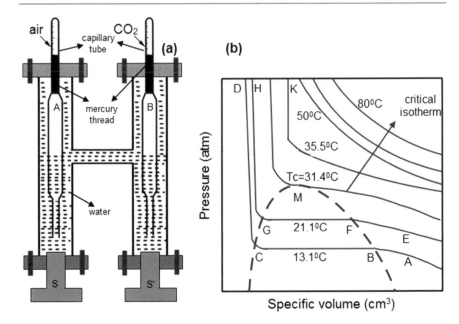

Fig. 7.5 a Schematic description of Andrew's experiment. b Illustration of isotherms from Andrew's experiment

be applied. Since the air in tube 'A' and CO_2 in tube 'B' are at the same pressure, from the known volume of air in tube 'A', one can apply Boyle's law to evaluate the pressure of CO_2 in tube 'B'. Tube 'B' is calibrated in such a way so that volume of CO_2 is directly read from the calibrated scale on tube 'B'. Usually, air is kept at a fixed temperature, and it is possible to maintain the temperature of CO_2 between 0 °C to 100 °C. Andrew's actual experiment consists of isotherms at 13.1 °C, 21.5 °C, 31.4 °C and 35.5 °C. These isotherms are shown in Fig. 7.5b.

Figure 7.5b displays the qualitative variation of pressure with specific volume (volume per gram) at a fixed temperature. Such a curve between P and V_c at a fixed temperature is known as an isotherm. The figure displays a few isotherms at different temperatures. The following conclusions can be drawn from the graph.

(i) At 13.1 °C, the portion AB represents the gaseous state up to point B. From B to C, pressure remains constant whereas volume decreases. Between these points, the vapour and liquid states co-exist. At point C, the gas is completely liquified and a further increase in pressure results in very little change (Portion CD) in volume. This is because liquids are slightly compressible.

(ii) With further increase in temperature (21.1 °C), we obtain an identical isotherm except that the horizontal portion FG representing the liquefaction stage has become narrow.

(iii) For CO_2, at temperature 31.4 °C the horizontal portion corresponding to liquefaction just disappears. The obtained isotherm at 31.4 °C becomes flat at point M (point of inflection). This isotherm is called critical isotherm, point M is called critical point and corresponding temperature 31.4 °C at which this isotherm is obtained is called critical temperature. The pressure (P_c) and volume (V_c) corresponding to temperature T_c are known as critical pressure and volume, respectively. The dashed curve joins the points (B, F, etc.) where liquefaction begins and points (C, G, etc.) where it is completed.

(iv) Further, we note that for any temperature larger than critical temperature (T_c), none of the isotherms shows a horizontal portion. This implies that for $T > T_c$, whatever large applied pressure may be, the gas cannot be liquefied. On the other hand for $T < T_c$, the presence of a flat curve indicates that CO_2 gas can be liquified by applying pressure alone. Therefore, we conclude the above results as follows. The temperature at which it becomes feasible to liquefy a gas by simply increasing pressure (compression) is known as critical temperature (T_c). At T $= T_c$, a liquid and its saturated vapour possess identical properties. Further, at T_c, the pressure required to liquefy a gas is called critical pressure, P_c. Corresponding volume (one mole of a gas) is called critical volume V_c. Collectively, these three parameters are known as critical constants.

7.5.1   Continuity of Liquid and Gaseous States

Following the discussion in the preceding section, we noted two important points.
(i) A gas can be liquefied only if it is cooled up to or below its critical temperature T_c.
(ii) There exists a continuity of liquid and gaseous states. This means they represent two distinct stages of a continuous physical phenomenon.
This can be understood in a better way by considering two isotherms, one below T_c and another above T_c. Consider two points H and K [Fig. 7.5b] for $P > P_c$. At point K, the substance is a gas ($T > T_c$) whereas at point H ($T < T_c$) it is purely a liquid. This implies that at a fixed pressure, gradually reducing the temperature (crossing T_c), the substance passes from gaseous state to liquid state.

7.6   Critical Constants of a Gas

The Van der Waals equation of state for one mole of a gas can be written as

$$P = \frac{RT}{(V - b)} - \frac{a}{V^2} \tag{7.5}$$

If we plot pressure against volume for a real gas (say CO_2 in Andrew's experiment), we obtain the curves shown below in Fig. 7.6.

These curves resemble those obtained in Andrew's experiment [Fig. 7.5b]. The only difference is that instead of a straight line portion BC, here we have a curve with one maxima (at B) and one minima (at C). In this region, the process of liquefaction takes place and the Van der Waals equation may not be strictly valid in this region. The dotted curve in Fig. 7.6 joins the points of maxima on one side to points of minima on the other side. This dotted curve corresponds to the one shown in Fig. 7.5b. With an

Fig. 7.6 Illustration of isotherms obtained from Van der Waals equation

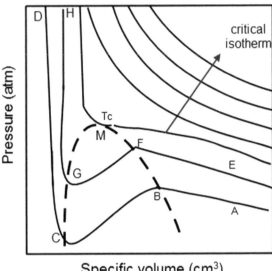

increase in temperature, we notice that maxima and minima come closer to each other and merge at point M. Temperature corresponding to this isotherm is denoted by T_c. Point M is known as the critical point. With reference to this critical temperature $T = T_c$, we define other parameters $V = V_c$, $P = P_c$, known as critical volume and critical pressure, respectively. Collectively, these parameters are known as critical constants of a gas. One can easily obtain their value mathematically as below. Using the Van der Waals equation of state, the point of inflection can be found by equating $(\frac{\partial P}{\partial V})_T = 0$. Therefore, we can write

$$\left(\frac{\partial P}{\partial V}\right)_T = -\frac{RT}{(V-b)^2} + \frac{2a}{V^3} = 0$$

or

$$T = \frac{2a(V-b)^2}{RV^3} \tag{7.6}$$

Now, the locus of the maxima and minima of all isotherms (i.e., the equation of the dotted curve shown in Fig. 7.6) can be obtained by eliminating temperature T between Eqs. 7.6 and 7.5. Therefore, we obtain

$$P = \frac{R}{(V-b)} \cdot \frac{2a(V-b)^2}{RV^3} - \frac{a}{V^2}$$

$$= \frac{2a(V-b)}{V^3} - \frac{a}{V^2} \tag{7.7}$$

Thus, Eq. (7.7) represents the dotted curve joining the points of maxima and minima. Corresponding to the maximum of this curve (i.e., critical point M), we can set

$\frac{dP}{dV} = 0$. This occurs at a certain critical volume $V = V_c$. That is from Eq. 7.7, corresponding to critical point M,

$$\frac{dP}{dV} = \frac{2a}{V^3} - \frac{2a(V-b) \times 3}{V^4} + \frac{2a}{V^3} = 0$$

After rearranging the terms, we obtain

$$\frac{4a}{V_c^3} = \frac{6a(V_c - b)}{V_c^4}$$

or

$$V_c = 3b \tag{7.8}$$

Substituting the value of V_c in Eq. 7.7, we obtain

$$P_c = \frac{2a(3b-b)}{27b^3} - \frac{a}{9b^2} = \frac{a}{9b^2}\left[\frac{4}{3} - 1\right]$$

$$P_c = \frac{a}{27b^2} \tag{7.9}$$

The third critical constant, i.e., critical temperature is obtained by substituting values of V_c and P_c in the Van der Waals equation of state (7.5).

$$T_c = \frac{1}{R}\left[P_c + \frac{a}{V_c^2}\right](V_c - b)$$

$$= \frac{1}{R}\left[\frac{a}{27b^2} + \frac{a}{9b^2}\right](3b - b)$$

$$T_c = \frac{8a}{27Rb} \tag{7.10}$$

7.6.1   Critical Coefficient

The Van der Waals equation, to some extent, fits well the experimental data but could not provide a perfect match with data for fixed values of a and b. Further, this equation predicts that the ratio

$$\frac{RT_c}{P_c V_c} = R\frac{8a}{27Rb} \cdot \frac{27b^2}{a} \cdot \frac{1}{3b}$$

$$= \frac{8}{3} = 2.66$$

This ratio is called the *critical coefficient* and its value is obtained by substituting the values of critical constants (P_c, T_c, V_c). As we see, the value of the critical coefficient is independent of a and b and hence should be the same for all gases. However, the experimentally obtained values for this ratio vary from 3.13 (Helium) to 3.48 (CO_2).

Table 7.1 Experimental values of critical coefficient $\dfrac{RT_c}{P_c V_c}$. Note that here V_c = molecular weight* specific volume

Gas	T_c	P_c	Specific volume	$\dfrac{RT_c}{P_c V_c}$
	in K	in atm		
Helium	5.1	2.25	15.4	3.13
Hydrogen	33.1	12.8	32.2	3.28
Nitrogen	125.9	33.5	3.21	3.42
Oxygen	154.2	49.7	2.32	3.42
CO_2	304	72.8	2.17	3.48

The experimental values for the critical coefficient for a few gases are given in Table 7.1. Experimentally obtained values of the critical coefficient are greater than the theoretically obtained value of 2.667. A more accurate equation of state can be obtained by using an actual form of the intermolecular force for dealing with the molecular collisions.

7.7 Van der Waals Equation and Boyle Temperature

One can expand the Van der Waals equation of state in terms of a Virial expansion and find the Boyle temperature in terms of the critical temperature. Rewrite the Van der Waals equation as

$$PV = \frac{RT}{V - b} + \frac{a}{V^2}$$
$$= \frac{RT}{V}\left[1 - \frac{b}{V}\right]^{-1} + \frac{a}{V^2}$$

Since a and b are small constants and V is a large quantity, the term (a/V^2) can be neglected. Further, using the binomial expansion, the term in brackets can be expanded into a series. This results in

$$\frac{PV}{RT} = 1 + \frac{1}{V}\left[b - \frac{a}{RT}\right] + \left(\frac{b}{V}\right)^2 + \left(\frac{b}{V}\right)^3 + ..$$

This equation is in the same form as the Virial expansion in Equation given below.

$$\frac{PV}{RT} = 1 + \frac{B}{V} + \frac{C}{V^2} + ...$$

Therefore, after comparison

$$B(T) = b - \frac{a}{RT}$$

Fig. 7.7 Illustration of PV
isotherms below, above and
at T_B

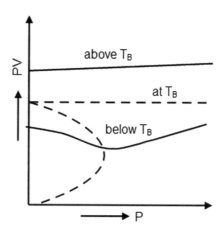

Boyle's temperature is defined as $B(T) = 0$, such that

$$T_B = \frac{a}{Rb} \tag{7.11}$$

Therefore, using Eq. 7.10, we obtain

$$T_B = \frac{27T_c}{8} \tag{7.12}$$

Equation 7.11 gives Boyle's temperature for all gases obeying the Van der Waals equation of state. As we see, T_B depends upon constants 'a' and 'b' which have different values for different gases. A PV versus V graph gives a description of the behaviour of gases close to Boyle's temperature. The following conclusions can be drawn:

(i) At Boyle's temperature:
The product PV is practically constant over a wide range of pressure. See Fig. 7.7.

(ii) Below Boyle's temperature:
The product PV first decreases with pressure and attains a minimum value. Thereafter, it increases further with a rise in pressure. See curves in Fig. 7.7 for T< T_B.

(iii) Above Boyle's temperature:
The product PV increases with an increase in pressure [Fig. 7.7].

7.8   Limitations of Van der Waals Equation

Though the Van der Waals equation of state did well in predicting the behaviour of real gases, it has a few limitations.

(i) Values of 'a' and 'b': The Van der Waals theory assumes 'a' and 'b' to be constant, whereas their values obtained by different methods differ considerably. In fact, 'a' is found to be temperature-dependent. At extremely large temperature, $a \to 0$.

(ii) Critical volume: Further, according to theory $V_c = 3b$, but it is found to depend upon the nature of the gas. However, experimental results imply $V_c \approx 2b$.

(iii) Critical coefficient: The value of the critical coefficient obtained from theory is constant (2.667) independent of a and b for all gases. However, experimentally it is found to vary from one gas to another. Therefore, the critical coefficient appears to depend on the molecular structure of the gas. Therefore, none of the gases obeys the Van der Waals equation of state in the vicinity of the critical point.

(iv) Boyle's temperature: As we noted in the previous section, $T_B = 3.37 T_c$. However, experimentally observed values of Boyle's temperature lie between $2.5 T_c$ and $3.7\ T_c$. Physically, these limitations are connected with the fact that we have ignored the repulsive forces between the molecules close to fluid phase boundaries. Despite these limitations, the Van der Waals equation is commonly employed due to its simple form.

7.9 Reduced Equation of State and Law of Corresponding States

For different gases, the size of the molecules (which controls b in the Van der Waals model) and the strength of the intermolecular interactions (which controls a in the Van der Waals model) will vary, and hence their phase diagrams will be different. For example, the critical temperatures and pressures for different gases are different. However, the phase diagram of substances should be the same when plotted in reduced coordinates, which can be obtained by dividing a quantity by its value at the critical point. Hence, if we replace the quantities P, V and T by their reduced coordinates P_r, V_r and T_r defined as

$$P_r = \frac{P}{P_c}$$

$$V_r = \frac{V}{V_c}$$

$$T_r = \frac{T}{T_c}$$

these relations together with Eqs. 7.8, 7.9 and 7.10 give

$$P = \frac{a}{27 b^2} P_r$$
$$V = 3 b V_r$$
$$T = \frac{8a}{27 R b} T_r$$

Substituting these values in Eq. 7.5, we obtain

$$\left[\frac{a}{27 b^2} P_r + \frac{a}{9 b^2 V_r^2} \right] (3 b V_r - b) = R \times \frac{8a}{27 R b} T_r$$

Multiplying both sides by $\dfrac{9b}{a}$, we obtain

$$\left[P_r + \frac{3}{V_r^2}\right]\left[V_r - \frac{1}{3}\right] = \frac{8}{3}T_r \qquad (7.13)$$

This equation can be arranged as

$$\left[P_r + \frac{3}{V_r^2}\right] = \frac{8T_r}{3V_r - 1} \qquad (7.14)$$

This equation is known as the reduced equation of state. This equation is independent of constants a and b and hence should be valid for all gases. In fact, this equation connects the new parameters P_r, T_r and V_r. From here, we conclude that if for any two gases, two out of three quantities P_r, T_r and V_r have the same values, then the third quantity must be the same for both. This is also known as *law of corresponding states*.

7.10   Multiple Choice Type Questions with Explanations

Q.1 **The condition under which a real gas can closely obey the ideal gas equation is**

(A) low pressure and high temperature
(B) high pressure and low temperature
(C) high pressure and temperature
(D) low pressure and temperature

Q.2 **At high pressure, the real gas behaviour resembles more closely with**

(A) Van der Waals equation   (C) both A and B
(B) ideal gas equation   (D) none of these

Q.3 **A reduced property of a substance means**

(A) ratio of critical property to an existing property of the same substance
(B) existing property of a substance minus critical property of the same substance
(C) critical property of a substance minus existing property of the same substance
(D) ratio of existing property to a critical property of the same substance

Q.4 **Two gases have the same reduced temperature and reduced pressure. The gas having a smaller molecular weight**

(A) shall have a larger reduced volume
(B) shall have a smaller reduced volume
(C) shall have reduced volume similar to gas with a larger molecular weight
(D) none of these

Q.5 **In the Van der Waals equation of state, the value of constant 'a' increases with**

(A) temperature   (C) volume
(B) pressure   (D) intermolecular force

Q.6 **According to the Van der Waals equation of state, the value of V_c is**

(A) $V_c = 2b$ (C) $V_c = b$
(B) $V_c = 3b$ (D) $V_c = 4b$

Q.7 **According to the Van der Waals equation of state, the critical coefficient $\frac{RT_c}{P_c V_c}$ is equal to**

(A) 3/8   (C) 3
(B) 8/3   (D) 8.3

Q.8 **The Van der Waals equation of state for one mole of a real gas is**

(A) $\left(P + \dfrac{a}{V^2}\right)(V - b) = RT$ (C) $\left(P - \dfrac{a}{V^2}\right)(V + b) = RT$
(B) $\left(P - \dfrac{a}{V^2}\right)(V - b) = RT$ (D) $P(V - b) = RT$

Q.9 **The dimensions of constant 'a' in the Van der Waals gas equation are that of**

(A) Pressure   (C) Pressure/Volume
(B) Volume   (D) Volume/Pressure

Q.10 **The dimensions of constant 'b' in the Van der Waals gas equation are that of**

(A) Pressure   (C) Pressure/Volume
(B) Volume   (D) Volume/Pressure

Q.11 **The Van der Waals equation explains the behaviour of**

(A) real gases   (C) vapour
(B) ideal gases   (D) non-real gases

Q.12 **At high pressure, the gases deviate from ideal gas behaviour because their molecules**

   (A) possess negligible volume
   (B) are polyatomic
   (C) do not attract each other
   (D) have forces of attraction between them

Q.13 **In the Van der Waals equation of state, the term that accounts for intermolecular forces is**

   IIT-1988

   (A) $(V - b)^{-1}$ (C) $P + \dfrac{a}{V^2}$
   (B) $(RT)^{-1}$
                                                 (D) RT

Q.14 **The temperature at which the second Virial coefficient of a real gas is zero is called**

   (A) boiling point   (C) critical temperature
   (B) temperature of inversion   (D) Boyle's temperature

Q.15 **The deviation from the ideal gas equation PV = nRT is more at**

   (A) high T and low P   (C) high T and high P
   (B) low T and high P   (D) low T and low high P

Q.16 **The Van der Waals equation reduces to the ideal gas equation at**

   (A) high P and high T   (C) low P and low T
   (B) low P and high T   (D) high P and low T

Q.17 **A gas is said to behave like an ideal gas when the relation $PV/T =$ constant. When do you expect a real gas to behave like an ideal gas?**

   [IIT-1999]

   (A) When the temperature is low
   (B) When the temperature is high and pressure is low
   (C) When both the temperature and pressure are high
   (D) When both the temperature and pressure are low

Q.18 **The temperature at which real gases obey the ideal gas laws over a wide range of pressure is called**

   [IIT-1981, 1994]

(A) critical temperature   (C) inversion temperature
(B) Boyle temperature   (D) reduced temperature

Q.19 **At high temperature and low pressure, the Van der Waals equation is
reduced to**

(A) $\left(P + \dfrac{a}{V^2}\right) V = RT$ (C) $P\,(V - b) = RT$

(B) $PV = RT$ (D) $\left(P + \dfrac{a}{V^2}\right)(V - b) = RT$

Q.20 **At high pressure, the Van der Waals equation reduces to**

(A) $PV = RT$ (C) $PV = RT - \dfrac{a}{V}$

(B) $PV = RT + \dfrac{a}{V}$ (D) $PV = RT + Pb$

Q.21 **At low pressure, Van der Waals equation reduces to**

(A) $\dfrac{PV}{RT} = 1 - \dfrac{a}{V\,RT}$ (C) $PV = RT$

(B) $\dfrac{PV}{RT} = 1 + \dfrac{bP}{RT}$ (D) $\dfrac{PV}{RT} = 1 - \dfrac{a}{RT}$

Q.22 **The critical temperature of a gas is 31 °C. The room temperature is 27 °C.
Then the gas in a room**

(A) can be liquefied   (C) will be solidified
(B) cannot be liquefied   (D) none of these

Q.23 **The Boyle temperature T_B is given by**

(A) $T_B = \dfrac{27}{8} T_c$ (C) both

                                         (D) none of these
(B) $T_B = \dfrac{a}{bR}$

Q.24 **The science of production and use of low temperature is called**

(A) refrigeration   (C) regelation
(B) cryogenics   (D) freezing

Q.25 **Let T_B be Boyle's temperature and T_C be the critical temperature then**

(A) $T_B > T_C$ (C) $T_B = T_C$
(B) $T_B < T_C$ (D) $T_B = 2.5 \times T_C$

Q.26 **The Van der Waals equation for one mole of a gas is**

$$\left(P + \frac{a}{V^2} \right)(V - b) = RT$$

The corresponding equation of state for n moles of this gas at pressure P, volume V and temperature T is

[IIT-1992, JRF-NET June 2018]

(A) $\left(P + \dfrac{an^2}{V^2} \right)(V - nb) = nRT$

(B) $\left(P + \dfrac{a}{V^2} \right)(V - nb) = nRT$

(C) $\left(P + \dfrac{a}{n^2V^2} \right)(V - nb) = nRT$

(D) $\left(P + \dfrac{an^2}{V^2} \right)(V - b) = nRT$

Q.27 **The Van der Waals equation for 0.5 mol gas is**

(A) $\left(P + \dfrac{a}{4V^2} \right)\left(\dfrac{V - b}{2} \right) = \dfrac{2RT}{2}$

(B) $\left(P + \dfrac{a}{4V^2} \right)(2V - b) = RT$

(C) $\left(P + \dfrac{a}{4V^2} \right)(2V - 4b) = RT$

(D) $\left(P + \dfrac{a}{4V^2} \right) = \dfrac{2RT}{2(V - b)}$

Q.28 **The Van der Waals equation of state for a gas is given by**

$$\left(P + \frac{a}{V^2} \right)(V - b) = RT$$

where V, P, and T represent the pressure, volume and temperature, respectively, and a and b are constant parameters. At the critical point, the volume is given by

[JRF-NET June 2014]

(A) $\dfrac{a}{9b}$

(B) $\dfrac{a}{27b^2}$

(C) $\dfrac{8a}{27bR}$

(D) $3b$

Keys and hints to MCQ type questions

Q.1 A	Q.6 B	Q.11 A	Q.16 B	Q.21 A	Q.26 A
Q.2 A	Q.7 B	Q.12 D	Q.17 B	Q.22 A	Q.27 B
Q.3 D	Q.8 A	Q.13 C	Q.18 B	Q.23 C	Q.28 D
Q.4 C	Q.9 A	Q.14 D	Q.19 B	Q.24 B	
Q.5 D	Q.10 B	Q.15 B	Q.20 D	Q.25 A	

Hint.12 The molecules of real gases have the force of intermolecular attraction. As a result, their impact on the walls of the container is diminished. This results in a decrease in the pressure of a real gas by a factor of a/V^2. Therefore, real gases deviate from ideal gas behaviour at high pressure and low temperature. The CORRECT option is **D**.

Hint.20 For one mole of a gas, the Van der Waals equation is

$$\left(P + \frac{a}{V^2}\right)(V - b) = RT$$

At high pressure, $\left(P + \frac{a}{V^2}\right) \approx P$, therefore, the above equation reduces to

$$P(V - b) = RT$$
$$PV = RT + Pb$$

Hint.21 For one mole of a gas, the Van der Waals equation is

$$\left(P + \frac{a}{V^2}\right)(V - b) = RT$$
$$PV = RT + Pb - \frac{a}{V} + \frac{ab}{V^2}$$
$$\frac{PV}{RT} = 1 + \frac{Pb}{RT} - \frac{a}{RTV} + \frac{ab}{RTV^2}$$

In the last equation, the second and last terms on the left can be neglected (P is small, b is small), therefore,

$$\frac{PV}{RT} = 1 - \frac{a}{VRT}$$

Hint.22 As $T_c > 27\,^\circ$C, the gas can be liquefied for any temperature below T_c. Therefore, for given conditions the gas in the room can be liquefied.

Hint.23 Boyle's temperature is given by

$$T_B = \frac{27}{8}T_c = \frac{27}{8}\frac{8a}{27Rb} = \frac{a}{bR}$$

Hint.25 Since

$$T_B = \frac{27}{8} T_C = 3.37 T_C > T_C$$

Hint.27 The equation of state for n moles of gas molecules is

$$\left(P + \frac{an^2}{V^2}\right)(V - nb) = nRT$$

Here, n = 1/2. Therefore, the desired equation of state is

$$\left(P + \frac{a}{4V^2}\right)(2V - b) = RT$$

Hint.28 At critical point $\dfrac{\partial P}{\partial V} = 0$ and $\dfrac{\partial^2 P}{\partial V^2} = 0$; by comparing, we get $V_c = 3b$.

7.11 Exercises

1. Describe Andrew's experiments on CO_2. Discuss the results obtained. Show that liquid and gaseous states are two different stages of a continuous phenomenon.

2. In what way does a real gas differ from an ideal gas?

3. Define the following parameters pertaining to a real gas.
   (i) Critical temperature T_c
   (ii) Critical pressure P_c
   (iii) Critical volume V_c

4. Give the reasons for the modification of a perfect gas equation.

5. Develop and discuss the Van der Waals equation of state of a gas. Compare the results with Andrew's experimental curves with CO_2.

6. What are Virial coefficients? What is the value of the first Virial coefficient? How does the second Virial coefficient change with temperature?

7. Distinguish between an ideal gas and a real gas. Write an expression for the Van der Waals equation of state, and explain each term in it. Further, use this equation to obtain expressions for critical constants in terms of constants of the Van der Waals equation.

8. Give four limitations of the Van der Waals equation of state.

9. Define the critical coefficient of a gas. Is it the same for all gases? Do the experimental values agree with the theoretical values?

10. Define critical constants of a gas. Obtain the values of these constants in terms of the constants of the Van der Waals equation of state.

11. Define Boyle's temperature. Discuss the behaviour of real gases above and below Boyle's temperature.

12. Explain, if Boyle's temperature is different for different gases?

13. Distinguish between a perfect gas and real gas. Derive the Van der Waals equation of state and use it to obtain expressions for the critical constants in terms of constants of the Van der Waals equation.

14. What are the limitations of the Van der Waals equation?

15. What are critical coefficients of a gas? Write and explain each term in the Van der Waals equation of state. Calculate the Van der Waals constants a and b in terms of V_c, P_c and T_c.

16. Develop the reduced equation of state for a gas starting from the Van der Waals equation of state. Show that if the two gases have the same reduced pressure and volume, they also have the same reduced temperature.

17. Starting from some basic ideas, set up the Van der Waals equation of state and show how the equation can be expressed in terms of the critical constants. What is the experimental bearing of this equation?

18. Explain the corrections introduced by Van der Waals in the gas equation. Show that for a gas obeying the Van der Waals equation

$$\frac{RT_c}{P_c V_c} = \frac{8}{3}$$

19. Deduce the Van der Waals equation of state. How far does it agree with Andrew's experimental results on CO_2? Discuss the importance of Andrew's experiment in the problem of liquefaction of gases?

Applications to Some Irreversible Changes, Cooling of Real Gases

8

In previous chapters, we learned that the thermodynamic variables (such as T, P, V and S) of a system can only be defined when the system is in equilibrium state. When the system passes through various non-equilibrium states (i.e., when it is away from equilibrium state), one cannot obtain well-defined values for these variables. Further, if these non-equilibrium states are well connected with initial and final equilibrium states, and sufficient information is given about the constraints subjected to the irreversible changes, then it is often possible to obtain useful information for such systems. The *Joule and Joule–Kelvin expansions* of gas are examples of these kinds of processes. We will examine these two processes in detail in this chapter. Let us begin with the Joule expansion.

8.1   The Joule Expansion

Consider a double-sectioned container (Fig. 8.1), with gas in the left part. The state of the gas in the left part is characterized by well-defined values of the thermodynamic variables P, V and T. Let the container is surrounded by an adiathermal wall so that heat exchange between the double-sectioned container and the reservoir (surrounding) is absent.

We also assume that the air from right-hand part is removed and there is a vacuum. Let the volumes of each part is equal to V and the intervening partition is removed so that the gas molecules rush into the right side before eventually settling down to a new equilibrium state. This process is known as a *free expansion*. We wish to calculate the work done by the gas in this expansion. An immediate application of the equation $W = - \int_{V_i}^{V_f} P \, dV$ (which holds for a reversible change) would give a finite number, as the volume is changing during the expansion. But the relation $W = - \int_{V_i}^{V_f} P \, dV$

© The Author(s) 2022
S. Sharma, *Thermal and Statistical Physics*,
https://doi.org/10.1007/978-3-031-07685-5_8

adiathermal wall

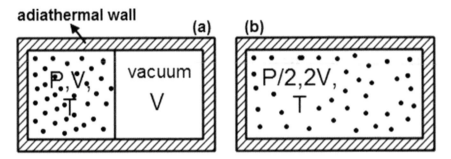

Fig. 8.1 Schematic illustration of the process of Joule expansion. **a** A gas is kept to one part of a vessel by a partition, the other part being evacuated. **b** The partition is removed, and the gas expands irreversibly to fill the entire vessel

cannot be applied here as this process is irreversible. In this special case, the work done by the gas on surrounding is zero, as there is no interaction between the system (confined chamber) and surrounding. The following two points should be clear while dealing with the thermodynamic processes.

a. It is important to be clear as to what is the system. Here, it is the chamber as a whole and not just the left-hand part initially containing all the gas.

2. $W = -\int_{V_i}^{V_f} P dV$ is applicable only to reversible processes and to those special irreversible processes where there is no finite pressure drop across the piston.

If a system, say a gas, expands in such a way that heat exchanges are absent (adiabatic process) and also no work is done by or on the system, then the expansion is called the '*free expansion*'. For this case applying the first law of thermodynamics, we get

$$U_f - U_i = Q + W = 0 + 0$$

This implies that $U_f = U_i$ and $T_f = T_i$, i.e., the initial and the final internal energies as well as temperatures are equal in free expansion. Therefore, the internal energy and the temperature of the gas do not change during the free expansion.

Now we can use this constraint to find the consequences of a Joule expansion. For instance, suppose that we wish to calculate the temperature change during Joule expansion of a gas (any gas). We would proceed as follows:

$$\left(\frac{\partial T}{\partial V}\right)_U = -\left(\frac{\partial U}{\partial V}\right)_T \left(\frac{\partial T}{\partial U}\right)_V \tag{8.1}$$

Although the process of free expansion is irreversible, we may always use the first law in the modified form if initial and final states can be connected by suitable reversible path. Thus,

$$dU = TdS - PdV \tag{8.2}$$

$$\left(\frac{\partial U}{\partial T}\right)_V = T\left(\frac{\partial S}{\partial T}\right)_V = C_V \quad \text{also} \tag{8.3}$$

$$\left(\frac{\partial U}{\partial V}\right)_T = T\left(\frac{\partial S}{\partial V}\right)_T - P \tag{8.4}$$

Equations above lead to

$$\left(\frac{\partial T}{\partial V}\right)_U = -\frac{1}{C_V}\left[T\left(\frac{\partial P}{\partial T}\right)_V - P\right]$$

$$= -\frac{T^2}{C_V}\frac{\partial}{\partial T}\left(\frac{P}{T}\right)_V \tag{8.5}$$

In this equation, the quantity

$$\mu_J = \left(\frac{\partial T}{\partial V}\right)_U = -\frac{1}{C_V}\left[T\left(\frac{\partial P}{\partial T}\right)_V - P\right] \tag{8.6}$$

is called as Joule coefficient.

Further, for a finite change in volume from V_1 to V_2, the total temperature change is found by integrating the above equation:

$$\Delta T = \int_{V_1}^{V_2} \mu_J dV = -\int_{V_1}^{V_2}\frac{1}{C_V}\left[T\left(\frac{\partial P}{\partial T}\right)_V - P\right]dV \tag{8.7}$$

8.1.1 Isothermal Expansion

From Eq. 8.2, we can write

$$\left(\frac{\partial U}{\partial V}\right)_T = T\left(\frac{\partial S}{\partial V}\right)_T - P \tag{8.8}$$

By using Maxwell's thermodynamic equation, it becomes

$$\left(\frac{\partial U}{\partial V}\right)_T = T\left(\frac{\partial P}{\partial T}\right)_V - P \tag{8.9}$$

So that the internal energy change during isothermal expansion of any gas is

$$\Delta U = \int_{V_1}^{V_2}\left[T\left(\frac{\partial P}{\partial T}\right)_V - P\right]dV \tag{8.10}$$

Internal Energy Change for a Perfect Gas

For an ideal gas, the equation of state is $PV = RT$. Thus $\left(\dfrac{\partial P}{\partial T}\right)_V = R/V$ and hence $T\left(\dfrac{\partial P}{\partial T}\right)_V - P = 0$. Therefore, for an ideal gas, during isothermal expansion, $\Delta U = 0$.

Internal Energy Change for a Real Gas

Similarly, by using the equation of state for a real gas, one can easily show that internal energy change during isothermal expansion of a real gas is

$$\Delta U = \int_{V_1}^{V_2} \frac{a}{V^2} dV = a\left[\frac{1}{V_1} - \frac{1}{V_2}\right] \tag{8.11}$$

Note that U depends on a, not b (it is influenced by the intermolecular interactions but does not 'care' that they have non-zero size). Note also that for large volumes, U becomes independent of V and one recovers the ideal gas limit.

8.1.2   Entropy Change Accompanying Joule Expansion

The entropy change accompanying the Joule expansion follows immediately from Eq. 8.2 and condition of constant U: For a perfect gas $dU = 0$, therefore Eq. 8.2 allows us to write

$$\Delta S = \int_{V_1}^{V_2} \frac{P}{T} dV = R \ln \frac{V_2}{V_1} \tag{8.12}$$

which agrees with the general expression discussed earlier in chapter dealing with "entropy" (free expansion). Since, for the perfect gas, there is no change in temperature. As $\Delta Q = 0$, the increase in entropy is entirely associated with the irreversibility of the expansion.

8.1.3   The Joule Coefficient for an Ideal Gas

We have, for one mole of an ideal gas,

$$PV = RT \quad \text{so that} \quad \left(\frac{\partial P}{\partial T}\right)_V = \frac{R}{V}$$

This gives $\mu_J = 0$. This was required, since we defined the perfect gas as one for which U is a function of T only. Thus Eq. 8.1 vanishes in this case.

8.1.4 The Joule Coefficient for Real Gas

The equation of state for real gas is

$$P = \frac{RT}{V - b} - \frac{a}{V^2}$$

so that

$$\left(\frac{\partial P}{\partial T}\right)_V = \frac{R}{V - b}$$

This allows us to write

$$\mu_J = -\frac{1}{C_V}\left[\frac{RT}{V - b} - \frac{RT}{V - b} + \frac{a}{V^2}\right] = -\frac{a}{C_V V^2} \tag{8.13}$$

and corresponding temperature change when gas expands from V_1 to V_2 in this case is given by

$$\Delta T = \int_{V_1}^{V_2} \mu_J dV = -\int_{V_1}^{V_2} \frac{1}{C_V}\left[T\left(\frac{\partial P}{\partial T}\right)_V - P\right]dV$$

$$= -\frac{a}{C_V}\int_{V_1}^{V_2} \frac{dV}{V^2} = -\frac{a}{C_V}\left[\frac{1}{V_1} - \frac{1}{V_2}\right] < 0 \tag{8.14}$$

$\Delta T < 0$ as $C_V > 0$ and $V_2 > V_1$ as it is an expansion. For real gases, therefore, a Joule expansion always (except at extremely high pressure) results in cooling.

8.1.5 Why Real Gases Produce Cooling on Expansion?

The expansion of real gases always results in cooling. The physical reason for this may be visualized as follows. In Eq. 8.1, the second term on the right is an inverse heat capacity which is necessarily a positive quantity. And the first term on the right represents the change in internal energy with volume when the temperature is kept constant. According to kinetic theory of gases, when temperature is held fixed, all contributions to the total energy from kinetic terms (or from degrees of freedom that contribute to kinetic energy or which are internal to the molecules) remain unchanged. Therefore, when gas expands (volume changes) at constant temperature, the internal energy can only change by virtue of the change in potential energy of constituent molecules of the gas. The latter can happen when the distance between the molecules changes (intermolecular interactions changes). This is illustrated in more detail in the Figure given below. The figure shows a plot of potential energy between two molecules as a function of separation (Fig. 8.2).

At large distances, there are weak attractive forces and the potential energy increases with separation. As the separation is decreased the forces eventually

Fig. 8.2 The potential energy of two molecules as a function of their separation r

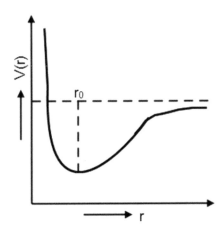

become repulsive and the potential energy again increases. The point of lowest potential energy, r_0, is the equilibrium separation at absolute zero, and corresponds to the density of the solid. Gases always have an average intermolecular separation greater than this, and the potential energy increases with volume, $T\left(\dfrac{dU}{dV}\right)$ is positive. Therefore, gases always cool in a Joule expansion.

8.2   Joule–Kelvin Expansion

The Joule–Kelvin (or Joule–Thomson) expansion is a steady flow process in which a gas is forced through a porous plug or a throttle valve under the constraint of thermal isolation from the surroundings, see Fig. 8.3. This is another example of an irreversible process. The throttling process is represented schematically in the Figure. Due to friction between the gas molecules and the walls of the pores in the plug, such process exhibits internal mechanical irreversibility. In other words, the gas passes through various dissipative non-equilibrium states on its way from the initial equilibrium state (P_1, V_1) to the final equilibrium state (P_2, V_2). These inter-

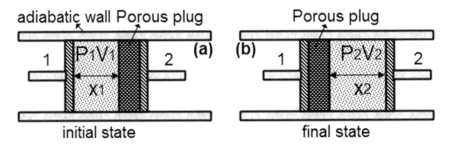

Fig. 8.3 Joule–Kelvin expansion of a gas through a porous plug. **a** The gas is entirely contained in chamber on left side **b** gas after passing through the porous plug and contained in chamber 2 at low pressure P_2

mediate non-equilibrium states cannot be described by thermodynamic coordinates, but an interesting conclusion can be drawn about the initial and final equilibrium states, which are described by thermodynamic coordinates. For throttling process, the modified form of the first law reads:

$$(U_2 - U_1) = W + Q \tag{8.15}$$

As the throttling process occurs under adiabatic conditions, i.e., $Q = 0$. The gas while passing through the porous plug performs external work at the cost of its internal energy. Let us evaluate the work done by the pistons (moving by distance x_1) on the gas to cause its flow across the porous plug. Consider a volume V_1 of gas on the high-pressure (P_1) side of the container. Let its internal energy is U_1. To push the gas through the porous plug, the high-pressure gas behind it has to do work on it.

The work done by the piston to move the gas molecules through porous plug is

$$W_1 = - \int_{V_1}^{0} P_1 dV = P_1 V_1 \tag{8.16}$$

(since the pressure P_1 is maintained on the high-pressure side of the porous plug). While passing through the porous plug the gas expands and now occupies a volume V_2 which is larger than V_1. As a result, the pressure drops from P_1 to P_2. Now the gas has to do work on the low-pressure gas in front of it which is at pressure P_2 and hence this work is

$$W_2 = - \int_{0}^{V_2} P_2 dV = -P_2 V_2 \tag{8.17}$$

Since the pressures on either side of the porus plug are kept constant, for instance, by pistons moving in cylinders at the appropriate rates, the net work done during throttling process is

$$W = W_1 + W_2 = P_1 V_1 - P_2 V_2 \tag{8.18}$$

During expansion, the gas may change its temperature, and hence its new internal energy is U_2. Then, the modified first law reads

$$U_2 - U_1 = P_1 V_1 - P_2 V_2$$

$$U_1 + P_1 V_1 = U_2 + P_2 V_2$$

$$H_1 = H_2 \tag{8.19}$$

This is an example of a steady flow process in which no external work is done during the expansion (the gas moves from higher pressure region to lower pressure region through porous plug), no heat is exchanged with the surroundings and in which kinetic and potential energies may normally be ignored, the constraint becomes simply that the enthalpy, H, of the gas (or fluid) is conserved. Note that it is not the internal energy but *enthalpy* (H) is conserved during this process.

The temperature drop can then be calculated as below

$$\left(\frac{\partial T}{\partial P}\right)_H = -\left(\frac{\partial T}{\partial H}\right)_P \left(\frac{\partial H}{\partial P}\right)_T \tag{8.20}$$

Also

$$dH = TdS + VdP \tag{8.21}$$

Therefore,

$$\left(\frac{\partial H}{\partial T}\right)_P = T\left(\frac{\partial S}{\partial T}\right)_P = C_P \tag{8.22}$$

and

$$\left(\frac{\partial H}{\partial P}\right)_T = T\left(\frac{\partial S}{\partial P}\right)_T + V = -T\left(\frac{\partial V}{\partial T}\right)_P + V \tag{8.23}$$

Here, the last step is written after making use of one of the Maxwell relations. Substituting the values of derivatives in Eq. 8.20

$$\mu_{JK} = \left(\frac{\partial T}{\partial P}\right)_H = \frac{1}{C_P}\left[T\left(\frac{\partial V}{\partial T}\right)_P - V\right]$$
$$= \frac{T^2}{C_P}\frac{\partial}{\partial T}\left(\frac{V}{T}\right)_P \tag{8.24}$$

For a finite change in pressure, one can evaluate the change in temperature as

$$\Delta T = \int_{P1}^{P2} \frac{1}{C_P}\left[T\left(\frac{\partial V}{\partial T}\right)_P - V\right]dP \tag{8.25}$$

Therefore, we have expressed the Joule–Thomson coefficient μ_{JK} in terms of the volume V of the gas, its heat capacity C_P and its expansion coefficient. All of these quantities are fairly easy to measure. The Joule–Kelvin expansion can result in heating or cooling in case of real gases. This is indeed determined by the sign of μ_{JK}.

It is convenient to consider when μ_{JK} changes sign, and this will occur when $\mu_{JK} = 0$, i.e., when

$$T\left(\frac{\partial V}{\partial T}\right)_P - V = 0$$

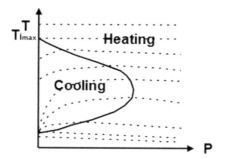

Fig. 8.4 Isenthalps (lines of constant enthalpy) for a real gas (obeying van der Waal equation of state) are shown as dotted line. When the gradients of the isenthalps, i.e., $\left(\dfrac{\partial T}{\partial P}\right)_H$ on this diagram are positive, then cooling can be obtained by reducing the pressure at constant enthalpy (i.e., in a Joule–Kelvin expansion). The solid line is the inversion curve

or equivalently when

$$\left(\frac{\partial V}{\partial T}\right)_P = \frac{V}{T} \tag{8.26}$$

This equation defines the so-called inversion curve in the T-P plane and is shown in Fig. 8.4. That is, the locus of all points at which the Joule–Thomson coefficient is zero (the locus of the maxima of the isenthalpic curves) is known as the inversion curve. This is plotted for the van der Waals gas in Fig. 8.4 below as a heavy solid line. The lines of constant enthalpy are also shown as dotted and their gradients change sign when they cross the inversion curve. The region inside the inversion curve, where μ_{JK} is positive, is called the region of cooling, that is, the final temperature of the gas is less than the initial temperature; whereas outside the inversion curve, where μ_{JK} is negative, it is called the region of heating, that is, the final temperature is more than the initial temperature. The maximum inversion temperature is a crucial parameter, below which the Joule–Kelvin expansion can result in cooling. In the case of helium, the maximum inversion temperature is 43 K, so helium gas must be cooled to below this temperature by some other means before it can be liquefied using the Joule–Kelvin process.

8.2.1   The Joule–Kelvin Coefficient for an Ideal Gas

A perfect gas obeys PV = RT equation. This gives us

$$\left(\frac{\partial V}{\partial T}\right)_P = \frac{R}{P}$$

$$T\left(\frac{\partial V}{\partial T}\right)_P = \frac{R}{P}T = V, \;\; or \;\; T\left(\frac{\partial V}{\partial T}\right)_P - V = 0$$

Therefore, for a perfect gas, Eq. 8.24 gives $\mu_{JK} = 0$.

8.2.2 The Joule–Kelvin Coefficient for a Real Gas

For a real gas obeying Van der Waals equation of state

$$\left(P + \frac{a}{V^2}\right)(V - b) = RT \tag{8.27}$$

After differentiating w.r.t T and keeping P constant, we obtain

$$\left(P + \frac{a}{V^2}\right)\left(\frac{\partial V}{\partial T}\right)_P - \frac{2a}{V^3}\left(\frac{\partial V}{\partial T}\right)_P (V - b) = R$$

$$\left(\frac{\partial V}{\partial T}\right)_P = \frac{R}{\left(P + \frac{a}{V^2}\right) - \frac{2a}{V^3}(V - b)}$$

$$= \frac{R}{\frac{RT}{V - b} - \frac{2a}{V^3}(V - b)} = \frac{R(V - b)}{RT - \frac{2a}{V^3}(V - b)^2}$$

$$T\left(\frac{\partial V}{\partial T}\right)_P = \frac{RT(V - b)}{RT - \frac{2a}{V^3}(V - b)^2}$$

Now making use of the fact that $b \ll V$, we can neglect the terms involving b and b^2 and can write V^2 in place of $(V - b)^2$. Therefore,

$$T\left(\frac{\partial V}{\partial T}\right)_P = \frac{RT(V - b)}{RT - \frac{2a}{V}} = \frac{V - b}{1 - \frac{2a}{VRT}} = (V - b)\left[1 - \frac{2a}{VRT}\right]^{-1}$$

By making use of binomial expansion and neglecting higher order terms (because a, $b \ll V$), we obtain

$$T\left(\frac{\partial V}{\partial T}\right)_P = (V - b)\left[1 + \frac{2a}{VRT}\right] = V - b + \frac{2a}{RT}$$

$$T\left(\frac{\partial V}{\partial T}\right)_P - V = \frac{2a}{RT} - b$$

Substituting back this expression in Eq. 8.24, we obtain

$$\mu_{JK} = \left(\frac{\partial T}{\partial P}\right)_H = \frac{1}{C_P}\left[\frac{2a}{RT} - b\right] \tag{8.28}$$

Above equation gives the Joule–Kelvin coefficient for the real gas. This equation implies that if $\frac{2a}{RT} > b$ or equivalently $T < \frac{2a}{Rb}$, in that case $\left(\frac{\partial T}{\partial P}\right)_H$ is positive.

Since, in present scenario ∂P is a negative quantity (the direction in which gas emerges through porous plug, is lower compared to the other side), it makes ∂T also negative. Therefore, in this case, the gas will be cooled after passing through the porous plug. In the second case, when $\dfrac{2a}{RT} < b$ or equivalently $T > \dfrac{2a}{Rb}$, then $\left(\dfrac{\partial T}{\partial P}\right)_H$ is negative and hence ∂T will be positive, i.e., gas will be heated as it passes through the porous plug. In a third case, when $T = \dfrac{2a}{Rb}$, $\left(\dfrac{\partial T}{\partial P}\right)_H = 0$, that is when gas passes through porous plug, its temperature does not change. This temperature at which Joule–Kelving coefficient (μ_{JK}) changes sign is known as *temperature of inversion* and denoted as

$$T_i = \frac{2a}{Rb} \tag{8.29}$$

Note that for most of the gases, the ordinary working temperature lies below the *inversion temperature* as a result they display a cooling effect. Whereas in case of hydrogen and helium gases, the inversion temperature are much below the ordinary temperature. As a result, these gases display heating effect at these ordinary temperature. However, if these gases are precooled below their inversion temperature before allowing them to undergo Joule–Kelvin expansion, they will also display cooling effect.

8.2.3 Relation Between Boyle Temperature, Temperature of Inversion and Critical Temperature

Equation 8.29 defines the inversion temperature. Boyle temperature and critical temperature are, respectively,

$$T_B = \frac{a}{Rb}, \quad T_c = \frac{8a}{27Rb}$$

Equations 8.29 and 8.30 imply

$$\begin{aligned} T_i &= 2T_B, \quad \text{and} \\ \frac{T_i}{T_c} &= \frac{2a}{Rb} \times \frac{27Rb}{8a} = \frac{27}{4} \end{aligned} \tag{8.30}$$

Since this ratio $\dfrac{T_i}{T_c}$ is a constant number independent of any kind of variable, it is *independent of nature of gas*. Note that experimental value for $\dfrac{T_i}{T_c}$ has been found to be less than 6. This means temperature of inversion is much higher than the critical temperature. For instance, for hydrogen $T_c = 33$ K, $T_i = 190$ K. Since, $T_i \gg T_c$, the methods which employ regenerative cooling are preferred over those employing initial cooling of gas below critical temperature.

8.2.4   Entropy Change Accompanying Joule–Kelvin Expansion

The entropy change during Joule–Kelvin expansion can be directly evaluated from Eq. 8.21. Since dH=0, this allows us to write

$$\left(\frac{\partial S}{\partial P}\right)_H = -\frac{V}{T}$$

$$\Delta S = -\int_{P_1}^{P_2} \frac{V}{T} dP = R \ln \frac{P_1}{P_2} > 0 \qquad (8.31)$$

Thus, for a perfect gas Joule–Kelvin coefficient is zero and entropy change is given by the above equation. Again, this is entirely associated with the irreversibility of the expansion.

Ex:8.1   **Show that isothermal curve for a perfect gas shown on a P-V diagram is also isoenthalpic (same enthalpy).**

Sol: The internal energy of a perfect gas depends solely on temperature and its internal energy remains constant, i.e., $U_f = U_i$. Further, an ideal gas under isothermal condition satisfy, $P_i V_i = P_f V_f$. Therefore,

$$U_i + P_i V_i = U_f + P_f V_f$$
$$H_i = H_f$$

That is enthalpy remains constant along an isothermal curve. In other words, one can say that curve is isoenthalpic.

8.3   Distinction Between Adiabatic Expansion, Joule Expansion and Joule–Kelvin Expansion

Having discussed all three processes which produce cooling after expansion, now we can compare them to figure out the similarities and differences among them. The system is thermally isolated (dQ = 0) from surrounding in these processes. However, in Joule expansion (i.e., free expansion), the system is also mechanically isolated from surrounding and net work done by system or on the system is zero. In contrast, in a Joule–Kelvin expansion the system is in contact with surrounding and both pistons can adjust their position to keep pressures P_1 and P_2 constant (Fig. 8.3). In this case, the net work done during throttling process is given by Eq. 8.18 as

$$dW = P_1 V_1 - P_2 V_2$$

In an adiabatic expansion, the system stays in contact with surrounding, i.e., not isolated mechanically from surrounding. Pressure of gas may also change ($PV^{\gamma} =$ constant). The work done by the expanding gas against external pressure is given by

$$dW = - \int P dV$$

The internal energy of the gas is utilized to perform this external work. Consequently, temperature of the gas falls. Applying the first law of thermodynamics ($dU = dQ + dW$) to these processes, we obtain

$$dU = 0, \quad \text{Joule expansion}$$
$$dU = dW = P_1 V_1 - P_2 V_2 \quad \text{Joule–Kelvin expansion}$$
$$dU = - \int P dV \quad \text{adiabatic expansion}$$

The largest fall in temperature takes place in case of adiabatic expansion, as the external work is performed at the cost of internal energy of the expanding gas. In Joule–Kelvin expansion, cooling or heating take place depending upon size of $P_1 V_1$ or $P_2 V_2$. In the third case, i.e., Joule expansion, it is the work done by the gas against molecular interaction and hence reduction in internal energy of the (real) gas undergoing free expansion. This reduction in internal energy is accompanied with cooling of gas.

8.4   Multiple Choice Questions

Q.1 **An ideal gas undergoes a Joule–Kelvin expansion. The temperature of an ideal gas**

(A)   increases
(B)   decreases
(C)   first increases and then decreases isenthalpically.
(D)   no change in temperature

Q.2 **The region in which Joule–Kelvin coefficient is negative is called as**

(A)   cooling region   (C)   negative region
(B)   heating region   (D)   positive region

Q.3 **The region in which Joule–Kelvin coefficient is positive is called as**

    (A)   cooling region   **(C)**   negative region

    (B)   heating region   **(D)**   positive region

Q.4 **At inversion temperature in Joule–Kelvin expansion of a gas, the value of Joule–Kelvin coefficient is**

    (A)   positive   **(C)**   zero

    (B)   negative   **(D)** ∞

Q.5 **In order to obtain maximum cooling effect in the Joule–Kelvin expansion, the initial temperature of the gas should be**

    (A)   less than the temperature corresponding to inversion point of isenthalpic curve of the expansion of gas

    (B)   more than the temperature corresponding to inversion point of isenthalpic curve of the expansion of gas

    (C)   equal to the temperature corresponding to inversion point of isenthalpic curve of the expansion of gas

    (D)   None of these

Q.6 **For achieving cooling effect in Joule–Kelvin expansion, the initial temperature of the gas must be**

    (A)   above the maximum inversion temperature

    (B)   below the maximum inversion temperature

    (C)   equal to the maximum inversion temperature

    (D)   there is no correlation between initial gas temperature and maximum inversion temperature

Q.7 **The inversion curve in Joule–Kelvin expansion or isenthalpic expansion is the locus of all points at which Joule–Kelvin coefficient**

    (A)   positive   **(C)**   zero

    (B)   negative   **(D)**   +ve, -ve or zero

Q.8 **An isolated box is divided into two equal compartments by a partition. One compartment contains a van der Waals gas while the other compartment is empty. The partition between the two compartments is now removed. After the gas has filled the entire box and equilibrium has been achieved, which of the following statement(s) is (are) correct?**

                                                                                          [JAM-2017]

(**A**)   Internal energy of the gas has not changed
(**B**)   Internal energy of the gas has decreased
(**C**)   Temperature of the gas has increased
(**D**)   Temperature of the gas has decreased

Keys and Hints to MCQ Type Questions

Q.1   D   Q.3   A   Q.5   C   Q.7   C
Q.2   B   Q.4   C   Q.6   B   Q.8   A, D

8.5   Exercises

1. What is Joule expansion? How it is different from Joule–Kelvin expansion?

2. What is Joule–Thomson (Joule–Kelvin) effect? Deduce expression for Joule Thomson cooling.

3. What is Joule–Thomson effect? Obtain an expression for the change in temperature for gas in Joule–Thomson expansion. What is inversion temperature?

4. The Joule–Kelvin coefficient $\mu_{JK} = \left(\dfrac{\partial T}{\partial P}\right)_H$ provides a measure of the result of a Joule–Kelvin expansion. Show that

$$\mu_{JK} = \frac{1}{C_P}\left[T\left(\frac{\partial V}{\partial T}\right)_P - V\right]$$

Further show that entropy change accompanying a Joule–Kelvin expansion is

$$\Delta S = -\int_{P_1}^{P_2} \frac{V}{T}dP = R\ln\frac{P_1}{P_2} > 0$$

5. What is Joule expansion? Is it similar/different from free expansion. Show that Joule coefficient μ_J for a gas undergoing Joule expansion is

$$\mu_J = -\frac{1}{C_V}\left[T\left(\frac{\partial P}{\partial T}\right)_V - P\right]$$

6. Show that Joule coefficient and Joule–Kelvin coefficient for an ideal gas is equal to zero.

7. Show that for a van der Waal gas the Joule–Kelvin coefficient is given by

$$\mu_{JK} = \left(\frac{\partial T}{\partial P}\right)_H = \frac{1}{C_P}\left[\frac{2a}{RT} - b\right]$$

8. What is the inversion temperature? Obtain an expression for it when a real gas is subjected to Joule–Kelvin expansion.

9. In what aspect, Joule expansion and Joule–Kelvin expansion are different?

10. Define Joule–Thomson effect. Obtain thermodynamically an expression for the cooling produced in a van der Waals gas and explain the reason why hydrogen and helium show heating effects at ordinary temperatures while other gases do not.

11. In Joule–Kelvin effect, gas is cooled by expansion through an insulated throttle a simple but inefficient process with no moving parts at low temperature. Explain why enthalpy is conserved in this process. Deduce that

$$\left(\frac{\partial T}{\partial P}\right)_H = \frac{1}{C_P}\left[T\left(\frac{\partial V}{\partial T}\right)_P - V\right]$$

Q.12 Describe Joule–Kelvin effect. Obtain an expression for Joule–Kelvin coefficient for a real gas and hence define the inversion temperature. Explain its significance.

Q.13 In what aspects adiabatic expansion, Joule expansion and Joule–Kelvin expansion are same and differ.

Q.14 Prove that the ratio of inversion temperature and critical temperature does not depend upon the nature of gas.

Q.15 What is inversion temperature? Show that inversion temperature for a Van der Waals gas is given by

$$T_i = \frac{2a}{Rb}$$

Theory of Radiation

9

This chapter is all about the properties of this thermal radiation. Readers will learn about a blackbody and its radiation spectrum. The chapter focuses on the various classical developments in understanding the blackbody radiation spectrum and how the limitations encountered by Wien's and Rayleigh–Jeans distribution law were addressed by Planck's hypothesis. The chapter concludes with the discussion that Planck's law is the most general law that explains all features of blackbody spectra and all other laws are contained in it.

9.1   Transfer of Thermal Energy

From elementary classes, it is known that thermal energy is transferred from one place to another by three processes: conduction, convection and radiation. In conduction, thermal energy is transferred by interactions among atoms or molecules, though there is no transport of the atoms or molecules themselves. For example, if one end of a solid bar is heated, the atoms in the heated end vibrate with greater energy than those at the cooler end. Because of the interaction of the more energetic atoms with their neighbours, this energy is transported along the bar. In convection, heat is transported by direct mass transport. For example, warm air in a room expands and rises because of its lower density. Thermal energy is thus transported upward along with the mass of warm air. In radiation, thermal energy is transported through space in the form of electromagnetic waves that move at the speed of light. Thermal radiation, light waves, radio waves, television waves and X-rays are all forms of electromagnetic radiation that differ from one another only in their wavelengths and frequencies.

© The Author(s) 2022
S. Sharma, *Thermal and Statistical Physics*,
https://doi.org/10.1007/978-3-031-07685-5_9

9.2   Thermal Radiation

Maxwell pointed out that light is an electromagnetic wave whose speed 'c' can be expressed in terms of fundamental constants ϵ_0 and μ_0 (permittivity and permeability of free space, respectively) taken from the theories of electricity and magnetism. The relation connecting these constant with speed of light can be written as

$$c = \frac{1}{\mu_0 \epsilon_0} \tag{9.1}$$

Later, Planck pointed out that light behaved not only like a wave but also like a particle. In the language of quantum mechanics, electromagnetic waves can be quantized as a set of particles which are known as *photons*. Energy possessed by a photon is given by

$$E = \hbar \omega = h \nu$$

where $\omega = 2\pi \nu$ is the angular frequency, $\hbar = h/2\pi$ and ν is the frequency. Further, the momentum associated with a photon can be written in terms of wave vector $k = 2\pi/\lambda$, λ being the wavelength of light.

Any substance at non-zero temperature emits electromagnetic radiation. This is known as *thermal radiation*. One never notices this effect from an object at room temperature, because the frequency of the electromagnetic radiation is low and most of the emission is in the infrared region of the electromagnetic spectrum. Our eyes being sensitive to electromagnetic radiation in the visible region cannot see the radiation whose frequency lies in infrared region of spectrum. However, you may have noticed that a piece of metal in a furnace glows 'red hot' so that, for such objects at higher temperature, your eyes are able to pick up some of the thermal radiation. Such type of thermal radiation has the following properties:

(i) Thermal radiation are electromagnetic in nature and can travel through empty space with the speed of light.

(ii) Similar to light waves, thermal radiation also travel in straight line.

(iii) Thermal radiation exhibit the phenomenon of reflection such as refraction, interference, diffraction and polarization.

(iv) They do obey the inverse square law.

Therefore, thermal radiation have same nature as that of light waves. The wavelength of thermal radiation is greater than red color. Their wavelength lies in infrared (IR) region of the electromagnetic spectrum. Thermal radiation cannot be detected by human eyes. These can be detected either by a **bolometer** or a **thermopile**.

9.3   A Few Important Definitions

9.3.1   Total Energy Density (u)

The number of Joules stored in a cubic metre of cavity is known as energy density u of electromagnetic radiation. In other words, the quantity u denotes the total radiant energy for all wavelength ranging from 0 to ∞ per unit volume. The units for u are Jm^{-3}.

9.3.2   Spectral Energy Density (u_λ)

Now, we want to specify that in which frequency ranges this energy is stored. To understand this, consider two containers, each in contact with thermal reservoirs at temperature T and joined to one another by a tube, as illustrated schematically in Fig. 9.1. This composite system is allowed to come to equilibrium.

Both thermal reservoirs are at the same temperature T. A simple application of the second law of thermodynamics implies that there can be no net heat flow from either one of the bodies to the other. Hence, there can be no net energy flux along the tube. This implies that the energy flux from the soot-lined cavity along the tube from left to right must be balanced by the energy flux from the mirror-lined cavity along the tube from right to left. Equation 23.12 thus tells us that each cavity must have the same energy density u. This argument can be repeated for cavities of different shapes and sizes as well as different coatings. Hence, we conclude that u is independent of shape, size or material of the cavity. But maybe one cavity might have more energy density than the other at certain wavelengths, even if it has to have the same energy density overall? This is not the case, as we shall now prove. First, we make a definition of spectral energy density u_λ as a measure of energy per unit volume per unit wavelength. Therefore, $u_\lambda d\lambda$ is the energy density due to those photons which have wavelengths ranging from λ to $\lambda + d\lambda$. Total energy density is then given by

$$u = \int u_\lambda d\lambda \qquad (9.2)$$

Further, we wish to discuss how effectively a particular surface of a cavity can absorb or emit electromagnetic radiation of a particular wavelength or frequency. We therefore make the following additional definitions

Fig. 9.1 Two cavities maintained at temperature T. One of them is lined with soot and the other one with a mirror coating

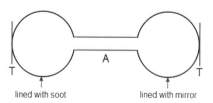

lined with soot   lined with mirror

9.3.3   The Spectral Absorptivity (α_λ)

It is defined as the fraction of the incident energy absorbed per unit surface area per second at wavelength λ.

9.3.4   Emissivity (e)

Emissivity or total emissive power 'e' is defined as the total radiant energy of all wavelengths from 0 to ∞ which is emitted per second per unit surface area of a body. Units for emissivity are $Jm^{-2}s^{-1}$ or Wm^{-2}.

9.3.5   Relative Emittance (ϵ)

It is defined as ratio of emittance of a surface to the emittance of a blackbody. Therefore, we can write

$$\epsilon = \frac{e'}{e} \tag{9.3}$$

where e' and e are emittance of a given surface and blackbody, respectively. For a perfect blackbody, e' = e, so that $\epsilon = 1$.

9.3.6   The Spectral Emissive Power (e_λ)

It signifies the radiant energy per second (power emitted) per unit surface area per unit range of wavelength. Therefore, $e_\lambda d\lambda$ is the power emitted per unit area by the electromagnetic radiation having wavelengths between λ and $\lambda + d\lambda$. One can evaluate the emissivity as

$$e = \int_0^\infty e_\lambda d\lambda \tag{9.4}$$

9.4   Kirchhoff's Law

Now, we will discuss how well a particular surface absorb or emit electromagnetic radiation of a particular frequency or wavelength. For this purpose, we will make use of the definitions of spectral absorptivity α_λ and spectral emissive power e_λ. We note that e_λ dλ is the power emitted per unit area by the electromagnetic radiation with wavelengths lying in the range of λ to $\lambda + d\lambda$.

For the incident spectral density $u_\lambda \, d\lambda$, the power per unit area ($\mathbb{P}'$) absorbed by the surface is

$$\mathbb{P}' = \left(\frac{1}{4}u_\lambda d\lambda c\right)\alpha_\lambda \tag{9.5}$$

Similarly, the power per unit area emitted ($\mathbb{P}''$) by a surface is given by

$$\mathbb{P}'' = e_\lambda d\lambda \tag{9.6}$$

In equilibrium, power emitted per unit area will be equal to power absorbed per unit area, i.e., $\mathbb{P}' = \mathbb{P}''$, giving

$$\frac{e_\lambda}{\alpha_\lambda} = \frac{c}{4}u_\lambda \tag{9.7}$$

Equation 9.7, expresses the Kirchhoff's law, which states that the ratio of $\dfrac{e_\lambda}{\alpha_\lambda}$ is a universal function of T and λ. Therefore, if we fix T and λ, the ratio is automatically fixed and $e_\lambda \propto \alpha_\lambda$. In other words, we can say that "*good absorbers are good emitters*" and "*bad absorbers are bad emitters*".

9.5 Blackbody

The quantities 'e' and 'e_λ' characterize the body as an emitter, whereas the quantity α_λ describes the absorbing properties of the surface. All these quantities depend upon temperature of the body and not on the nature of surface. Whenever a radiation of particular wavelength falls on any randomly chosen surface, in may be partially reflected, partially absorbed and partially transmitted. If we denote r_λ and t_λ as reflection and transmission coefficients of the body at wavelength λ, then we can write

$$r_\lambda + \alpha_\lambda + t_\lambda = 1$$

Further, if $r_\lambda = t_\lambda = 0$ and $\alpha_\lambda = 1$ for all wavelengths, i.e., *when a body absorbs all radiation incident on it, the body is said to be perfectly black*. A perfectly blackbody neither reflects nor transmits any radiation, and it appears black. It behaves like a perfect absorber.

9.5.1 A Blackbody in Practice

A perfectly blackbody neither reflects nor transmits any radiation, and it appears black. It behaves like a perfect absorber. Note that in practice no body strictly satisfies the ideal definition of a blackbody. For instance, for lamp black and platinum black, α_λ is less than unity. These two have absorbency of 96 and 98%, respectively. Below we discuss one example of a body that has absorption close to a perfectly blackbody.

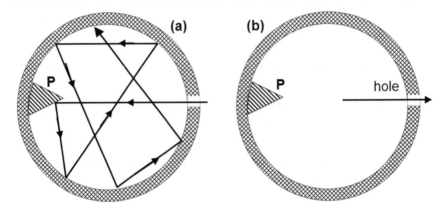

Fig. 9.2 Blackbody **a** absorber **b** emitter

Fery's Blackbody

It consists of a double-walled hollow sphere having a small opening O on one side and a conical projection P just opposite to it (Fig. 9.2). Its inner surface is coated with lamp black. Any radiation entering the body through the opening suffers multiple reflections at its inner wall and about 97% of it is absorbed by lamp black at each reflection.

9.5.2   Energy Distribution in a Blackbody Radiation Spectrum

A blackbody placed inside an isothermal enclosure starts emitting radiation. These radiation are independent of nature of substance, wall of enclosure its shape and even presence of any other body in surrounding. It simply depends upon temperature. A perfect blackbody absorbs all radiation (irrespective of wavelength) which falls on it. The radiation emitted by the body are also independent of the nature of substance, but purely depend upon temperature. *Such radiation spectra obtained in a uniform temperature enclosure is known as blackbody radiation spectrum.*

The spectral energy density of blackbody at different temperature was investigated by Lummer and Pringesheim. Their experimental blackbody was a small aperture of an electrically heated chamber whose temperature was monitored through a thermocouple (bolometer). They measured the emissive power at different temperatures, and a qualitative description of the results is given in Fig. 9.3. The following points are noted:

(i) At fixed temperature, energy is non-uniformly distributed over the spectrum.

(ii) At a given temperature, intensity of heat radiation increases with increase in wavelength, attains a maxima at certain wavelength (λ_m) and decreases further with rise in wavelength.

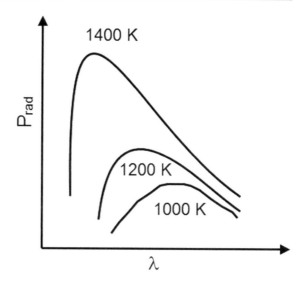

Fig. 9.3 Blackbody spectral density at different temperatures

(iii) With rise in temperature, λ_m, the wavelength at which emitted energy is maximum, shifts towards lower wavelength so that the product λ_m T = constant. This is also known as *Wien's displacement* law.

(iv) With increase in temperature, emitted energy density also increases.

(v) At particular temperature, total area under a curve gives total energy emitted by the body over a range of wavelengths. This area increases with increase in temperature of the body. It is found that increased area is directly proportional to the fourth power of temperature and hence power radiated can be written as P$\propto T^4$.

9.6   Radiation Emitted from a Blackbody and Other Objects (Stefan–Boltzmann Law)

Consider a box with a hole and a black object at same initial temperature T are facing each other as shown in Fig. 9.4. Each object emits photons and some of the emitted photons are absorbed by the other object. If both objects have same size, each of them will absorb same fraction of radiation emitted by other. Now suppose that blackbody does not emit same amount of power as that of hole (in a box). Let us assume that it emits less power. Then more energy will flow from hole to the blackbody than from the blackbody to the hole. Thus, blackbody will gradually become hotter!!! This process would violate the second law of thermodynamics. In the other case, if blackbody emits more radiation than the hole, then blackbody will gradually cool off and box with a hole gets hotter. Again this cannot happen. So the total power emitted by the blackbody per unit area at any given temperature must be same as that emitted by a hole in box. Further, if we imagine that a filter is inserted between these two objects so that it allows only certain range of wavelengths to pass through between hole and blackbody. Again if one of the objects emits more radiation at these wavelengths than the other, its temperature will decrease while other's temperature

Fig. 9.4 A thought experiment illustrating that a perfectly black surface emits radiation identical to that emitted by a hole in a box of thermal photons (radiation)

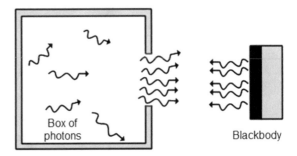

Box of photons

Blackbody

increases, violating the second law again. Thus, we conclude that entire spectrum of radiation emitted by a blackbody must be identical as for hole.

A perfect blackbody absorbs all heat radiation (of any wavelength) incident on it. It neither reflects ($r_\lambda = 0$) nor transmits ($t_\lambda = 0$) any of the incident radiation and therefore appears to be black for all kinds of incident radiation. However, if an object is not black, then it reflects some of the photons instead of absorbing all of them. This makes the analysis even more complicated. Let us assume that out of every five incident photons (at certain wavelength λ) that hit the object it reflects two photons back and absorbs other three. In order to remain in thermal equilibrium with the hole it only needs to emit three photons, which join the (two) reflected photons on its way back. In general, if e is the fraction of photons absorbed (here e = 3/5), then e is also the fraction emitted in comparison to a blackbody. This number is called *emissivity* of the material. Emissivity equals 1 for a perfectly blackbody, whereas it equals zero for a perfectly reflecting body. Thus, a good reflector is a poor emitter and vice-versa. In general, emissivity e depends upon the wavelength of the incident radiation. By taking a weighted average of e over relevant wavelengths the total power radiated (energy radiated per second) by any object can be written as

$$Power = Ae\sigma T^4 \tag{9.8}$$

where A is surface area of the object, e is emissivity of the object, a number between 0 and 1 that depends on composition of the surface of the object, and σ is Stefan's constant whose value is $5.67 \times 10^{-8} \mathrm{Wm^{-2}K^{-4}}$.

For a perfect blackbody, e = 1, so that power radiated by a blackbody with surface area A is

$$Power = A\sigma T^4 \tag{9.9}$$

This result, found empirically by Josef Stefan in 1879 and derived theoretically by Ludwig Boltzmann about five years later, is called the **Stefan–Boltzmann law**. According to this law,

The rate at which an object radiates energy is proportional to the area of the object and to the fourth power of its absolute temperature.

Let T_0 be the temperature of the surroundings. Then the rate at which an object absorbs radiation from the surrounding is given by

$$P_{abs} = Ae\sigma T_0^4 \tag{9.10}$$

Let the temperature of the object is T. Then the power radiated by an object at temperature T (if T>T_0) is

$$P_{rad} = Ae\sigma T^4 \tag{9.11}$$

The net power radiated by an object at temperature T in an environment at temperature T_0 is

$$P_{net} = P_{rad} - P_{abs} = Ae\sigma \left(T^4 - T_0^4\right) \tag{9.12}$$

When an object is in thermal equilibrium with its surroundings, i.e., T = T_0, the object emits and absorbs radiation at the same rate and P_{net}=0. Here, one should note that P_{abs} and P_{rad} are simply the absorbed and radiated powers not the power densities which are obtained after dividing P_{abs} or P_{rad} with surface area of the object involved.

9.6.1   Newton's Law of Cooling from Stefan's Law

Stefan's law just discussed is applicable for all temperatures of a hot object. On the other hand, when the temperature T of an object is not too different from the surrounding temperature T_0, a radiating object obeys Newton's law of cooling. We can see this by writing Eq. 9.12

$$\begin{aligned}
P_{net} &= Ae\sigma \left(T^4 - T_0^4\right) \\
&= Ae\sigma \left[T^2 + T_0^2\right]\left[T^2 - T_0^2\right] \\
&= Ae\sigma \left[T^2 + T_0^2\right][T + T_0][T - T_0]
\end{aligned}$$

When T-T_0 is small, we can replace T_0 by T in the sums with little change in the result. Then

$$\begin{aligned}
P_{net} &= Ae\sigma \left(T^4 - T_0^4\right) \\
&\approx Ae\sigma \left[T^2 + T^2\right][T + T][T - T_0] \\
&= 4Ae\sigma T^3 \Delta T
\end{aligned} \tag{9.13}$$

This equation implies that net power radiated is approximately proportional to the temperature difference, in agreement with Newton's law of cooling and it is true even when the temperature difference ΔT is small.

This result can also be obtained by computing the differential dP for a small change in temperature dT. We have

$$P_{rad} = Ae\sigma T^4$$

So that

$$dP_{rad} = Ae\sigma(4T^3)dT = 4Ae\sigma T^3 dT$$

9.7 Pressure Exerted by Radiation

Similar to light, thermal radiation also exerts a small but definite pressure on a surface. For normal incident this pressure is equal to the energy density u, i.e., $P = u$. In case of diffuse radiation, where photons move in all directions the resulting pressure can be obtained by applying simple result obtained in kinetic theory of gases. In kinetic theory, the pressure P of a gas of molecules (particles) was $P = \frac{1}{3}nm < v >^2$. Here, we replace $< v >^2$ with c^2, velocity of photon and noting that the photon energy is mc^2, the pressure exerted by the photon gas is

$$P = \frac{1}{3}u \qquad\qquad (9.14)$$

9.8 Classical Thermodynamics of Electromagnetic Radiation

Since, radiation exerts pressure, one can apply the laws of thermodynamics to obtain a few important results belonging to them. Here, we will discuss the thermodynamical properties of electromagnetic radiation confined to an enclosed cavity by following the classical approach. Consider, photons confined to a cavity of volume V and are in thermodynamic equilibrium with the wall of the cavity. We also assume that the cavity wall is diathermal (i.e., it can transmit heat between photons and surrounding). If n (N/V) photons per unit volume are available in the cavity, then the energy density u of the gas may be written as

$$u = \frac{U}{V} = n\hbar\omega \qquad\qquad (9.15)$$

where $\hbar\omega$ denotes the mean energy of the photons. As discussed in the previous section, the pressure exerted by the photon gas is

$$P = \frac{1}{3}u$$

Also from kinetic theory, the photon flux, i.e., number of photons falling unit area of the wall per second is given by

$$\phi = \frac{1}{4}nc \tag{9.16}$$

here c is the speed of light. The power striking per unit area (i.e., power density) of cavity wall, due to the photons, is

$$\mathbb{P} = \hbar\omega\phi = \hbar\omega\frac{1}{4}nc = \frac{1}{4}uc \tag{9.17}$$

We will make use of this relation to develop well known Stefan–Boltzmann law, which relates its temperature and energy flux emanating from a body in the form of radiation. We can develop this using the first law of thermodynamics (dU=TdS-PdV) as below

$$\left(\frac{\partial U}{\partial V}\right)_T = T\left(\frac{\partial S}{\partial V}\right)_T - P = T\left(\frac{\partial P}{\partial T}\right)_V - P \tag{9.18}$$

here, in the last step we have made use of Maxwell's relation. Noting that left side in Eq. 9.18 is energy density, and making use of Eq. 9.14 we obtain

$$u = \frac{1}{3}T\left(\frac{\partial u}{\partial T}\right)_V - \frac{u}{3} \tag{9.19}$$

After rearranging we obtain

$$4u = T\left(\frac{\partial u}{\partial T}\right)_V, \quad \text{giving} \frac{du}{u} = 4\frac{dT}{T} \tag{9.20}$$

while writing the last step we have made use of the fact that we are left with a relation between u and T only, so partial differential can be replaced with direct differential. This equation can be readily integrated to obtain

$$\ln u = 4\ln T + \ln\alpha = \ln aT^4 \tag{9.21}$$

After taking antilog, this yields

$$u = aT^4 \tag{9.22}$$

Here, a is constant of integration, independent of surface. We can make use of Eq. 9.17 to obtain power incident per unit area (i.e., energy flux)

$$\mathbb{P} = \frac{1}{4}uc = \frac{1}{4}aT^4c = \left(\frac{1}{4}ac\right)T^4 = \sigma T^4 \tag{9.23}$$

The term in the brackets, i.e., $\left(\frac{1}{4}ac\right) = \sigma$ is Stefan's constant that is determined from experiment. Equation 9.23 is known as Stefan–Boltzmann law.

9.8.1 Isothermal Expansion of Blackbody Radiation

Again, we consider a cylinder of volume V fitted with a movable piston and having perfectly reflecting walls. Latter ensures that energy exchange between walls and blackbody radiation is prevented. Let it be filled with thermal radiation with energy density u at temperature T. We assume that these radiation are in equilibrium with a small black piece of matter whose thermal capacity is negligible. If we assume the entire system is in equilibrium, we can define radiation temperature as that of the wall of cylinder and black matter inside. One can write equilibrium pressure exerted by radiation in terms of temperature T as

$$P = \frac{1}{3}u = \frac{aT^4}{3} \tag{9.24}$$

Suppose, the entire system is expanded isothermally and reversibly. The amount of heat absorbed from an external source (to keep the temperature of cavity constant) can be evaluated using first TdS equation.

$$TdS = C_V dT + T \left(\frac{\partial P}{\partial T}\right)_V dV \tag{9.25}$$

As the internal energy U = uV = aT^4V, the heat capacity $C_V = \left(\frac{\partial U}{\partial T}\right)_V = 4aT^3V$.

Same way using Eq. 9.24 we obtain $\left(\frac{\partial P}{\partial T}\right)_V = \frac{4}{3}aT^3$. Therefore, for an isothermal expansion of cavity, Eq. 9.25 gives

$$TdS = \Delta Q = \frac{4}{3}aT^4 dV$$

During isothermal expansion, if volume changes from V_i to V_f, the heat exchange is given by

$$Q = \int_{V_i}^{V_f} \Delta Q = \frac{4}{3}aT^4 \left(V_f - V_i\right) \tag{9.26}$$

9.8.2 Adiabatic Expansion of Blackbody Radiation

Again we will consider the same system just described in the previous section. But now, we assume reversible adiabatic expansion of the radiation inside cavity. As cylinder walls are perfectly reflecting, there is no absorption or emission of radiation from walls. The expansion and hence work on surroundings is achieved at the cost of internal energy of radiation. Consequently, the energy density of radiation will drop from some initial value u_1 to u_2. This results in drop in temperature of radiation. Also note that radiation are always in equilibrium with small blackbody placed inside the cavity, its temperature will also fall. In order to evaluate, final temperature inside cavity, we put dS = 0 (dQ = 0, adiabatic process) in TdS equation to obtain

$$4aT^3 V dT + \frac{4}{3}aT^4 dV = 0, \quad or$$

$$4aT^3 V dT = -\frac{4}{3}aT^4 dV$$

$$\left[\frac{dV}{V}\right] = -3\left[\frac{dT}{T}\right]$$

Last equation after integration gives

$$\ln V = -3 \ln T + \ln C \quad or$$

$$[VT^3]_S = \text{constant} \tag{9.27}$$

This equation tells us about the change in temperature when radiation expand adiabatically. For instance, if volume changes eight times corresponding decrease in temperature will be one-half initial value of temperature. Note that radiation does not change due to adiabatic expansion (or compression), it is only the temperature and corresponding energy density that is reduced (increased) at the end of process.

9.9   Wien's Displacement Law

As we discussed earlier, an object that absorbs all the radiation incident upon it has an emissivity equal to 1 and is called a blackbody. A blackbody is also an ideal radiator. Similar to a blackbody, other objects at non-zero temperature emit radiation whose wavelength depend upon temperature of the object. Most of the radiation emitted by an object at temperatures below about 600 °C is concentrated at wavelengths much longer than those of visible light. As an object is heated, the rate of energy emission increases, and the energy radiated extends to shorter and shorter wavelengths. Between about 600 and 700° C, enough of the radiated energy is in the visible spectrum for the body to glow a dull red. Figure 9.5 shows the power radiated (energy per second) by a blackbody as a function of wavelength for several different temperatures. The wavelength at which the power is a maximum varies inversely with the temperature.

In other words, **Wien's shows that the product of the wavelength λ_m corresponding to maximum power (or energy radiated/s) and the absolute temperature T is constant**. This result is known as **Wien's displacement law**. Mathematically, it is written as

$$\lambda_m T = 2.898 mm K = constant \tag{9.28}$$

Fig. 9.5 Radiated power,
i.e., energy radiated per
second versus wavelength
for radiation emitted by a
blackbody. The wavelength
of the maximum power
varies inversely with the
absolute temperature of the
blackbody

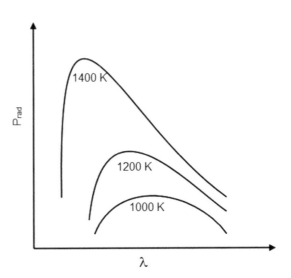

Wien's displacement law is used to determine the temperatures of stars from analyses of their radiation. It can also be used to map the variation in temperature over different regions of the surfaces of an object. Such a map is called a thermograph.

Wien's further extended the theory to explain the observed features of the black-body spectrum. According to Wien's distribution law, the $u_\lambda d\lambda$, the amount of energy du emitted by a blackbody at temperature T and associated with spectral region lying between λ and $\lambda + d\lambda$ is given by

$$u_\lambda d\lambda = K\lambda^{-5} e^{-b/\lambda T} d\lambda \qquad (9.29)$$

This equation is known as Wien's distribution law. However, it should be noted that it holds only in the region of shorter wavelength at lower temperatures. The law does not hold at higher wavelengths and higher temperatures.

9.10 Rayleigh–Jeans Law

As mentioned in the previous section, Wien's law does not comply with experimental results at longer wavelengths and high temperatures. These limitations of the thermodynamics stimulated Lord Rayleigh to look for an alternative explanation for blackbody radiation spectra. He analysed the problem by applying the principle of equipartition of energy to the electromagnetic (thermal) radiation inside a cavity. The numerical error he made was subsequently corrected by Jeans in 1906. To begin with, Rayleigh considered that blackbody radiation inside a cavity suffer multiple reflections and give rise to the formation of standing wave. The next problem was to determine the total number of modes corresponding to the standing waves inside cavity. To determine the total number of modes, we consider radiation enclosed in a cubic cavity of side L and volume V. To make the problem simpler, we consider the formation of standing waves in one dimension and extend the problem into three dimensions. We recall that wave equation in one dimension is given by

$$\frac{\partial^2 v}{\partial^2 t^2} = c^2 \frac{\partial^2 v}{\partial^2 x^2} \tag{9.30}$$

Here, x is direction of propagation of wave and v is the displacement perpendicular to x. In three dimensions v can be written as v(x, y, z). Note that here v is not the velocity, rather it represents vertical displacement of wave perpendicular to the direction of propagation. Here, c represents the wave velocity (here thermal radiation, hence velocity of light). Note that in a one-dimensional cavity extending from x = 0 to x = L, these points will act as nodal points. The wave equation for radiation inside cavity is

$$\frac{\partial^2 v}{\partial^2 x^2} + \frac{\partial^2 v}{\partial^2 y^2} + \frac{\partial^2 v}{\partial^2 z^2} = \frac{1}{c^2} \frac{\partial^2 v}{\partial^2 t^2} \tag{9.31}$$

Solutions to this equation are of the type

$$v(x, y, z) = A e^{-i\omega t} \sin\left(\frac{n_x \pi x}{L}\right) \sin\left(\frac{n_y \pi y}{L}\right) \sin\left(\frac{n_z \pi z}{L}\right) \tag{9.32}$$

Here, n_x, n_y, n_z are integers subjected to the condition that v(x,y,z) vanishes at x = y = z = 0 and x = y = z = L and ω is the angular frequency.

Now, v(x,y,z) has to satisfy Eq. 9.31. Putting the value of v(x,y,z) and after simplifications, we obtain relation between n_x, n_y, n_z and ω.

$$n_x^2 + n_y^2 + n_z^2 = \frac{\omega^2 L^2}{\pi^2 c^2} = \left(\frac{2vL}{c}\right)^2 = \left(\frac{2L}{\lambda}\right)^2 \tag{9.33}$$

Here, we have used $\lambda = c/v$ as the wavelength of standing wave inside cavity. Above equation gives the allowed frequencies or modes that are subjected to the condition that n_x, n_y and $n_z > 0$. This equation represents an equation of sphere whose radius is $n = 2L/\lambda$ and written as

$$n_x^2 + n_y^2 + n_z^2 = n^2 \tag{9.34}$$

The fact that n_x, n_y and $n_z > 0$ and each of these integers changes by one unit, implies that only 1/8 of the total volume of the sphere gives the possible number of modes inside cavity. In other words, only the positive octant will determine the allowed modes inside cavity. Therefore, the allowed number of modes is

$$N = \frac{1}{8}\left[\frac{4}{3}\pi n^3\right] = \frac{1}{8}\left[\frac{4}{3}\pi \left(\frac{2L}{\lambda}\right)^3\right] = \frac{4\pi L^3}{3 \lambda^3}$$

Therefore, the number of modes in the wavelength range between λ and $\lambda + d\lambda$ is given by

$$N_\lambda d\lambda = \frac{4}{3}\pi V \left|\left(-\frac{3}{\lambda^4}\right)\right| d\lambda = \frac{4\pi V}{\lambda^4}d\lambda \tag{9.35}$$

Recall that thermal radiation are electromagnetic (transverse) in nature and have two polarization states for a given wavevector $k = 2\pi/\lambda$. Therefore, the total number of modes in a cavity of volume V is

$$N_\lambda d\lambda = \frac{8\pi V}{\lambda^4}d\lambda \tag{9.36}$$

Since, $\lambda = c/v$, implying that $d\lambda = -c/v^2 dv$. After substituting the value of λ and $d\lambda$ in Eq. 9.36, we obtain an expression for number of modes in the frequency range v to $v + dv$ as below

$$N_v dv = \frac{8\pi V}{c^3}v^2 dv \tag{9.37}$$

Rayleigh and Jeans further made use of principle of equipartition of energy and assigned an energy equal to kT to each mode of vibration. Therefore, total energy inside a cavity in the frequency ranging from v to $v + dv$ is

$$E_v dv = \frac{8\pi V k_B T}{c^3}v^2 dv \tag{9.38}$$

so that the energy density in the frequency ranging from v to $v + dv$ is given by

$$u_v dv = \frac{8\pi k_B T}{c^3}v^2 dv \tag{9.39}$$

In terms of wavelength, we can rewrite the previous equation (using $v = c/\lambda$) as

$$u_\lambda d\lambda = u_v dv = \frac{8\pi k_B T}{\lambda^4}d\lambda \tag{9.40}$$

This equation gives the Rayleigh–Jeans law. Like Wien's law, this law was capable of reproducing the experimentally observed blackbody spectra only at higher wavelengths and higher temperature. The law failed at lower wavelengths and temperatures [Fig. 9.6a].

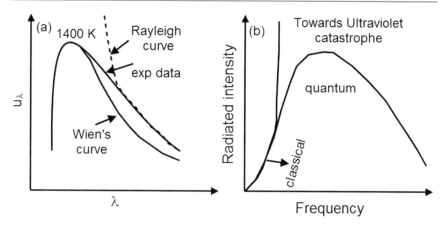

Fig. 9.6 **a** Comparing Wien's law and Rayleigh–Jeans law with blackbody spectra **b** qualitative depiction of ultraviolet catastrophe

9.10.1 Ultraviolet Catastrophe

As discussed above, the Rayleigh law agree well with emission spectra at lower frequencies but failed in agreement at higher frequencies. Equation 9.40 can be written in terms of frequency giving energy density as

$$du = \int u_\lambda d\lambda = \frac{8\pi k_B T}{c^4} \times v^4 \times \frac{c}{v^2} dv = \frac{8\pi v^2 k_B T}{c^3} dv \qquad (9.41)$$

Here, we have used $\lambda = \frac{c}{v}$, $d\lambda = \frac{c}{v^2} dv$ This equation implies that energy density will continuously increase with increase in frequency and approaches infinity as v approaches infinity. Also at any temperature T, total energy density is given by

$$u = \int_0^\infty \frac{8\pi v^2 k_B T}{c^3} dv \qquad (9.42)$$

As we see, this also approaches infinity. A comparison between Wien's distribution and Rayleigh–Jeans law is given in Fig. 9.6b. The discrepancy in observed spectrum and predicted increase in energy density at higher frequency is known as Ultraviolet catastrophe. This was one of the major failures of classical physics. The problem was later resolved by Planck, who came up with radiation law that considered the discrete nature of radiation.

9.11   Planck's Law

Recall that Rayleigh–Jeans law is based on equipartition of energy, which considers energy as infinitely divisible continuous variable. But, as we noted, the final obtained result for energy density was not able to address the experimental data over full range of wavelength. To tackle this issue, Planck imagined that blackbody cavity not only contains the radiation but also molecules of a perfect gas, which were capable of exchanging energy with radiation through matter-radiation interaction. These molecules act like resonators (oscillators) after absorbing energy from radiation and may transfer it to other molecules when they collide with them. This way equilibrium can be maintained inside the cavity. With these ideas Planck came up with the following points:

(i) Energy exchange between matter (cavity wall or resonators) and radiation (inside cavity) could take place only in bundles (quantum) of certain minimum size.

(ii) The quantum of energy exchange is directly proportional to frequency. That is energy of an oscillator having frequency v is given by hv, where h is known as Planck's constant. These oscillators cannot radiate or absorb energy continuously, but energy is absorbed or emitted in the form of packets known as quanta.

With these points in hand, the energy density inside a cavity can be obtained if the number of available modes and energy associated with a single resonator or oscillator is known. Assuming a cubic cavity of volume V, total number of modes in the frequency range v to $v + dv$ will be

$$N_v dv = \frac{8\pi V}{c^3} v^2 dv \qquad (9.43)$$

Next, if we assume that there are N oscillators (molecules) inside cavity and U be their total energy, so that average energy per oscillator is U/N. Let, N oscillators are further subdivided into $N_0, N_1, N_2, ...N_r...$ various groups of resonator with energies $0, \epsilon, 2\epsilon, 3\epsilon, ...r\epsilon, ...$, respectively. Then, according to Maxwell's law of molecular motion,

$$
\begin{aligned}
N &= N_0 + N_1 + N_2 + ... + N_r + ... \\
&= N_0 + N_0 e^{-\epsilon/k_B T} + N_0 e^{-2\epsilon/k_B T} + ... + N_0 e^{-r\epsilon/k_B T} + .. \\
&= N_0 \left[1 + e^{-\epsilon/k_B T} + e^{-2\epsilon/k_B T} + .. e^{-r\epsilon/k_B T} + .. \right] \\
&= \frac{N_0}{1 - e^{-\epsilon/k_B T}} \qquad (9.44)
\end{aligned}
$$

Similarly, total energy of Planck's resonators can be found as below

$$
\begin{aligned}
U &= 0 \times N_0 + \epsilon \times N_1 + 2\epsilon \times N_2 + \ldots + r\epsilon \times N_r + \ldots \\
&= 0 + \epsilon N_0 e^{-\epsilon/k_B T} + 2\epsilon N_0 e^{-2\epsilon/k_B T} + \ldots + r\epsilon N_0 e^{-r\epsilon/k_B T} + \ldots \\
&= N_0 \epsilon \left[e^{-\epsilon/k_B T} + \ldots + r0 e^{-r\epsilon/k_B T} + \ldots \right] \\
&= N_0 \epsilon \frac{e^{-\epsilon/k_B T}}{\left(1 - e^{-\epsilon/k_B T}\right)^2}
\end{aligned}
\tag{9.45}
$$

Therefore, average energy per oscillator comes out to be

$$
\bar{u} = \frac{U}{N} = \frac{N_0 \epsilon \dfrac{e^{-\epsilon/k_B T}}{\left(1 - e^{-\epsilon/k_B T}\right)^2}}{\dfrac{N_0}{1 - e^{-\epsilon/k_B T}}} = \frac{\epsilon}{e^{\epsilon/k_B T} - 1}
\tag{9.46}
$$

Hence, energy density in the frequency range v to $v + dv$ can be written as

$$
\begin{aligned}
u_v dv = n_v dv &= \left[\frac{8\pi h v^2}{c^3} dv \right] \left[\frac{hv}{e^{hv/k_B T} - 1} \right] \\
&= \frac{8\pi h v^3}{c^3} \frac{1}{e^{hv/k_B T} - 1} dv
\end{aligned}
\tag{9.47}
$$

This equation is known as Planck's radiation law. This can be changed in terms of wavelength by making use of $v = c/\lambda$ and $dv = |- \frac{c}{\lambda^2} d\lambda|$ we obtain

Planck's law in terms of wavelength reduces to

$$
\begin{aligned}
u_\lambda d\lambda = u_v dv \\
&= \frac{8\pi h}{c^3} \left[\left(\frac{c}{\lambda}\right)^3 \left(\frac{1}{e^{\frac{hc}{\lambda k_B T}} - 1} \right) \right] \left[-\frac{c}{\lambda^2} d\lambda \right] \\
&= \frac{8\pi hc}{\lambda^5} \left[\frac{1}{e^{\frac{hc}{\lambda k_B T}} - 1} \right] d\lambda
\end{aligned}
\tag{9.48}
$$

This equation reproduces essential features of blackbody spectrum in entire range of wavelengths. It represents a complete law and therefore the idea of discrete energy (quantum of energy) came into final form. In fact this concept led to the developments in photoelectricity, spectral emission and gave rise to the new field of quantum mechanics.

9.11.1 Deduction of Rayleigh–Jeans Law

In previous section, we concluded that Planck's law provides us a general description of blackbody radiation. This law can be used to deduce all other laws pertaining to blackbody radiation spectrum. In the region where $\lambda >> hc/k_B T$, the exponential term is Eq. 9.48 can be approximated as

$$e^{\frac{hc}{\lambda k_B T}} = 1 + \frac{hc}{\lambda k_B T} + \ldots$$

so that the denominator reduces to

$$e^{\frac{hc}{\lambda k_B T}} - 1 = \frac{hc}{\lambda k_B T}$$

Therefore, for longer wavelength ($\lambda >> hc/k_B T$), the energy density in the range λ to $\lambda + d\lambda$ is given by

$$u_\lambda d\lambda = \frac{8\pi hc}{\lambda^5} \times \left[\frac{\lambda k_B T}{hc} \right] d\lambda = \frac{8\pi k_B T}{\lambda^4} d\lambda \qquad (9.49)$$

This equation represents the Rayleigh–Jeans law. Therefore, for longer wavelengths, Rayleigh's law holds.

9.11.2 Deduction of Wien's Distribution Law

For smaller wavelength, i.e., ($\lambda << hc/k_B T$), the exponential term in Eq. 9.48 will be much greater than unity. Therefore, ignoring one in the denominator, Eq. 9.48 reduces to

$$u_\lambda d\lambda = \frac{8\pi hc}{\lambda^5} e^{-\left(\frac{hc}{\lambda k_B T} \right)} d\lambda \qquad (9.50)$$

This equation represents Wien's distribution law. Therefore, for shorter wavelengths, Wien's law holds. We can therefore say that Wien's law and Rayleigh's law are a special case of Planck's law.

9.11.3 Deduction of Wien's Displacement Law

Refering to Eq. 9.48, Planck's law in terms of wavelength is

$$u_\lambda d\lambda = \frac{8\pi hc}{\lambda^5} \left[\frac{1}{e^{\frac{hc}{\lambda k_B T}} - 1} \right] d\lambda \qquad (9.51)$$

This equation gives the energy density due to radiation which have wavelength between λ and $\lambda + d\lambda$. The wavelength ($\lambda = \lambda_m$) at which this function acquires a maximum value can be obtained from the above equation as below

$$\left[\frac{\partial u_\lambda}{\partial \lambda}\right]_{\lambda_m} = 0 \tag{9.52}$$

Therefore, we obtain

$$\frac{\partial u_\lambda}{\partial \lambda} = \frac{1}{e^{hc/\lambda k_B T} - 1} \times 8\pi hc \left(\frac{-5}{\lambda^6}\right) + \frac{8\pi hc}{\lambda^5} \times \frac{\frac{hc}{\lambda^2 k_B T} e^{hc/\lambda k_B T}}{\left(e^{hc/\lambda k_B T} - 1\right)^2}$$

$$\dot{} = -\frac{40\pi hc}{\lambda^6} \times \frac{1}{e^{hc/\lambda k_B T} - 1} + \frac{8\pi hc}{\lambda^5} \times \frac{hc}{\lambda^2 k_B T} \times \frac{e^{hc/\lambda k_B T}}{\left(e^{hc/\lambda k_B T} - 1\right)^2}$$

Applying the condition Eq. 9.52 and rearranging the terms in above equation, we obtain

$$\frac{8\pi hc}{\lambda^6 \left(e^{hc/\lambda k_B T} - 1\right)} \left[-5 + \frac{hc}{\lambda k_B T} \times \frac{e^{hc/\lambda k_B T}}{e^{hc/\lambda k_B T} - 1}\right] = 0$$

OR

$$-5 + \frac{hc}{\lambda k_B T} \times \frac{e^{hc/\lambda k_B T}}{e^{hc/\lambda k_B T} - 1} = 0 \tag{9.53}$$

Let us put $hc/\lambda k_B T = x$, in Eq. 9.53 so that we obtain

$$-5 + \frac{xe^x}{e^x - 1} = 0$$

$$OR$$

$$\frac{xe^x}{e^x - 1} = 5 \tag{9.54}$$

Equation 9.54 represents a transcendental equation and can be solved numerically. This way a root in the neighbourhood of 5 is expected. Applying the method of approximation, one can obtain a value of x = 4.965. Therefore, we can write

$$x = \frac{hc}{\lambda k_B T} = 4.965$$

Here, we can write $\lambda = \lambda_m$, which represents the wavelength at which energy density (u_λ) is maximum. Therefore, in terms of λ_m we have

$$\frac{hc}{\lambda_m k_B T} = 4.965$$

OR

$$\lambda_m T = \frac{hc}{4.965 k_B} = constant \qquad (9.55)$$

This equation represents Wien's displacement law.

9.11.4 Deduction of Stefan's Law

Again, according to Planck's law, the energy density due to radiation which have wavelength between λ and $\lambda + d\lambda$ is

$$u_\lambda d\lambda = \frac{8\pi hc}{\lambda^5} \left[\frac{1}{e^{\frac{hc}{\lambda k_B T}} - 1} \right] d\lambda \qquad (9.56)$$

Therefore, total energy density for radiation of wavelength ranging from 0 to ∞ is

$$u(T) = \int_0^\infty u_\lambda d\lambda$$
$$= 8\pi hc \int_0^\infty \frac{1}{\lambda^5 \left[e^{hc/\lambda k_B T} - 1 \right]} \qquad (9.57)$$

To evaluate this integral, we put $x = hc/\lambda k_B T$, thereby giving $\lambda = hc/x k_B T$ and $d\lambda = -(hc/x^2 k_B T)dx$. With these changes the limit of integration changes to $-\infty$ to 0. Therefore, we obtain

$$u(T) = 8\pi hc \int_{-\infty}^0 \frac{\left[-\dfrac{hc}{x^2 k_B T} \right] dx}{\left[\dfrac{hc}{x k_B T} \right]^5 [e^x - 1]}$$
$$= \frac{8\pi k_B^4 T^4}{c^3 h^3} \int_0^\infty \frac{x^3 dx}{e^x - 1}$$
$$= \frac{8\pi^5 k_B^4}{15 c^3 h^3} T^4 = a T^4 \qquad (9.58)$$

where $a = \dfrac{8\pi^5 k_B^4}{15 c^3 h^3}$. Since, $ac/4 = \sigma$, where σ is Stefan's constant. We finally write

$$u(T) = \sigma T^4 \qquad (9.59)$$

Here, $\sigma = \dfrac{2\pi^5 k_B^4}{15 c^2 h^3}$ is Stefan's constant. Equation 9.59 gives Stefan's law. Hence, Planck's law is complete in itself and reproduces main features of blackbody spectra.

9.12   Solved Problems

Q.1 **The total power emitted by a spherical blackbody of radius R at a temperature T is P_1. Let P_2 be the total power emitted by another spherical blackbody of radius $R/2$ kept at temperature 2T. The ratio, P_1/P_2 is ....(Give your answer up to two decimal places).**

[GATE 2016]

Sol: The power density is

$$P \propto AT^4$$

Therefore,

$$\frac{P_1}{P_2} = \frac{R_1^2 T_1^4}{R_2^2 T_2^4}$$

$$= \frac{R^2 T^4}{\left(\frac{R}{2}\right)^2 (2T)^4} = \frac{4}{16} = 0.25$$

Q.2 **Calculate the number of modes in a cavity of volume 55 c.c. in the frequency range from 4×10^{14} and 4.01×10^{14} s^{-1}.**

Sol: The number of modes per c.c.

$$N_\nu d\nu = \frac{8\pi \nu^2}{c^3} d\nu$$

$$= \frac{8 \times 3.14 \times (4 \times 10^{14})^2}{(3 \times 10^{10})^3} \cdot (4.01 - 4) \times 10^{14}$$

$$= 7.5 \times 10^{12}$$

Q.3 **Calculate the number of modes per unit volume in cavity of volume 100 c.c. in the wavelength range 500 nm to 500,2 nm.**

Sol: Number of modes per unit volume in wavelength range between λ and $\lambda + d\lambda$ is

$$n(\lambda) = \frac{8\pi}{\lambda^4} d\lambda$$

$$= \frac{8 \times 3.14}{(5 \times 10^{-5}\,\text{cm})^4} \cdot (2 \times 10^{-8}\,\text{cm})$$

$$= 8 \times 10^{10}$$

Therefore, number of modes in volume 100 c.c. is $n(\lambda)V = 8 \times 10^{12}$.

Q.4 A spherical blackbody with radius of 4 cm is maintained at 227 K. Evaluate the power radiated by it.

Sol: Power radiated is

$$P_{rad} = A\sigma T^4$$
$$= 4\pi(16 \times 10^{-4}\,\text{m}^2)(5.67 \times 10^8\,\text{Wm}^{-2}\text{K}^{-4} \times (500\,\text{K})^4)$$
$$\approx 71W$$

Q.5 In Q.4 above, what is the wavelength at which maximum energy is radiated?

Sol: Using Wien's displacement law

$$\lambda_m = \frac{2898 \times 10^{-6}\,\text{mK}}{500\,\text{K}} = 5.79 \times 10^{-6} m$$

9.13   Multiple Choice Questions

Q.1 Which one of the following statements is NOT TRUE about thermal radiation ?

(A)   All bodies emit thermal radiation at all temperatures
(B)   thermal radiation are electromagnetic waves
(C)   thermal radiation are not reflected by from a mirror
(D)   thermal radiation travels in free space with speed of light

Q.2 The amount of energy radiated by a body depends upon

(A)   nature of its surface   (C)   temperature of the surface
(B)   area of the surface   (D)   all of the above

Q.3 The wavelength of radiation emitted by a body depends upon

(A)   nature of its surface   (C)   temperature of the surface
(B)   area of the surface   (D)   all of the above

Q.4 A solid sphere and a hollow sphere of same material and size are heated to same temperature and allowed to cool in the same surroundings. If the temperature difference between the surroundings and each sphere is T, then

(A)   the hollow sphere will cool at a faster rate for all values of T
(B)   the solid sphere will cool at a faster rate for all values of T
(C)   both spheres will cool at a faster rate for all values of T
(D)   both spheres will cool at a faster rate only for small values of T

Q.5 **An ideal blackbody at room temperature is thrown into a furnace. it is observed that**

(A)   initially it is the darkest body and at later times the brightest
(B)   it is the darkest body at all times
(C)   it cannot be distinguished at all
(D)   initially it is the darkest body and at later times it cannot be distinguished

Q.6 **A sphere, a cube and a thin circular plate have the same mass and are made from the same material. All of them are heated to the same temperature. The rate of cooling is**

(A)   the maximum for the sphere and minimum for the plate
(B)   the maximum for the sphere and minimum for the cube
(C)   the maximum for the plate and minimum for the sphere
(D)   same for all

Q.7 **In which of the following processes, is the heat transfer primarily via radiation?**

(A)   boiling of water
(B)   land and sea breezes
(C)   heating of a metal rod placed over flame
(D)   heating of the glass surface of an electric bulb due to current in its filament

Q.8 **A spherical body of emissivity ϵ, placed inside a perfectly blackbody (emissivity = 1), is maintained at absolute temperature T. The energy radiated by a unit area of the body per second will be (σ be Stefan's constant).**

(A) σT^4 (C) $(1-\epsilon)\sigma T^4$
(B) $\epsilon\sigma T^4$ (D) $(1+\epsilon)\sigma T^4$

Q.9 **When the temperature of a blackbody is doubled, the maximum value of its spectral energy density, with respect to that at initial temperature, would become**

[JAM 2012]

(A) $\dfrac{1}{16}$ times

(B) 8 times

(C) 16 times

(D) 32 times

Q.10 **A blackbody at temperature T emits radiation at a peak wavelength λ. If the temperature of the blackbody becomes 4T, the new peak wavelength is**

[JAM 2013]

(A) $\dfrac{\lambda}{256}$

(B) $\dfrac{\lambda}{64}$

(C) $\dfrac{\lambda}{16}$

(D) $\dfrac{\lambda}{4}$

Q.11 **Wien's displacement law is**

(A) $\lambda_m T = \text{constant}$

(B) $\dfrac{\lambda_m}{T} = \text{constant}$

(C) $\dfrac{T}{\lambda_m} = \text{constant}$

(D) $\lambda_m^2 T = \text{constant}$

Q.12 **Stefan's law for a perfect blackbody is (here u is the energy density)**

(A) $u = \sigma T^4$

(B) $u = \sigma(T^4 - T_0^4)$

(C) $u = \sigma(T^4 - T_0)$

(D) $u = e\sigma(T^4 - T_0^4)$

Q.13 **A blackbody radiation cavity is made to undergo an adiabatic change at a pressure of $1.5 \times 10^{-2} Nm^{-2}$. What will be the final pressure in cavity if its volume is increased by 8 times?**

(A) $1.5 \times 10^{-2}\,\text{Nm}^{-2}$

(B) $3 \times 10^{-2}\,\text{Nm}^{-2}$

(C) $15 \times 10^{-2}\,\text{Nm}^{-2}$

(D) $24 \times 10^{-2}\,\text{Nm}^{-2}$

Q.14 **The velocity of thermal radiation in vacuum is**

(A) equal to that of light

(B) less than velocity of light

(C) greater than that of light

(D) equal to that of sound

Q.15 **If the absolute temperature of a blackbody is increased to twice its value, the rate of emission of energy per unit area will become**

(A) 2 times

(B) 4 times

(C) 8 times

(D) 16 times

Q.16 Radiation is passing through a transparent medium, then

  (A)   temperature of medium increases
  (B)   temperature of medium decreases
  (C)   temperature of medium does not change
  (D)   none of these

Q.17 A hot body will radiate heat most rapidly if its surface is

  (A)   white and polished   **(C)**   black and polished
  (B)   white and rough   **(D)**   black and rough

Q.18 Good absorbers of heat are

  (A)   poor emitters   **(C)**   good emitters
  (B)   non-emitters   **(D)**   highly polished

Q.19 A blackbody emits

  (A)   line spectrum   **(C)**   continuous spectrum
  (B)   band spectrum   **(D)**   mixed spectrum

Q.20 Two bodies are at temperature $27\,^\circ$C and $927\,^\circ$C. The heat energy radiated by them will be in the ratio

  (A)   1:4   **(C)**   1:64
  (B)   1:16   **(D)**   1:256

Q.21 A polished metal plate with a rough black spot on it is heated to about 1400 K and quickly taken into a dark room. Which one of the following statements will be true?.

  (A)   the spot will appear brighter than the plate
  (B)   the spot will appear darker than the plate
  (C)   the spot and plate will appear equally bright
  (D)   the spot and plate will not be visible in the dark room

Q.22 Electromagnetic radiation is emitted

  (A)   only by radio and TV transmitting antennas
  (B)   only by bodies at temperature higher than surrounding
  (C)   only by red hot bodies
  (D)   by all bodies at all temperatures

Q.23 **In the radiation emitted by a blackbody, the ratio of the spectral densities at frequencies 2 ν and ν will vary with ν as (k is Boltzmann constant):**

[JAM 2017]

(A) $\left[e^{h\nu/kT} - 1 \right]^{-1}$ (C) $\left[e^{h\nu/kT} - 1 \right]$

(B) $\left[e^{h\nu/kT} + 1 \right]^{-1}$ (D) $\left[e^{h\nu/kT} + 1 \right]$

Q.24 **A red star having radius R_R at a temperature T_R and a white star having radius R_W at a T_W temperature, radiate the same total power. If these stars radiate as perfect blackbodies, then**

[JAM 2019]

(A) $R_R > R_W$ and $T_R > T_W$ (C) $R_R > R_W$ and $T_R < T_W$
(B) $R_R < R_W$ and $T_R > T_W$ (D) $R_R < R_W$ and $T_R < T_W$

Q.25 **The blackbody at a temperature of 6000 K emits a radiation whose intensity spectrum peaks at 600 nm. If the temperature is reduced to 300 K, the spectrum will peak at**

[JEST 2015]

(A) 120μm   (C) 22mm
(B) 12μm   (D) 120mm

Q.26 **For blackbody radiation in a cavity, photons are created and annihilated freely as a result of emission and absorption by the walls of the cavity. This is because**

[GATE 2015]

(A)   the chemical potential of the photons is zero
(B)   photons obey Pauli exclusion principle
(C)   photons are spin-1 particles
(D)   the entropy of the photons is very large

Q.27 **For an energy state E of a photon gas, the density of states is proportional to**

(A) E (C) E^2
(B) $\sqrt{E}$ (D) $E^{2/3}$

Q.28 **"Good absorbers are good emitters"—The statement is called:**

(A)   Prevost's law   (C)   Stefan's law
(B)   Kirchoff's law   (D)   Wien's law

Q.29 **For a perfect blackbody, the absorptive power is**

(A)   1   (C)   0
(B)   0.5   (D)   ∞

Q.30 **Two solid spheres A and B have the same emissivity. The radius of A is four times the radius of Band temperature of A is twice the temperature of B. The ratio of the rate of heat radiated from A to that from B is ....**
                                                                        [GATE 2018]

Q.31 **A cavity contains blackbody radiation in equilibrium at temperature T. The specific heat per unit volume of the photon gas in the cavity is of the form**

$$C_V = \gamma T^3$$

where γ is a constant. The cavity is expanded to twice its original volume and then allowed to equilibrate at the same temperature T. The new internal energy per unit volume is
                                                                  [NET-JRF June-2011]

(A) $4\gamma T^4$ (C) γT^4
(B) $2\gamma T^4$ (D) $\dfrac{\gamma T^4}{4}$

Q.32 **Consider blackbody radiation contained in a cavity whose walls are at temperature T. The radiation is in equilibrium with the walls of the cavity. If the temperature of the walls is increased to 2T and the radiation is allowed to come to equilibrium at the new temperature, the entropy of the radiation increases by a factor of**
                                                                  [NET-JRF June-2012]

(A)   2   (C)   8
(B)   4   (D)   16

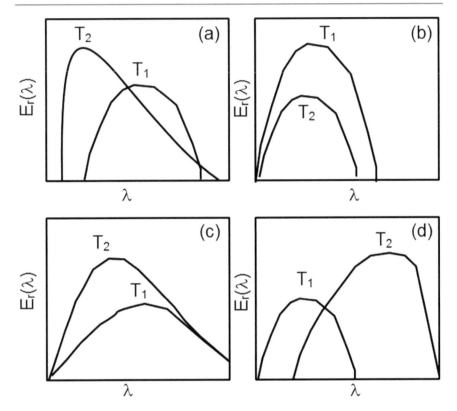

Fig. 9.7 Options for Q.33

Q.33 **Which of the graphs below in Fig. 9.7, gives the correct qualitative behaviour of the energy density $E_r(\lambda)$ of blackbody radiation of wavelength λ at two temperatures T_1 and T_2 s.t $T_2 > T_1$.**

[NET-JRF June-2014]

Q.34 **A gas of photons inside a cavity of volume V is in equilibrium at temperature T. If the temperature of the cavity is changed to 2T, the radiation pressure will change by a factor of**

[NET-JRF June-2017]

(A) 2   (C) 8
(B) 16   (D) 4

Q.35 **The maximum intensity of solar radiation is at the wavelength of $\lambda \approx$ 5000Å and corresponds to its surface temperature $T_{Sun} = 10^4 K$. If the wavelength of the maximum intensity of an X-ray star is 5Å, its surface temperature is of the order of**

[NET-JRF June-2018]

(A) $10^{16} K$ (C) $10^{10} K$

(B) $10^{14} K$ (D) $10^7 K$

Q.36 **The tungsten filament of an electric lamp has a surface area A and a power rating P. If the emissivity of the filament is ϵ and σ is Stefan's constant, the steady temperature of the filament will be**

(A) $T = \left(\dfrac{P}{A\epsilon\sigma}\right)^2$ (C) $T = \left(\dfrac{P}{A\epsilon\sigma}\right)^{1/2}$

(B) $T = \left(\dfrac{P}{A\epsilon\sigma}\right)$ (D) $T = \left(\dfrac{P}{A\epsilon\sigma}\right)^{1/4}$

Q.37 **If the temperature of a blackbody is increased from $7°C$ to $287°C$, then the rate of energy radiated increases by**

(A) $\left(\dfrac{287}{7}\right)^4$ (C) 4

(B) 16   (D) 2

Q.38 **Two metallic spheres S_1 and S_2 are made from the same material and have identical surface finish. The mass of sphere S_1 is three times that of S_2. Both spheres are heated to the same high temperature and placed in the same room having low temperature but are thermally insulated from each other. The ratio of the initial rate of cooling of S_1 to that of S_2 is**

(A) $\dfrac{1}{3}$ (C) $\dfrac{\sqrt{3}}{1}$

(B) $\dfrac{1}{\sqrt{3}}$ (D) $\left(\dfrac{1}{3}\right)^{1/3}$

Q.39 **A spherical blackbody of radius 12 cm radiates 450 W power at 500 K. If radius of sphere is halved and temperature is doubled, the power radiated in Watt would be.....**

(A) 225   (C) 900

(B) 450   (D) 1800

Q.40 **A blackbody is at temperature of 2880 K. The energy of radiation emitted by this body between wavelengths 499 nm and 500 nm is U_1, between 900 nm and 1000 nm is U_2 and between 1499 nm and 1500 nm is U_3. Given Wien's constant $b = 2.88 \times 10^6 nmK$, then**

(A) $U_1 = 0$ (C) $U_1 > U_2$
(B) $U_3 = 0$ (D) $U_2 > U_1$

Q.41 **A cavity contains blackbody radiation in equilibrium at temperature T.**
The specific heat per unit volume of the photon gas in the cavity is of the
form $C_v = \gamma T^3$ where, γ is a constant. The cavity is expanded to twice its
original volume and then allowed to equilibrate at the same temperature
T. The new internal energy per unit volume is

[JRF-NET June 2011]

(A) $4\gamma T^4$ (C) γT^4

(B) $2\gamma T^4$ (D) $\dfrac{\gamma T^4}{4}$

Q.42 **The Helmholtz free energy for a photon gas is given by $A = -\dfrac{a}{3}VT^4$,**
where a is a constant. The pressure of the photon gas is?

(A) $-\dfrac{a}{3}T^4$ (C) $-\dfrac{4a}{3}VT^3$

(B) $\dfrac{a}{3}T^4$ (D) $\dfrac{4a}{3}VT^3$

Q.43 **The Helmholtz free energy for a photon gas is given by $A = -\dfrac{a}{3}VT^4$,**
where a is a constant. The entropy S of the photon gas is?

(A) $-\dfrac{a}{3}T^4$ (C) $-\dfrac{4}{3}aVT^3$

(B) $\dfrac{a}{3}T^4$ (D) $\dfrac{4a}{3}VT^3$

Q.44 **The Helmholtz free energy for a photon gas is given by $A = -\dfrac{a}{3}VT^4$,**
where a is a constant. The chemical potential μ of the photon gas is ?

(A)   0   (C) $\dfrac{a}{3}VT^3$

(B) $\dfrac{4}{3}aVT^3$ (D) aVT^4

Q.45 **The entropy of a photon gas is proportional to**

(A)   T   (C) T^3
(B) T^2 (D) T^5

Q.46 **Consider a radiation cavity of volume V at temperature T. Specific heat** C_V **for photon gas is proportional to**

(A)   T   (C) T^3

(B) T^2 (D) T^5

Q.47 **Consider a radiation cavity of volume V at temperature T. The average number of photons in equilibrium inside the cavity is proportional to**

(A)   T   (C) T^3

(B) T^2 (D) T^4

Q.48 **Consider a radiation cavity of volume V at temperature T. The density of states at energy E of the quantized radiation (photons) is**

(A) $\dfrac{8\pi V}{c^3 h^3} E^2$ (C) $\dfrac{8\pi V}{c^2 h^3} E$

(B) $\dfrac{8\pi V}{c^3 h^3} E^{3/2}$ (D) $\dfrac{8\pi V}{c^2 h^3} E^{1/2}$

Q.49 **The density of available photon states in the energy interval E to E+dE in a cubical cavity of volume** $V = l^3$ **are**

(A) $\dfrac{8\pi V}{ch}$ (C) $\dfrac{8\pi V E^2}{c^3 h^3}$

(B) $\dfrac{8\pi V E^2}{ch}$ (D) $\dfrac{8\pi V E^3}{c^2 h^2}$

Q.50 **A radiation cavity of volume V at temperature T has total energy U. The energy density in the cavity is proportional to**

(A)   T   (C) T^3

(B) T^2 (D) T^4

Q.51 **A radiation cavity of volume V at temperature T has total energy U. Let P be the pressure exerted by the radiation. Then which one of the following is true for radiation inside the cavity?**

(A) $P = \dfrac{2}{3}\dfrac{U}{V}$ (C) $P = \dfrac{3}{5}\dfrac{U}{V}$

(B) $P = \dfrac{1}{3}\dfrac{U}{V}$ (D) $P = \dfrac{3}{2}\dfrac{U}{V}$

Q.52 **A radiation cavity of volume V at temperature T has total energy U. The pressure exerted by the radiation inside the cavity is proportional to**

(A) T (C) T^3

(B) T^2 (D) T^4

Q.53 **The temperature of a radiation cavity of fixed volume is doubled. Which one of the following is true for blackbody radiation inside the cavity?**

(A) Its energy and number of photons both increase 8 times

(B) its energy increases 8 times and number of photons increase 16 times

(C) its energy increases 16 times and the number of photons increases 8 times

(D) both energy and number of photons increase 16 times

Q.54 **The spectrum of radiation emitted by a blackbody at a temperature 1000 K peaks in the**

(A) visible range of frequencies

(B) infrared range of frequencies

(C) ultraviolet range of frequencies

(D) microwave range of frequencies

Q.55 **Consider blackbody radiation in a cavity maintained at 2000K. If the volume of the cavity is reversibly and adiabatically increased from 10 cc to 640 cc, the temperature of the cavity changes to**

(A) 800K   (C) 600K

(B) 700K   (D) 500K

Q.56 **Which one of the following statements is true for blackbody radiation?**

(A) As the temperature increases, peak of energy distribution curve decreases in height and shifts to longer wavelength

(B) As the temperature increases, peak of energy distribution curve increases in height and shifts to lower wavelength

(C) As the temperature increases, peak of energy distribution curve increases in height and shifts to longer wavelength

(D) As the temperature increases, peak of energy distribution curve decreases in height and shifts to lower wavelength

Keys and hints to MCQ type questions

Q.1 C	Q.11 A	Q.21 A	Q.31 D	Q.41 D	Q.51 B
Q.2 D	Q.12 A	Q.22 D	Q.32 C	Q.42 B	Q.52 D
Q.3 C	Q.13 D	Q.23 B	Q.33 C	Q.43 B	Q.53 C
Q.4 C	Q.14 A	Q.24 C	Q.34 B	Q.44 A	Q.54 D
Q.5 A	Q.15 D	Q.25 B	Q.35 D	Q.45 C	Q.55 D
Q.6 C	Q.16 C	Q.26 A	Q.36 D	Q.46 C	Q.56 B
Q.7 D	Q.17 D	Q.27 C	Q.37 B	Q.47 C	
Q.8 B	Q.18 C	Q.28 B	Q.38 D	Q.48 A	
Q.9 C	Q.19 C	Q.29 A	Q.39 D	Q.49 C	
Q.10 D	Q.20 D	Q.30 -	Q.40 D	Q.50 D	

Hint.7 Here, **D** is correct. Because heat transfer in boiling of water and in land and sea breezes is primarily due to convection. A metal rod placed over a flame is heated via conduction. Heat transfer via convection and conduction takes place through a medium (or matter), whereas transfer via radiation is feasible in vacuum. This is the situation in case of **D**. A bulb is evacuated. Hence, heat transfer is not via conduction or convection.

Hint.8 The energy radiated from a hot body per second per unit area (i.e., power radiated)

$$E = A\epsilon\sigma T^4, \quad P = \frac{E}{A} = \epsilon\sigma T^4$$

The energy radiated per second, per unit area by a hot body is $\epsilon\sigma T^4$ independent of surrounding.

Hint.9 Since $U \propto T^4$.

$$\frac{U_1}{U_2} = \left[\frac{T_1}{T_2}\right]^4, \text{ giving, } U_2 = 16U_1$$

Hint.13 When a blackbody radiation cavity undergoes reversible adiabatic change,

$$PV^{4/3} = constant$$

So that,

$$\frac{P_2}{P_1} = \left[\frac{V_1}{V_2}\right]^{4/3}$$

OR

$$P_2 = P_1\left[\frac{V_1}{V_2}\right]^{4/3} = 1.5 \times 10^{-2}\left[\frac{1}{8}\right]^{4/3} = 24 \times 10^{-2} Nm^{-2}$$

Hint.23 Spectral density is proportional to number of photons available at that particular energy. Therefore,

$$\frac{n_{2\nu}}{n_\nu} = \frac{\frac{1}{e^{2h\nu/kT}-1}}{\frac{1}{e^{h\nu/kT}-1}} = \frac{e^{h\nu/kT}-1}{e^{2h\nu/kT}-1} = \frac{e^{h\nu/kT}-1}{(e^{h\nu/kT}-1)(e^{h\nu/kT}+1)}$$

$$= \frac{1}{(e^{h\nu/kT}+1)} = \left[(e^{h\nu/kT}+1)\right]^{-1}$$

Hint.24 Given that stars behave like perfectly black bodies, i.e., $\epsilon = 1$. Therefore, energy (E) radiated per second, i.e., power P is

$$P = A\epsilon\sigma T^4 = A\sigma T^4$$

Because, both stars radiate the same power, therefore,

$$4\pi R_W^2 T_W^4 = 4\pi R_R^2 T_R^4, \quad T_W = T_R \left[\frac{r_R}{r_W}\right]^2 \tag{9.60}$$

Therefore, **C** is satisfied, i.e., $R_R > R_W$ implies $T_R < T_W$.

Hint.25 Wien's displacement law implies

$$\lambda_2 T_2 = \lambda_2 T_2, \quad \lambda_2 = \frac{\lambda_2 T_2}{T_1} = \frac{600 \times 6000}{300} = 12\mu m$$

Hint.30 The rate of heat radiated per second from a hot body is the power density

$$P = A\epsilon\sigma T^4$$

Therefore, for same emissivity (σ being Stefan's constant) and $R_A = 4R_B$, $T_A = 2T_B$

$$\frac{P_A}{P_B} = \frac{4\pi R_A^2 T_A^4}{4\pi R_B^2 T_B^4} = \frac{(4R_B)^2(2T_B)^4}{(R_B)^2(T_B)^4} = 256$$

Hint.31

$$U = \int dU = \int C_V dT = \int \gamma T^3 dT = \frac{\gamma T^4}{4}$$

Hint.32 For a blackbody the free energy is given by

$$A = -kT \ln Z = \frac{8\pi^5 V}{45c^3 h^3}(kT)^4$$

The entropy is

$$S = -\left(\frac{\partial A}{\partial T}\right)_V = \frac{32\pi^5 k^4}{45c^3 h^3} V T^3 \propto T^3$$

therefore, when temperature is doubled, the entropy of the radiation increases by a factor of 8.

Hint.34 For a photon gas inside a cavity of volume V, the pressure P is linked with T as

$$P \propto T^4$$

Therefore,

$$\frac{P_2}{P_1} = \left[\frac{2T}{T}\right]^4 = 16$$

Hint.35 According to Wien's displacement law

$$\lambda T = constant$$

So that

$$\lambda_2 T_2 = \lambda_1 T_1 \quad T_2 = \frac{\lambda_1 T_1}{\lambda_2} = \frac{5000 \times 10^4}{5} = 10^7 K$$

Hint.36 The power radiated is

$$P = A\epsilon\sigma T^4, \quad T = \left(\frac{P}{A\epsilon\sigma}\right)^{1/4}$$

Hint.37 According to Stefan's law, energy radiated per second is

$$E = A\epsilon\sigma T^4 \propto T^4$$

Therefore,

$$\frac{E_2}{E_1} = \left[\frac{T_2}{T_1}\right]^4 = \left[\frac{273 + 287}{273 + 7}\right]^4 = 16$$

Hint.38 The rate of heat loss is given by

$$mc\frac{d\theta}{dt} = (4\pi r^2)\sigma T^4$$

c is specific heat. The initial rate of cooling is

$$\frac{d\theta}{dt} \propto \frac{r^2}{mc}$$

as $m = \frac{4}{3}\pi r^3\rho$, implying $r \propto m^{1/3}$. Therefore,

$$\frac{d\theta}{dt} \propto \frac{m^{2/3}}{m} \propto \frac{1}{m^{1/3}}$$

Giving

$$\frac{(\frac{d\theta}{dt})_{S_1}}{(\frac{d\theta}{dt})_{S_2}} = \left[\frac{m_{S_2}}{m_{S_1}}\right]^{1/3} = \left[\frac{1}{3}\right]^{1/3}$$

Hint.39 The power radiated by a hot body is

$$P = A\epsilon\sigma T^4 = (4\pi r^2)\epsilon\sigma T^4 \propto r^2 T^4$$

$$\frac{P_2}{P_1} = \left[\frac{r_2}{r_1}\right]^2 \left[\frac{T_2}{T_1}\right]^4 = \left[\frac{1}{2}\right]^2 \left[\frac{2}{1}\right]^4 = 4, \quad P_2 = 4P_1 = 4 \times 450$$
$$= 1800W$$

Hint.40 According to Wien's displacement law, the wavelength λ_m at which the maximum radiation is emitted by a hot body at temperature T is

$$\lambda_m T = b, \quad \lambda_m = \frac{b}{T} = \frac{2.88 \times 10^6 nm K}{2880K} = 1000nm$$

Hence, U_2 is the maximum. Since a blackbody emits radiation at all wavelengths, implying $U_1 \neq 0$ and $U_3 \neq 0$. Therefore, **D** is correct.

Hint.41 Answer is **D**. $dU = C_v dT = \gamma T^3 dT \Rightarrow u = \frac{\gamma T^4}{4}$

Hint.43 Answer is **B**. $S = -\frac{\partial A}{\partial T} = \frac{4a}{3}V T^3$.

Hint.51 For relativistic case

$$P = \frac{1}{3}\frac{U}{V}$$

Therefore, **B** is correct. Note that for non-relativistic case

$$P = \frac{2}{3}\frac{U}{V}$$

Hint.52 Since for a photon gas

$$P = \frac{1}{3}\frac{U}{V} \propto T^4$$

Hint.54 using Wien's displacement law,

$$\lambda T = 2.898 \times 10^{-3}$$

Gives

$$\lambda = \frac{2.898 \times 10^{-3}}{1000} = 2898\text{nm}$$

Therefore, the wavelength lies in microwave range (1 mm to 1 m) of frequencies.

Hint.55 For a radiation cavity undergoing reversible adiabatic process

$$VT^3 = \text{constant}$$

$$\frac{T_2}{T_1} = \left[\frac{V_1}{V_2}\right]^{1/3}$$

$$T_2 = T_1 \left[\frac{V_1}{V_2}\right]^{1/3} = 2000 \left[\frac{10}{640}\right]^{1/3} = 500 \text{ K}$$

9.14 Exercises

1. Define thermal radiation.

2. Explain the terms emissive power and absorptive power.

3. What is a perfect blackbody? Discuss, how the idea of a perfect blackbody can be achieved in practice?

4. What is blackbody radiation? Explain temperature dependence of blackbody radiation spectra.

5. Show that Planck's law reduces to Wien's law for shorter wavelengths and Rayleigh–Jeans law for longer wavelengths.

6. Derive Wien's displacement law and Stefan's law from Planck's radiation law.

7. What is a perfect blackbody? Draw curves for distribution of energy in spectra of a blackbody for two different temperatures. Explain important features of the spectra.

8. Show that Planck's law and Rayleigh–Jeans law become identical if the size of quantum is allowed to vanish or if temperature becomes too high.

9. Explain ultraviolet catastrophe according to Rayleigh–Jeans distribution law.

10. Explain the terms emissive power and absorptive power. Prove that at any temperature the ratio of the emissive power to the absorptive power of a substance is constant and is equal to the emissive power of a perfect blackbody.

11. Show that number of modes of vibrations per unit volume of a cavity in the frequency range from v to $v + dv$ is given by

$$n_v dv = \frac{8\pi v^2}{c^3} dv$$

where terms have their usual meaning.

12. Show that average energy of a Planck's oscillator of frequency v in thermal equilibrium with a heat reservoir at temperature T is given by

$$< E > = \frac{hv}{e^{hv/k_B T} - 1}$$

Elementary Statistical Mechanics

10

In this chapter, we will give a brief introduction to three kinds of statistics. The concept of microstates, macrostates and thermodynamic probability is introduced in this chapter. This is followed by the development of probability distribution functions for three kinds of statistics. A few examples are discussed to understand the distribution of particles according to three different types of statistics.

10.1   Probability

Let us consider the example of tossing a dice. As we know, a dice has six faces and when thrown any one of six faces may turn up. That is, out of six possibilities, only one will be true. We say that the probability of a face with six dots is 1/6. Therefore, we define the probability P as the number of cases in which the event occurs to the total number of cases.

10.1.1 Probability of Two Independent Events

Sometimes, we come across random events whose occurrences are independent of each other. For instance, consider two different molecules to be distributed in certain region of space of volume V. Let the probability of molecule 1 to be in volume V_1 be $P_1 = V_1/V$ and that of molecule 2 to be in volume $V_2 = V - V_1$ be equal to $P_2 = V_2/V$. These two events are independent of each other and their joint probability of occurrence is given by

$$P = P_1 \times P_2$$

© The Author(s) 2022
S. Sharma, *Thermal and Statistical Physics*,
https://doi.org/10.1007/978-3-031-07685-5_10

10.1.2 Principle of Equal a Priori Probability

Suppose we have to throw a ball into any one of two identical boxes. There are two possible outcomes, either ball can enter box 1 or it may enter box 2. It is evident that while throwing the ball, it may enter into any one of two boxes. The probability of entering into box one is 1/2. Similarly, the probability of entering into box two is also 1/2. Another best example illustrating this point is that of tossing a coin. While doing that either head or tail will come up. That is, the probability of occurrence of head or tail is equal (=1/2). *This principle of assuming equal probability for events that are equally likely to occur is known as the principle of equal a priori probability.*

10.1.3 Permutations and Combinations

Though, students might have learned about permutations and combinations in their elementary mathematical classes, we will discuss these concepts here in brief. We take the example of three distinguishable objects a, b and c. If we consider two objects at a time, possible six arrangements are: ab, ba, ac, ca, bc and cb. Here, two objects a and b can be arranged in two ways and we say that the number of permutations are 2 and the combinations (of two objects) is 1. Similarly, out of three objects, considering arrangements of two at a time gives 6 arrangements with three group combinations. We say that, in this case, the number of permutations are 6 and combinations are only 3. Mathematically, we can write it as below: Number of permutations of 3 objects taking two at a time are

$$^3P_2 = \frac{3!}{(3-2)!} = \frac{3 \times 2 \times 1}{1} = 6$$

And number of combinations (groups) are given by

$$^3C_2 = \frac{3!}{2!(3-2)!} = 3$$

In general, we can write

$$^nP_r = \frac{n!}{(n-r)!}$$

$$^nC_r = \frac{n!}{r!(n-r)!} = \frac{1}{r!}\frac{n!}{(n-r)!} = \frac{^nP_n}{r!}$$

10.2   Phase Space

Before we describe the phase space, let us take an example of a static particle whose position in space can be specified precisely by three position coordinates (x, y, z). *Such a three-dimensional space in which the position of an object is completely specified by three position coordinates is called position space.* We can represent a small volume element in position space as

$$dV = dxdydz$$

Now take an example of a dynamic system where particles are continuously moving. Consider a single particle of mass m freely moving in an open space. We can extend the problem to N particles as required in statistical mechanics. For a single moving particle, we require only three velocity coordinates (v_x, v_y, v_z) or precisely momentum coordinates (p_x, p_y, p_z), where $p_x = mv_x$, $p_y = mv_y$, $p_z = mv_z$. *The position of moving particle can be described by three mutually perpendicular momentum coordinates p_x, p_y and p_z in a space known as momentum space.* Small volume in momentum space is given by

$$d\Gamma = dp_x dp_y dp_z$$

When position space and momentum space are combined we get another space known as phase space which is six-dimensional. It consists of mutually perpendicular, three position and three momentum coordinates. In phase space coordinates of an object are specified by a point with six coordinates (x, p_x, y, p_y, z, p_z). A small element in phase space is denoted by $d\tau$ and given as

$$d\tau = dxdp_x dydp_y dzdp_z$$

10.2.1 Division of Phase Space into Cells

The phase space introduced above is six-dimensional and is purely a mathematical concept. It cannot be drawn. However, to understand the underlying approach, we consider one-dimensional (dx) space and associated momentum (dp_x) coordinate. Then the problem can be extended to three-dimensional position space, and hence to six- dimensional phase space. For the sake of convenience, consider the Fig. 10.1. Let dx be the interval size along x-axis and dp_x along p_x axis. The smallest cell in this two-dimensional plot will have an area $= dxdp_x = h_0$. Note that classically, there is no limit in choosing the value of h_0. It could be as small (large) as possible. One can distinguish between neighbouring cells even if h approaches zero. The end results in classical mechanics are independent of the size of h_0, and hence its size is immaterial in classical statistics. If a particle lies within a cell, its state is represented by a pair of coordinates (x, p_x).

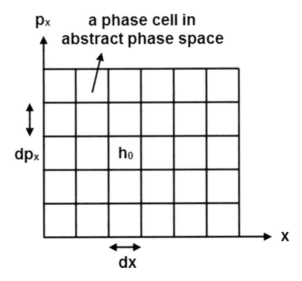

In quantum mechanics, the Heisenberg uncertainty principle fixes the limit on
the size of a cell. According to this principle, $dx dp_x \geq h$. Therefore, for a six-
dimensional phase space, the elemental volume of phase cell is

$$d\tau = dx dp_x dy dp_y dz dp_z \geq h^3 \qquad (10.1)$$

That is minimum possible volume that can be associated with a cell in phase space
is h^3. Now total number of cells in phase space is given by

$$\frac{\text{Total phase space volume}}{\text{Volume of one cell}} = \frac{\int \int \int dx dy dz \int \int \int dp_x dp_y dp_z}{d\tau} = \frac{V\Gamma}{d\tau}$$

10.2.2 Constraints on a System and Accessible States

In statistics, we deal with a system consisting of a large number of particles. While
considering the distribution of these N particles into a large number of energy states
(also called level) and further into various cells (or sublevels or quantum states or
single particle states), the system under consideration needs to satisfy a few constraints. The first constraint fixes the total number of particles in the system and
mathematically it is represented as

$$N = \sum_i n_i = \text{constant} \qquad (10.2)$$

The second constraint imposed on the system is related to total energy of the system
taken as conserved. That is we can write

$$E = \sum_i \epsilon_i n_i = \text{constant} \qquad (10.3)$$

Here, n_i represents number of particles in ith cell where energy of each particle is ϵ_i. These two constraint ensure that total number of particles and total energy of the system is constant. *From here, we can define accessible states as the number of states which are consistent with the given constraints of the system. Later, we will take a few examples to make it more clear.*

10.3   Macrostates and Microstates

Before we give a proper definition of a macrostate and microstate, let us first consider the distribution of four distinguishable particles in two identical boxes (or energy levels). This case is discussed below.

10.3.1 Distribution of Four Distinguishable Particles in Two Identical Boxes

Let us consider the distribution of four distinguishable particles, a, b, c and d in two different boxes. The possible five distributions are given in Table 10.1. These five box wise distributions i.e., (4,0), (3,1), (2,2), (1,3) and (0,4) are five different *macrostates*. That is for four distinguishable particles when distributed in two identical boxes have five distinct macrostates. In general, when we have n distinguishable particles to be distributed in 2 boxes, we will have (n + 1) distinct macrostates. Further, we note that each macrostate has various possible distributions for particles. For instance, the macrostate (4,0) has only one such distribution. Macrostate (3,1) has four, (2,2) has six such distributions. Each such distribution is called a *microstate* corresponding to a particular macrostate of the system, that is, to say the (2,2) macrostate has 6 microstates. Similarly, macrostate (3,1) has only four microstates. In total, system has 16 microstates. Let us evaluate the probability of a particular macrostate and microstate. Consider the (4,0) macrostate which has only one microstate. The probability for this microstate is 1/16 (see column 5, Table 10.1). This is also the probability for the macrostate (4,0). Now take another example of macrostate (3,1) which has four microstates. Each of the microstates has a probability of 1/16. But now the macrostates (3,1) has four equally probable microstates, its probability will be 4/16. Similarly, (2,2) has a probability of 6/16. That is to say, the system is going to spend the largest amount of time in the macrostate with the largest number of microstates. This macrostate is known as *most probable macrostate*. Note that within a macrostate, say (4,0) when we interchange particles, it doesn't give a new microstate.

Table 10.1 Distribution of four distinguishable particles, a, b, c and d in two different boxes

Macrostate	Possible arrangements		No of microstates	Probability of macrostates
	Box 1	Box 2		
(4,0)	abcd	0	1	1/16
(3,1)	abc	d	4	4/16
	dab	c		
	cda	b		
	bcd	a		
(2,2)	ab	cd	6	6/16
	ac	db		
	ad	bc		
	bc	da		
	bd	ac		
	cd	ab		
(1,3)	a	bcd	4	4/16
	b	cda		
	c	dba		
	d	abc		
(0,4)	0	abcd	1	1/16

Total no of microstates = 16

10.3.2 Distribution of Four Indistinguishable Particles in Two Identical Boxes

Now, we take an example of four identical or indistinguishable particles which have to be distributed in the same two boxes. The possible macrostates and microstates are given in Table 10.2. As we see, by imposing the constraint of indistinguishability, the number of accessible microstates has drastically reduced. Now, the system has only five microstates. Probability of each microstate being 1/5. Note that all microstates are *equally probable*. This means all have equal probability. This is also referred to as the principle of equal a priori probability. Therefore, we conclude the discussion with following points:

(i) All microstates in a system are equally probable, i.e., they have equal probability.

(ii) Constraints reduce the number of microstates significantly.

(iii) The probability of a microstate $= \dfrac{1}{\text{Total number of microstates}}$.

(iv) The probability of a macrostate $= \dfrac{\text{Number of microstates in it}}{\text{Total number of microstates in system}}$.

Table 10.2 Distribution of four indistinguishable particles, a, a, a and a in two different boxes

Macrostate	Possible arrangements		No of microstates	Probability of macrostates
	Box 1	Box 2		
(4,0)	aaaa	0	1	1/5
(3,1)	aaa	a	1	1/5
(2,2)	aa	aa	1	1/5
(1,3)	a	aaa	1	1/5
(0,4)	0	aaaa	1	1/5

Total no of microstates = 5

10.4 Static and Dynamic Systems

Let us take the following two examples of systems. When a coin is tossed, either head or tail will turn up. As long as the coin is not thrown again, its previous microstate cannot change itself. Similarly, when a ball is thrown into a box, it cannot change its state as long as it is removed from there and again thrown into some other box. Both of these cases represent a static system. *Therefore, a system whose constituent particles remain at rest in a particular microstate is called a static system.*

On the other hand, a gaseous system represents a dynamic system, where gas molecules always keep on moving from one microstate to another. These microstates may correspond to the same macrostate. *Therefore, a system will spend maximum time in a macrostate having largest number of microstates and the corresponding macrostate is known as the most probable microstate.*

10.5 Thermodynamic Probability

So far we have learned about probability, microstates and macrostates. Now, we will try to understand the physical meaning of thermodynamic probability. *In a given macrostate, the number of meaningful arrangements or microstates is known as thermodynamic probability.* Or simply we can say that thermodynamic probability is the number of microstates in a particular macrostate. It is denoted by either W or Ω. Note the word meaningful here. For instance, when we considered the arrangements or 2 particles out of three, we encountered arrangements like 'ab' and 'ba'. Here, these arrangements are identical and should be counted only once. Therefore, meaningful arrangements are usually less than total possible arrangements. The result is that we are supposed to divide the total arrangements by the number of occurrence of such type of arrangements that give identical distribution. The number of meaningful arrangements of n total particle out of which r are of one kind and (n − r) or another is

$$W_{(r,n-r)} = \frac{n!}{r!(n-r)!} = {}^n C_r \qquad (10.4)$$

10.5.1 Probability of a Macrostate

In Sect. 10.3, we noticed that a system of four distinguishable particles has five macrostates where each macrostate was further associated with different numbers of microstates. The sum of all microstates in different macrostates gives total number of microstates present in a system. From this information, one can define the probability of a macrostate as follows:

$$P_{macrostate} = \frac{\text{No of microstaes in a given macrostae}}{\text{Total no of microstaes in a system}}$$

10.6   Three Kinds of Statistics

Basically, there are two types of statistics, classical and quantum statistics. Quantum statistics further has two types: Fermi–Dirac (F–D) and Bose–Einstein (B–E) statistics. F–D statistics applies to half-integral spin particles such as electron, proton, neutron, etc., whereas B–E statistics applies to integral spin particles, for instance, photons, alpha-particles, etc. The basic differences among these are as below

(i) Classical statistics also known as Maxwell–Boltzmann statistics treats the particle as distinguishable, whereas in quantum statistics, particles are treated as indistin-

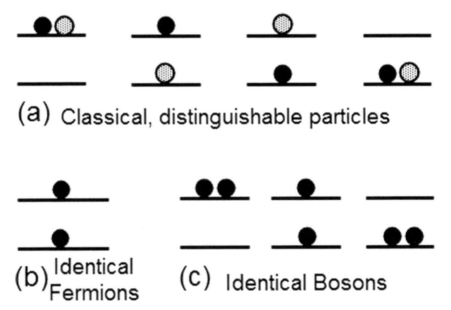

(a) Classical, distinguishable particles

(b) Identical Fermions   **(c)** Identical Bosons

Fig. 10.2 Distribution of two particles in two different energy states assuming **a** Classical statistics (distinguishable particles). **b** Fermi–Dirac statistics (identical particles). **c** Bose–Einstein statistics (identical particles)

guishable.

(ii) Classical statistics does not impose any constraint on occupancy of a particular level. That is, as many particles can occupy a particular energy level. Whereas, in quantum statistics, this is not the case. For instance, Fermi–Dirac statistics obey the Pauli exclusion principle and no more than one Fermion can occupy a single particle state. On the other hand, Bose–Einstein statistics does not impose any restriction on the occupancy of a particular energy state. That is, as many Bosons can occupy a particular energy state. Note that Fermions are spin half particles, whereas Bosons are integral spin particles.

(iii) In classical statistics, h_0 can be as small as possible, whereas in quantum case, uncertainty principle imposes a minimum limit on h (Fig. 10.2).

10.7 The Distribution Functions

So far necessary tools for statistical mechanics have been discussed in details. These tools can now be applied to understand various problems under classical (Maxwell–Boltzmann) or quantum statistics (Bose–Einstein) statistics. We will continue with classical statistics. First of all, we will develop an expression for thermodynamic probability (total number of meaningful arrangements of particles) followed by the determination of most probable distribution.

10.7.1 Maxwell–Boltzmann Distribution

Thermodynamic Probability, Ω_{MB}

Here, we consider a classical system of N distinguishable particles enclosed in a volume V and having total energy E. Though, there is no constraint on the energy a particle can have (single particle energy could be zero or E), for mathematical convenience, we assume that these N particles can be put into k discrete subgroups labelled as 1, 2, 3, 4...k − 2, k − 1, k, such that each subgroup contains $n_1, n_2, n_3, n_4, ...n_k$ particles, respectively. We make another assumption that, each subgroup is further subdivided into small cells. That is to say the above said groups have $g_1, g_2, g_3, ...g_k$ cells, respectively. The system under consideration is further subjected to the following constraint:

(i) Total number of particles in the system is conserved, i.e,

$$N = \sum_{i}^{k} n_i.$$

(ii) Total energy of the system is also conserved, i.e.,

$$E = \sum_{i}^{k} n_i \epsilon_i.$$

(iii) There is no constraint on the number of particles occupying a particular energy level.

Note that the above two constraints (i) and (ii) are satisfied by classical as well as quantum (Fermi–Dirac and Bose–Einstein) systems. Once constraints are known, next task is to determine the number of ways in which we can distribute N different particles in various groups which are further subdivided into cells. These different distributions or number of ways in principal equal to permutations of N objects out of which n_1 are of one kind, n_2 of second kind, n_3 of third kind and so on. Let us try to understand it as follows. First of all, we focus on the first group which contains n_1 particles. For the time being, we will not take into account that how these n_1 particles are further distributed in g_1 cells. The number of ways of distributing n_1 particles out of N in group 1 is

$$^N C_{n_1} = \frac{N!}{n_1!\,(N-n_1)!} \tag{10.5}$$

Now, we are left with $(N-n_1)$ particles. Out of these, n_2 particles will occupy group 2 and respective number of ways are

$$^{N-n_1} C_{n_2} = \frac{(N-n_1)!}{n_2!\,(N-n_1-n_2)!} \tag{10.6}$$

Note that distribution of particles in each group is independent of each other. Therefore, total number of ways in n_1 particles are arranged in group 1, n_2 in group 2, n_i in ith group and n_k in kth group is given by

$$W = \frac{N!}{n_1!\,(N-n_1)!} \frac{(N-n_1)!}{n_2!\,(N-n_1-n_2)!} \frac{(N-n_1-n_2)!}{n_3!\,(N-n_1-n_2-n_3)!}\cdots$$
$$\cdots \frac{(N-n_1-n_2-n_3-\ldots n_i)!}{n_i!\,(N-n_1-n_2-n_3\ldots n_i)!}\cdots\cdots$$
$$= \frac{N!}{n_1!n_2!n_3!\ldots n_i!\ldots} = \frac{N!}{\prod_i n_i!} \tag{10.7}$$

Our next step is to determine the number of ways (P_{MB}) in which particles belonging to a particular group can be distributed in respective cells in that group. Let us take example of group one, where we have to distribute n_1 particles in g_1 cells. Let us start with particle 1 out of n_1

Number of ways in which first particle can enter into g_1 cells $= g_1$
Number of ways in which second particle can enter into g_1 cells $= g_1$
$$\ldots\ldots$$

$$\ldots\ldots$$

Number of ways in which n_1th particle can enter into g_1 cells $= g_1$
Total number of ways in which n_1 particles can enter into g_1 cells $= g_1^{n_1}$

Hence, total number of ways in which n_1 particles may enter g_1 cells, n_2 particles may enter g_2 cells and n_i may enter g_i cells and so on, is given by

$$P_{MB} = g_1^{n_1} g_2^{n_2} g_3^{n_3} \dots g_k^{n_k} = \prod_i g_i^{n_i} \tag{10.8}$$

Note that while writing the above result, we have made use of the fact that distribution of particles in each group is independent of distribution in other groups. Hence, we can multiply respective probabilities. Therefore, total number of ways in which we can distribute N distinguishable particles into k groups and the particles in various groups can further be distributed into respective cells in each group is

$$\Omega_{MB} = \frac{N!}{\prod_i n_i} \prod_i g_i^{n_i} \tag{10.9}$$

Equation 10.9 gives the thermodynamic probability for distribution of N distinguishable particles in k groups which are further distributed in respective cells in each group.

In case of indistinguishable (identical) particles, the thermodynamic probability is obtained by dividing Eq. 10.9 by a factor of $N!$. This is because, in case of identical particles, the number of accessible microstates reduces significantly. This way we obtain expression for classical identical particles.

$$\Omega_{cl} = \prod_i \frac{g_i^{n_i}}{n_i!} \tag{10.10}$$

Note that in classical statistics, there is no constraint on number of particles that can enter into a particular cell. However, in quantum statistics, we will see that it is not the case.

10.7.2 Fermi–Dirac Distribution

Thermodynamic Probability, Ω_{FD}

It is known that Fermion are identical and half-integral spin particles and subjected to satisfy the Pauli exclusion principle. We will consider a Fermi system of N indistinguishable (identical) particles enclosed in a volume V and having total energy E. For mathematical convenience we assume that these N particles can be put into k discrete groups or levels marked as $1, 2, 3, 4 \dots k-2, k-1, k$, and have $n_1, n_2, n_3, n_4, \dots n_k$, particles, respectively. Further, we assume that each level is degenerate, i.e., each level is further having several states (also called as sublevels) of equivalent energy. We assume that k levels have degeneracy $g_1, g_2, g_3, \dots g_k$, respectively. Let us take an example of ith level (with degeneracy g_i) containing n_i particles (Fermions). Since, Fermions are subjected to the Pauli exclusion principle, only one Fermion can stay in a single particle state. Let us continue with first particle. It will have g_i options to enter into any single particle state out of total g_i states. The second particle will have only $g_i - 1$ and third will have only $g_i - 2$. This way, total number of ways in which n_i Fermions can occupy g_i states is equal to

$$g_i(g_i - 1)(g_i - 2)(g_i - 3)........(g_i - n_i + 1) = \frac{g_i!}{(g_i - n_i)!} \qquad (10.11)$$

Note that permutations among identical particles do not yield a different arrangement and such arrangements must be excluded from the above equation. This can be done by dividing the above equation by $n_i!$. Therefore, the number of ways in which n_i Fermions can be arranged in g_i sublevels are

$$\Omega_{FDi} = \frac{g_i!}{n_i!(g_i - n_i)!} \qquad (10.12)$$

This was the case for ith level with degeneracy g_i. Similarly, we can write expression for other levels (groups). Since, the distribution of Fermions in each level is independent of the other, we can write for distribution of N Fermions in k levels with degeneracy $g_1, g_2, g_3, ...g_k$, respectively

$$\Omega_{FD} = \prod_i \frac{g_i!}{n_i!(g_i - n_i)!} \qquad (10.13)$$

10.7.3 Bose–Einstein Distribution

Thermodynamic Probability, Ω_{BE}

Before we obtain a general result for a Bose system, we consider a system of n_i indistinguishable bosons, say photons. Our objective is to distribute n_i identical objects among g_i cells such that there is no constraint on number of objects that may enter into particular cell. For this purpose, we will take a simple example of $n_i = 7$ identical objects to be distributed into $g_i = 6$ cells. Some of the possible arrangements are shown in Fig. 10.3. We notice that, n_i objects and $(g_i - 1)$ partitions have to be arranged to obtain the final distributions. To do this, we consider $(n_i + g_i - 1)$ objects out of which n_i are of one kind and $(g_i - 1)$ are of other kinds. Therefore, the required number of ways in which n_i identical objects can be distributed in g_i cells or sublevels of ith state is

$$\frac{(n_i + g_i - 1)!}{n_i!(g_i - 1)!} \qquad (10.14)$$

Identical expressions can be found for other groups (or quantum states) and thermodynamic probability is given by

$$W_{BE} = \frac{(n_1 + g_1 - 1)!}{n_1!(g_1 - 1)!} \cdot \frac{(n_2 + g_2 - 1)!}{n_2!(g_2 - 1)!} \cdots \frac{(n_i + g_i - 1)!}{n_i!(g_i - 1)!} \cdots$$

$$= \prod_i \frac{(n_i + g_i - 1)!}{n_i!(g_i - 1)!} \qquad (10.15)$$

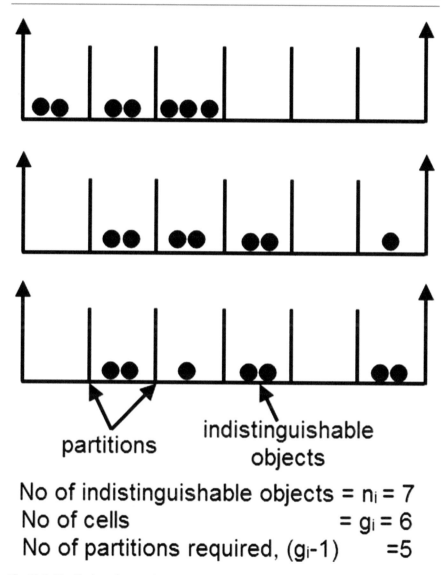

No of indistinguishable objects = n_i = 7
No of cells   = g_i = 6
No of partitions required, (g_i-1)   =5

Fig. 10.3 Distribution of $n_i = 7$ identical objects among $g_i = 6$ cells

10.8   Approach to Obtain the Most Probable Distribution in Different Statistics

Here, we will describe the common approach to find the most probable distribution for three kinds of statistics. Since, Ω is very large, it is convenient to maximize $\ln \Omega$

instead of Ω. So that we can set $\delta(\ln W) = 0$. Further, system will be subjected to the constraint that total number of particles and energy is conserved. That is

$$N = n_1 + n_2 + n_3 + \ldots\ldots n_k = \sum_i n_i = \text{constant}$$

$$E = n_1\epsilon_1 + n_2\epsilon_2 + \ldots\ldots + n_i\epsilon_i + \ldots = \sum_i n_i\epsilon_i = \text{constant}$$

In differential form, the above equations reduce to

$$\sum_i \delta n_i = 0 \tag{10.16}$$

$$\sum_i \epsilon_i \delta n_i = 0 \tag{10.17}$$

Further, if the total number of accessible microstates, i.e., Ω is known, then maximizing $\ln \Omega$ ($\because \Omega$ is very large), we get

$$\delta(\ln \Omega) = 0 \tag{10.18}$$

Any system, classical or quantum, is supposed to satisfy these three constraints simultaneously. In order to obtain a most probable distribution function, we use the method of Lagrange undetermined multiplier where we multiply Eq. 10.16 with α, Eq. 10.17 with β and then add these two into Eq. 10.18 to obtain

$$\delta(\ln \Omega) + \alpha\delta n_i + \beta\delta\epsilon_i = 0 \tag{10.19}$$

This is the general equation and can be solved for obtaining the most probable distribution for any type of statistics.

10.8.1 The Most Probable Distribution, Maxwell–Boltzmann Statistics

As mentioned in one of the previous sections, the thermodynamic probability for system obeying Maxwell–Boltzmann statistics is given by

$$\Omega_{MB} = \frac{N!}{\prod_i n_i} \prod_i g_i^{n_i} \tag{10.20}$$

Taking natural logarithm, we obtain

$$\ln \Omega_{MB} = \ln N! + \sum_i [n_i \ln g_i - \ln n_i!] \tag{10.21}$$

Since n is very large, we will make use of Stirling's approximation; $\ln n! = n \ln n - n$. As N, n_i and g_i are all large, we obtain

$$\ln \Omega_{MB} = N \ln N - N + \sum_i [n_i \ln g_i - n_i \ln n_i + n_i]$$

$$= N \ln N + \sum_i [n_i \ln g_i - n_i \ln n_i] \qquad (10.22)$$

where, we have written last step using $N = \sum_i n_i$. Further, we will make use of the fact that N is constant, g_i is not subjected to change but n_i can vary continuously. This gives $\delta \ln \Omega_{MB} = 0$. That is

$$\delta(\ln \Omega_{MB}) = \sum_i \delta [n_i \ln g_i - n_i \ln n_i]$$

$$= \sum_i \left[\delta n_i \ln g_i - n_i \left(\frac{1}{n_i} \right) \delta n_i - \delta n_i \ln n_i \right]$$

$$= - \sum_i \left[\ln \left(\frac{n_i}{g_i} \right) + 1 \right] \delta n_i = 0 \qquad (10.23)$$

The above equation together with Eqs. 10.16 and 10.17 needs to satisfy simultaneously. In order to find general result for Maxwell–Boltzmann distribution, we make use of method of Lagrange undetermined multipliers. In this method, we multiply Eq. 10.16 with α and Eq. 10.17 with β and add these three equations to get

$$\sum_i \delta n_i \left[\ln \left(\frac{n_i}{g_i} \right) + \alpha + \beta \epsilon_i \right] = 0 \qquad (10.24)$$

Note that here we have absorbed the integer 1 within α. In the above equation, the coefficients δn_i are arbitrary and cannot be zero. Therefore, the above result will hold if

$$\ln \left(\frac{n_i}{g_i} \right) + \alpha + \beta \epsilon_i = 0$$

$$\ln \left(\frac{n_i}{g_i} \right) = -(\alpha + \beta \epsilon_i)$$

This leads to

$$n_i = g_i exp \left[-(\alpha + \beta \epsilon_i) \right] = \frac{g_i}{exp \left[-(\alpha + \beta \epsilon_i) \right]} \qquad (10.25)$$

This equation is known as *Maxwell–Boltzmann distribution law*. From the above equation, we cane write

$$n_i \propto e^{-\beta \epsilon_i} = K e^{-\beta \epsilon_i} \qquad (10.26)$$

here, K is a constant. Probability (P_i) to find a particle in energy state ϵ_i is given by

$$P_i = \frac{n_i}{N} = \frac{K}{N}e^{-\beta\epsilon_i} = Ce^{-\beta\epsilon_i}$$

Here C is another constant as N is a constant. Normalization condition implies that $\sum_i P_i = 1$, i.e., $C\sum_i e^{-\beta\epsilon_i} = 1$ or $C = \frac{1}{e^{-\beta\epsilon_i}}$. Hence, the probability for a particle to have an energy ϵ_i is

$$P_i = \frac{e^{-\beta\epsilon_i}}{\sum_i e^{-\beta\epsilon_i}} = \frac{e^{-\beta\epsilon_i}}{\mathbb{Z}} \tag{10.27}$$

where $\mathbb{Z} = \sum_i e^{-\beta\epsilon_i}$ is known as *Partition function* and gives the distribution of particles over various energy states. Note that we have introduced two unknown constants α and β and haven't yet obtained their value. In order to know their precise value, rather than considering discrete values for energy, we consider continuous distribution of energies of particles. Let us assume an infinitesimally small energy interval from ϵ to $(\epsilon + d\epsilon)$. The energy of any particle which lies in this interval can be taken as ϵ. Let us denote $n(\epsilon)$ as the number of particles around ϵ in a unit energy interval between $\left(\epsilon - \frac{1}{2}\right)$ to $\left(\epsilon + \frac{1}{2}\right)$. Then $n(\epsilon)$ can be written as

$$n(\epsilon) = g(\epsilon)e^{-\alpha}e^{-\beta\epsilon} \tag{10.28}$$

In terms of momentum, $\epsilon = \frac{p^2}{2m}$ and above equation reduces to

$$n(p) = g(p)e^{-\alpha}e^{-\beta p^2/2m} \tag{10.29}$$

here $n(p)$ denotes number of particles in unit momentum interval between $\left(p - \frac{1}{2}\right)$ to $\left(p + \frac{1}{2}\right)$ and $g(p)$ corresponds to number of cells in phase space corresponding to this interval.

Number of particles that lie in momentum interval dp (i.e., between p to p + dp) = No of particles in unit momentum interval × size of momentum interval = n(p)dp. Therefore

$$n(p)dp = g(p)e^{-\alpha}e^{-\beta p^2/2m}dp \tag{10.30}$$

here g(p)dp denotes the number of cells in phase space corresponding to a momentum interval from p tp p + dp. That is

$$g(p)dp = \frac{\text{phase space volume between p to p + dp}}{h_0^3} \tag{10.31}$$

Fig. 10.4 A sphere of radius p in momentum space. Also, shown a spherical shell of thickness dp

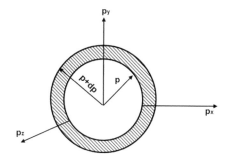

Number of Phase Space Cells

As noted number of phase space cells between p to p + dp are

$$g(p)dp = \frac{1}{h_0^3} \int \int \int dxdydz \int \int \int dp_x dp_y dp_z$$

$$= \frac{V}{h_0^3} \int \int \int dp_x dp_y dp_z \qquad (10.32)$$

Integrals can be separated as the terms are independent. Here V corresponds to volume occupied by gas molecules or particles in ordinary position space. The only problem left with us is the determination of the momentum integral $\int \int \int dp_x dp_y dp_z$ representing the volume in momentum space between interval p to p + dp. In momentum space, a sphere of radius p is drawn in Fig. 10.4. Then all points lying on this momentum sphere will satisfy the equation

$$p^2 = p_x^2 + p_y^2 + p_z^2 \qquad (10.33)$$

Similarly, for another sphere of radius p + dp, the all points which have momentum between p to p + dp will lie in a spherical shell of thickness dp. The volume enclosed between two momentum values p and p + dp is equal to volume of this spherical shell of thickness dp. If we denote this volume by V_{dp}, it is given by

$$V_{dp} = \text{surface area of sphere of radius } p \times \text{thickness of shell} = 4\pi p^2 dp$$

Therefore,

$$g(p)dp = \frac{V}{h_0^3} \int \int \int dp_x dp_y dp_z = \frac{4\pi V p^2 dp}{h_0^3} \qquad (10.34)$$

With this value of g(p)dp, the number of gas molecules in momentum interval p to p + dp are

$$n(p)dp = \frac{4\pi V p^2 dp}{h_0^3} e^{-\alpha} e^{-\beta p^2 / 2m} \qquad (10.35)$$

Now one can obtain total number of phase space cells in a system by integrating Eq. 10.34 from 0 to p_{max} as

$$\text{Total phase space cells} = \frac{4\pi V}{h_0^3} \int_0^{p_{max}} p^2 dp$$

$$= \frac{V}{h_0^3}\left[\frac{4}{3}\pi p_{max}^3\right] = \frac{V\Gamma}{h_0^3} \quad (10.36)$$

Here $\Gamma = \frac{4}{3}\pi p_{max}^3$ represents total volume in momentum space. Therefore, we see that Γ depends on p_{max} and hence on the temperature of the system through kinetic energy, which is a function of temperature.

10.8.2 Determination of Unknown Coefficients

Knowing the value of number of molecules in momentum interval dp, total number of molecules with momentum ranging from 0 to p_{max} is given by

$$N = \int_0^{p_{max}} n(p)dp$$

For $p > p_{max}$, this integral is zero. Therefore, we can replace the limits of integration from 0 to p_{max} with 0 to ∞. Therefore, integral reduces to

$$N = \int_0^{p_{max}} n(p)dp = \int_0^{\infty} \frac{4\pi V}{h_0^3} e^{-\alpha} e^{-\beta p^2/2m} dp$$

Using the standard integral, $\int_{-\infty}^{\infty} x^2 e^{-ax^2} dx = \frac{1}{2}\sqrt{\frac{\pi}{a^3}}$, we get

$$N = \frac{4\pi V}{h_0^3} e^{-\alpha} \frac{1}{4}\sqrt{\frac{\pi}{(\beta/2m)^3}}$$

Rearranging, we obtain

$$e^{-\alpha} = \frac{Nh_0^3}{V}(\beta/2\pi m)^{3/2}$$

Substituting this value in Eq. 10.35, a few terms cancel and we obtain

$$n(p)dp = 4\pi N \left[\frac{\beta}{2\pi m}\right]^{3/2} p^2 e^{-\beta p^2/2m} dp \quad (10.37)$$

Now, momentum p and energy ϵ are related as $p^2 = 2m\epsilon$, giving $2pdp = 2md\epsilon$, or

$$dp = \frac{m}{p}d\epsilon = \frac{m}{\sqrt{2m\epsilon}}d\epsilon = \sqrt{\frac{m}{2\epsilon}}d\epsilon \tag{10.38}$$

Substituting the value of d(p) in Eq. 10.34, we obtain

$$g(\epsilon)d\epsilon = \frac{4\pi V}{h_0^3}2m\epsilon\sqrt{\frac{m}{2\epsilon}}d\epsilon = \frac{4\sqrt{2}\pi V}{h_0^3}m^{3/2}\epsilon^{1/2}d\epsilon \tag{10.39}$$

Further, Eq. 10.37 in terms of energy gives

$$n(\epsilon)d\epsilon = 4\pi N\left[\frac{\beta}{2\pi m}\right]^{3/2}2m\epsilon e^{-\beta\epsilon}\sqrt{\frac{m}{2\epsilon}}d\epsilon$$
$$= \frac{2N}{\sqrt{\pi}}\beta^{3/2}\sqrt{\epsilon}e^{-\beta\epsilon}d\epsilon \tag{10.40}$$

Equation 10.40 can be used to evaluate total energy of gas molecules and using the fact that it is equal to $\frac{3}{2}Nk_BT$, we obtain

$$U = \frac{3}{2}Nk_BT = \int_0^\infty \epsilon n(\epsilon)d\epsilon$$
$$\frac{3}{2}Nk_BT = \frac{2N\beta^{3/2}}{\sqrt{\pi}}\int_0^\infty \epsilon^{3/2}e^{-\beta\epsilon}d\epsilon$$
$$= \frac{2N\beta^{3/2}}{\sqrt{\pi}}\frac{3}{4\beta^2}\frac{\sqrt{\pi}}{\sqrt{\beta}} = \frac{3}{2}N\frac{1}{\beta}$$

After eliminating common terms, we obtain $\beta = \frac{1}{k_BT}$. This value of β is common to all three statistics. Note that while evaluating the above integral, we have made use of standard integral

$$\int_0^\infty x^{3/2}e^{-ax}dx = \frac{3}{4a^2}\sqrt{\frac{\pi}{a}}$$

10.8.3 Maxwell–Boltzmann's Law of Distribution of Energy and Velocity

Having determined the value of β, we can now obtain the number of molecules having energy between ϵ and $(\epsilon + d\epsilon)$ from Eq. 10.40 as below

$$n(\epsilon)d\epsilon = \frac{2\pi N}{(\pi k_BT)^{3/2}}\sqrt{\epsilon}e^{-\epsilon/k_BT}d\epsilon \tag{10.41}$$

Similarly, Eq. 10.37 gives

$$n(p)dp = \frac{\sqrt{2}\pi N}{(\pi m k_B T)^{3/2}} p^2 e^{-p^2/2mk_B T} dp \qquad (10.42)$$

In terms of velocity v, (using p = mv) the above distribution reduces to

$$n(v)dv = \frac{\sqrt{2}\pi N (m)^{3/2}}{(\pi k_B T)^{3/2}} v^2 e^{-mv^2/2k_B T} dv \qquad (10.43)$$

Equations 10.41 and 10.43 defines Maxwell–Boltzmann's law of distribution of energies and velocity, respectively.

10.8.4 The Most Probable Distribution, Fermi–Dirac Statistic

Following similar approach, we can now develop the expression for the most probable distribution in case of a system obeying Fermi–Dirac statistics. From Eq. 10.13, we can write

$$\ln \Omega_{FD} = \sum_i [\ln g_i! - \ln n_i! - \ln(g_i - n_i)!]$$

Making use of Stirling relation, this equation reduces to

$$\ln \Omega_{FD} = \sum_i [g_i \ln g_i - g_i - n_i \ln n_i + n_i - (g_i - n_i) \ln(g_i - n_i) + (g_i - n_i)]$$

$$= \sum_i [g_i \ln g_i - (g_i - n_i) \ln(g_i - n_i) - n_i \ln n_i] \qquad (10.44)$$

As done in the previous case, we make use of the fact that N is constant and g_i is not subject to change. Therefore, differentiating the above equation w.r.t, n_i, we get

$$\delta(\ln \Omega_{FD}) = \sum_i \delta [g_i \ln g_i - n_i \ln n_i + (n_i - g_i) \ln(g_i - n_i)]$$

$$= \sum_i \left[\delta n_i \ln(g_i - n_i) + \frac{(n_i - g_i)}{g_i - n_i}(-\delta n_i) - n_i \frac{1}{n_i}\delta n_i - \delta n_i \ln n_i \right]$$

$$= \sum_i [\delta n_i \ln(g_i - n_i) + \delta n_i - \delta n_i - \delta n_i \ln n_i]$$

$$= \sum_i (-\delta n_i) [\ln n_i - \ln(g_i - n_i)]$$

$$= -\sum_i \left[\ln \left(\frac{n_i}{g_i - n_i} \right) \right] \delta n_i = 0 \qquad (10.45)$$

This equation, along with constraint Eqs. 10.16 and 10.17 can be solved by using the method of Lagrange undetermined multipliers. As done before

$$\sum_i \delta n_i \left[\ln \left(\frac{n_i}{g_i - n_i} \right) + \alpha + \beta \epsilon_i \right] = 0 \qquad (10.46)$$

Note that α and β are new constants. As argued earlier, δn_i are arbitrary and subject to change. Therefore, the above equation will be valid if

$$\ln \left(\frac{n_i}{g_i - n_i} \right) + \alpha + \beta \epsilon_i = 0 \qquad (10.47)$$

or we can transform this equation as

$$\frac{n_i}{g_i - n_i} = exp\left[-(\alpha + \beta \epsilon_i) \right], \quad or \quad \frac{g_i - n_i}{n_i} = exp\left[(\alpha + \beta \epsilon_i) \right]$$

After a few mathematical rearrangement of the terms, we get

$$n_i = \frac{g_i}{exp\left[(\alpha + \beta \epsilon_i) \right] + 1} \qquad (10.48)$$

This equation defines the *Fermi–Dirac distribution* function.

10.8.5 The Most Probable Distribution, Bose–Einstein Statistic

For Bose–Einstein statistics, with the help of Eq. 10.15, we can write

$$\ln \Omega_{BE} = \sum_i [\ln(n_i + g_i - 1)! - \ln n_i! - \ln(g_i - 1)!]$$

Using the fact that n_i and g_i are very large, and employing Stirling formula, we obtain

$$\ln \Omega_{BE} = \sum_i \left[(n_i + g_i - 1) \ln(n_i + g_i - 1) - (n_i + g_i - 1) \right.$$

$$\left. - n_i \ln n_i + n_i - (g_i - 1) \ln(g_i - 1) + (g_i - 1) \right]$$

As n_i and g_i are very large, we can approximate $(n_i + g_i - 1) \approx (n_i + g_i)$ and $(g_i - 1) \approx g_i$. Therefore, $\ln \Omega_{BE}$ reduces to

$$\ln \Omega_{BE} = \sum_i [(n_i + g_i) \ln(n_i + g_i) - (n_i + g_i) - n_i \ln n_i + n_i - g_i \ln g_i + g_i]$$

Now, as we have done earlier for the case M–B and F–D statistics, we recall that n_i changes continuously but g_i does not. Therefore, differentiating the above equation w.r.t. n_i, we obtain

$$\delta(\ln \Omega_{BE}) = \sum_i \delta \left[(n_i + g_i) \ln(n_i + g_i) - (n_i + g_i) \right.$$

$$\left. - n_i \ln n_i + n_i - g_i \ln g_i + g_i \right]$$

or we can write

$$\delta(\ln \Omega_{BE}) = \sum_i \left[\delta n_i \ln(g_i + n_i) + \frac{(n_i + g_i)}{g_i + n_i} \delta n_i - n_i \frac{1}{n_i} \delta n_i - \delta n_i \ln n_i \right]$$

$$= \sum_i [\delta n_i \ln(g_i + n_i) + \delta n_i - \delta n_i - \delta n_i \ln n_i]$$

$$= \sum_i \delta n_i [\ln(n_i + g_i) - \ln n_i] \tag{10.49}$$

The last equation after rearranging the terms and the fact that $\delta(\ln \Omega_{BE}) = 0$ gives

$$\delta(\ln \Omega_{BE}) = - \sum_i \left[\ln \left(\frac{n_i}{n_i + g_i} \right) \right] \delta n_i = 0 \tag{10.50}$$

This equation together combined with Eqs. 10.16 and 10.17 will determine the B–E distribution function. For this, we will make use of the method of Lagrange Undetermined multiplier. We multiply Eq. 10.16 with α and Eq. 10.17 with β and add these to Eq. 10.50 to obtain

$$\sum_i \delta n_i \left[\ln \left(\frac{n_i}{n_i + g_i} \right) + \alpha + \beta \epsilon_i \right] = 0 \tag{10.51}$$

As noted earlier, the coefficient δn_i are arbitrary and cannot be zero, therefore, the above equation will be satisfied if

$$\ln \left(\frac{n_i}{n_i + g_i} \right) + \alpha + \beta \epsilon_i = 0$$

We can rewrite this as

$$\frac{g_i + n_i}{n_i} = exp \left[\alpha + \beta \epsilon_i \right]$$

Finally, after rearranging the terms, it gives

$$n_i = \frac{g_i}{exp[\alpha + \beta \epsilon_i] - 1} \tag{10.52}$$

This equation gives the *Bose–Einstein distribution law.*

Fig. 10.5 A comparison between three different statistical distribution. Distribution function 'f' plotted against $\dfrac{(\epsilon - \mu)}{k_B T}$ for a fixed value of μ and T

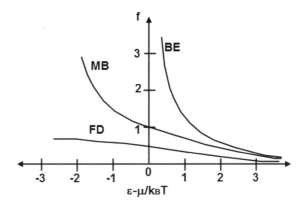

10.8.6   Comparison Between Three Statistics

After knowing the newly introduced constants $\alpha = -\mu/k_B T$ and the fact that $\beta = 1/k_B T$, three distribution can be combined into a single equation as below

$$\frac{n_i}{g_i} = \frac{1}{exp\,[(\epsilon_i - \mu)/k_B T] + a} \qquad (10.53)$$

where, $a = 0$ for Maxwell–Boltzmann statistics, $a = -1$ for Fermi–Dirac statistic and $a = +1$ for Bose–Einstein statistic. Despite several differences among the ways, these three statistics handle the particle distribution for entirely different systems, their distribution function looks identical except small difference in the value of parameter 'a'. Now we will discuss how three distribution function differ from each other and under what conditions they represent identical results. Three distribution functions are plotted in Fig. 10.5. We notice the following points: (i) The distribution functions differ significantly at low temperature. The Bosons distribution function is skewed towards low energies, whereas for Fermions, towards higher energy; implying that there is higher probability of finding Bosons in low energy states and Fermions in higher energy states.
(ii) At sufficiently high temperature, all three statistics give the same result. This means that at high temperature, the distinction between three distributions disappear, whereas at low temperature, the quantum and classical results will be different.

10.9   Solved Problems

Q.1 **Calculate the number of different arrangements of seven Bosons among three cells of equal a priori probability.**

Sol: Different arrangements of $n_i = 7$ Bosons in $g_i = 3$ cells are

$$W = \frac{(n_i + g_i - 1)!}{n_i!(g_i - 1)!} = \frac{(7 + 3 - 1)!}{7!(3 - 1)!} = \frac{9 * 8 * 7!}{7! * 2 * 1} = 36 \qquad (10.54)$$

Q.2 **Classify the following particles according to Fermi or Bose statistics:**
α **particles,** ^3He, H_2 **molecule, positron,** $^6Li^+$ **ion,** $^7Li^+$ **and** $^7_3Li^{++}$ **ion,** ^{12}C **atom,** $^{12}C^+$ **ion,** $^4He^+$ **ion, H$^-$ ion,** ^{13}C **atom, deuterium, deutron.**

Sol: Proton (P), neutron (N), electron (e^-), and positron (e^+) obey Fermi statistics. One can easily obtain the answer by seeing how many Fermi particles a given particle has:

α particle $= 2P + 2N$ (Bose particle),
$^3He = 2P + N + 2e^-$ (Fermi particle),
H_2 molecule $= 2(P + e^-)$ (Bose particle),
positron $= e^+$ (Fermi particle),
$^6Li^+$ ion $= 3P + 3N + 2e^-$ (Bose particle),
$^7Li^+$ ion $= 3P + 4N + 2e^-$ (Fermi particle),
$^7_3Li^{++}$ ion $= 3P + 4N + 1e^-$ (Bose particle), ^{12}C atom $= 6e^- + 6p + 6n$ (Bose particle),
$^{12}C^+$ ion $= 5e^- + 6p + 6n$ (Fermi particle)
$^4He^+$ ion $= 1e^- + 2p + 2n$ (Fermi particle)
H$^-$ ion $= 2e^- + 1p$ (Fermi particle)
^{13}C atom $= 6e^- + 6p + 7n$ (Fermi particle)
deuterium $= {_1}H^2 = 1e^- + 1P + 1N$ (Fermi particle)
deutron $=$ nucleus of deuterium, i.e., $1P + 1N$ (Bose particle)

Q.3 **There is no need to take into account the orbital angular momentum of the electrons in order to determine if an atom is a boson or a fermion, because**

Sol: The orbital angular momentum of an electrons is always an integer. Hence, the addition of the orbital angular momentum cannot change an integer spin into a half-integer spin and vice versa.

10.10   Multiple Choice Questions

Q.1 **The phase space is a**

    (A)   three-dimensional space   **(C)**   five-dimensional space
    (B)   four-dimensional space   **(D)**   six-dimensional space

Q.2 **The Maxwell–Boltzmann statistics is obeyed by particles that are**

    (A)   identical   **(C)**   photons
    (B)   distinguishable   **(D)**   B and C

Q.3 **Which one of the following shall obey Maxwell–Boltzmann statistics**

(A)   oxygen molecule   (C)   neutron
(B)   electron   (D)   photon

Q.4 **The one obeying Fermi–Dirac statistics is**

(A) α particle   (C) H_2 molecule
(B)   H   (D)   Phonon

Q.5 **Bosons have a spin value**

(A)   0   (C)   1/2
(B)   1   (D)   0 or 1

Q.6 **Fermions have a spin value**

(A)   0   (C)   1/2
(B)   1   (D)   0 or 1

Q.7 **The statistics obeyed by Photons is**

(A)   M.B   (C)   B.E
(B)   F.D   (D)   All

Q.8 **The energy at absolute zero cannot be zero. This fact is supported by which one of the following statistics ?**

(A)   M.B.   (C)   F.D
(B)   B.E   (D)   None

Q.9 **Which one of the following is a Boson ?**

(A) α-particle   (C)   proton
(B)   neutron   (D)   positron

Q.10 **Planck's radiation law can be obtained from**

(A)   B–E statistics   (C)   M–B statistics
(B)   F–D statistics   (D)   all of the above

Q.11 **The Bose-Einstein distribution is**

(A) $n_i = \dfrac{g_i}{e^{\alpha+\beta\epsilon_i} + 1}$ (C) $n_i = \dfrac{g_i}{1 - e^{\alpha+\beta\epsilon_i}}$

(B) $n_i = \dfrac{g_i}{e^{\alpha+\beta\epsilon_i} - 1}$ (D) $n_i = \dfrac{g_i}{e^{\alpha+\beta\epsilon_i}}$

Q.12 **The distribution function for a photon gas can be written as**

(A) $n_i = e^{-\frac{\epsilon_i}{kT}}$

(C) $n_i = \dfrac{1}{e^{\frac{\epsilon_i}{kT}} - 1}$

(B) $n_i = \dfrac{1}{e^{\frac{\epsilon_i}{kT}} + 1}$

(D) $n_i = \dfrac{1}{e^{-\frac{\epsilon_i}{kT}} - 1}$

Q.13 **The distribution function for an electron gas can be written as**

(A) $n = e^{-\frac{\epsilon}{kT}}$

(C) $n = \dfrac{1}{e^{\frac{\epsilon - \mu}{kT}} - 1}$

(B) $n = \dfrac{1}{e^{\frac{\epsilon - \mu}{kT}} + 1}$

(D) $n = \dfrac{1}{e^{-\frac{\epsilon}{kT}} - 1}$

Q.14 **The Bosons spin is**

(A) always a positive integer

(C) positive integral multiple of 1/2

(B) an integer

(D) any fraction

Q.15 **Which one of the following is a Fermion?.**

[GATE-2014]

(A) α particle

(C) Hydrogen atom $_1H^1$

(B) $_4Be^7$ nucleus

(D) Deutron

Q.16 **The number of ways in which N identical bosons can be distributed in two energy levels, is**

[JRF-NET Dec 2012]

(A) N+1

(C) $\dfrac{N(N + 1)}{2}$

(B) $\dfrac{N(N - 1)}{2}$

(D) N

Q.17 **Which of the following atoms cannot exhibit Bose–Einstein condensation, even in principle?**

[GATE 2010]

(A) $_1H^1$

(C) $_{11}Na^{23}$

(B) $_2H^4$

(D) $_{19}K^{30}$

Q.18 **The quantum statistics changes into classical statistics when**

(A) $\dfrac{n_i}{g_i} \gg 1$ (C) $\dfrac{n_i}{g_i} = 1$

(B) $\dfrac{n_i}{g_i} \ll 1$ (D) $\dfrac{n_i}{g_i} = 01$

Q.19 **Thermodynamic probability of a system in equilibrium is**

(A) 0   (C) maximum

(B) 1   (D) minimum

Q.20 **For a system in equilibrium**

(A) entropy and probability are zero

(B) entropy and probability are maximum

(C) entropy and probability are minimum

(D) entropy is maximum and probability is minimum

Q.21 **A system of identical particles obey Maxwell–Boltzmann statistics when**

(A) occupation index is zero   (C) occupation index is very low

(B) occupation index is unity   (D) occupation index is very high

Q.22 **Five Fermions are to be arranged in 8 equally likely energy states. The number of different ways are**

(A) 8^5 (C) $\dfrac{8!}{5!}$

(B) 8   (D) $\dfrac{8!}{5!3!}$

Q.23 **In Q.22, if particle are bosons then number of arrangements is**

(A) 8^5 (C) $\frac{12!}{5!7!}$

(B) 8   (D) $\frac{8!}{5!3!}$

Q.24 **Which of the following statements is INCORRECT ?**

[JAM 2006]

(A) Indistinguishable particles obey Maxwell–Boltzmann statistics

(B) All particles of an ideal Bosegas occupy a single energy state at $T = 0$

(C) The integral spin particles obey Bose–Einstein statistics

(D) Protons obey Fermi–Dirac statistics

Q.25 Let N_{MB}, N_{BE}, N_{FD}, denote the number of ways in which two particles can be distributed in two energy states according to Maxwell–Boltzmann, Bose–Einstein and Fermi–Dirac statistics, respectively. Then $N_{MB}{:}N_{BE}{:}N_{FD}$ is

[JAM 2013]

(A) 4:3:1   (C) 4:3:3
(B) 4:2:3   (D) 4:3:2

Q.26 The number of ways of distributing 11 indistinguishable Bosons in 3 different energy levels is

[NET-JRF June-2018]

(A) 3^{11}

(C) $\dfrac{(13)!}{2!(11)!}$

(B) 11^3

(D) $\dfrac{(11)!}{3!8!}$

Keys and hints to MCQ type questions

Q.1 D	Q.6 C	Q.11 B	Q.16 A	Q.21 C	Q.26 C
Q.2 B	Q.7 C	Q.12 C	Q.17 D	Q.22 D	
Q.3 A	Q.8 C	Q.13 B	Q.18 B	Q.23 C	
Q.4 B	Q.9 D	Q.14 B	Q.19 C	Q.24 A	
Q.5 D	Q.10 A	Q.15 B	Q.20 B	Q.25 A	

Hint.14 The Bosons spin is always an integer (including 0, because 0 is neither positive nor negative, it is simply a whole number). For instance, Higgs has zero, the gluon and photon have spin 1 and graviton is postulated to have 2 units of spin. Therefore, the correct option **B**.

Hint.16 Number of boson = N, Number of energy level = g So the number of ways to distribute N boson into g level is

$$W = \frac{(N + g - 1)!}{N!(g - 1)!} = \frac{(N + 2 - 1)!}{N!(2 - 1)!} = N + 1$$

Hint.17 Answer is **D**. For Bose–Einstein: Number of electron + number of proton + number of neutron = even
For, $_{19}K^{30}$ Number of proton = 19 ,Number of electron = 19 , Number of neutron = 21, Total half-integral particles, $19 + 19 + 21 = 49$ which is odd. So it will not exhibit Bose–Einstein condensation.

Hint.21 When occupation index is very low, i.e., $\frac{n_i}{g_i} <<< 1$, M.B stat gives same results as quantum statistics. Correct option (C).

Hint.22 The number of ways are

$$W = \frac{g!}{n!(g-n)!} = \frac{8!}{5!3!}$$

Hint.23 The number of ways are

$$W = \frac{(n+g-1)!}{n!(g-1)!} = \frac{12!}{5!7!}$$

Hint.25 Here, total number of particles $N = 2$ and $n_1 = 2$, energy states with equal a priori probability $g_1 = 2$. Then

$$N_{MB} = N!\frac{g_1^{n_1}}{n_1!} = 2!\frac{2^2}{2!} = 4$$

$$N_{BE} = \frac{(n_1+g_1-1)!}{n_1!(g_1-1)!} = \frac{(2+2-1)!}{2!(2-1)!} = 3$$

$$N_{FD} = \frac{g_1!}{n_1!(g_1-n_1)!} = \frac{2!}{2!(2-1)!} = 1$$

Therefore, $N_{MB}:N_{BE}:N_{FD}$ is 4:3:1

Hint.26 Number of boson n = 11, Number of energy levels g = 3. Therefore, number of ways to distribute 18 Bosons into 3 level is

$$W = \frac{(n+g-1)!}{n!(g-1)!} = \frac{(11+3-1)!}{(11)!(3-1)!} = \frac{(13)!}{2!(11)!}$$

10.11 Exercises

1. Compare Maxwell–Boltzmann, Fermi–Dirac and Bose–Einstein's statistics. Under what conditions, F–D and B–E statistics yield results similar to classical statistics.

2. What is thermodynamic probability? Obtain an expression for thermodynamic probability for a classical system obeying Maxwell–Boltzmann law.

3. Starting with the expression for thermodynamic probability for a F–D system, obtain Fermi–Dirac distribution law.

4. Show that thermodynamic probability for a system obeying B–E statistics is given by

$$W_{BE} = \prod_i \frac{(n_i + g_i - 1)!}{n_i!(g_i - 1)!}$$

Further, show that the corresponding Bose–Einstein distribution is given by

$$n_i = \frac{g_i}{exp[\alpha + \beta \epsilon_i] - 1}$$

5. Define the following terms:
   (i) Microstates (ii) Macrostates (iii) Thermodynamic probability

6. Two identical particles have to be distributed among three energy levels. Let r_B, r_F, r_C represent the ratios of probability of finding two particles to that of finding one particle in a given energy state. The subscripts B, F and C correspond to whether the particles are Bosons, Fermions and Classical particles, respectively. Find the ratio of $r_B : r_F : r_C$ of probabilities.

Index

© The Author(s) 2022
S. Sharma, *Thermal and Statistical Physics*,
https://doi.org/10.1007/978-3-031-07685-5

Printed in the United States
by Baker & Taylor Publisher Services